Sound Synthesis and Sampling

Series introduction

The Focal Press Music Technology Series is intended to fill a growing need for authoritative books to support college and university courses in music technology, sound recording, multimedia and their related fields. The books will also be of value to professionals already working in these areas who want either to update their knowledge or to familiarise themselves with topics that have not been part of their mainstream occupations.

Information technology and digital systems are now widely used in the production of sound and in the composition of music for a wide range of end uses. Those working in these fields need to understand the principles of sound, musical acoustics, sound synthesis, digital audio, video and computer systems. This is a tall order, but people with this breadth of knowledge are increasingly sought after by employers. The series will explain the technology and techniques in a manner which is both readable and factually concise, avoiding the chattiness, informality and technical woolliness of many books on music technology. The authors are all experts in their fields and many come from teaching and research backgrounds.

Dr Francis Rumsey
Series Editor

Music Technology Titles

Acoustics and Psychoacoustics, 2nd edition
David M. Howard and James Angus

The Audio workstation Handbook
Francis Rumsey

Composing Music with Computers
Eduardo Reck Miranda

Computer Sound Design: synthesis techniques and programming, 2nd edition
Eduardo Reck Miranda

Digital Audio CD and Resource Pack
Markus Erne

Digital Sound Processing for Music and Multimedia
Ross Kirk and Andy Hunt

MIDI Systems and Control, 2nd edition
Francis Rumsey

Network Technology for Digital Audio
Andrew Bailey

Sound and Recording: An introduction, 4th edition
Francis Rumsey and Tim McCormick

Sound Synthesis and Sampling CD-ROM
Martin Russ

Spatial Audio
Francis Rumsey

Sound Synthesis and Sampling

Martin Russ

OXFORD AMSTERDAM BOSTON LONDON NEW YORK PARIS
SAN DIEGO SAN FRANCISCO SINGAPORE SYDNEY TOKYO

Focal Press
An imprint of Elsevier Science
Linacre House, Jordan Hill, Oxford OX2 8DP
200 Wheeler Road, Burlington, MA 01803

First published 1996
Reprinted 1998, 1999, 2000 (twice), 2003

British Library Cataloguing in Publication Data
A catalogue record for this book is available from the British Library

Library of Congress Cataloguing in Publication Data
A catalogue record for this book is available from the Library of Congress

ISBN 0 240 51429 7

For information on all Focal Press publications
visit our website at www.focalpress.com

Printed and bound in Great Britain

Contents

Preface

This is a book about sound synthesis and sampling. It is intended to provide a reference guide to the many techniques and approaches that are used in both commercial and research sound synthesizers. The coverage is more concerned with the underlying principles, so this is not another 'build your own synthesizer' type of book. Instead it aims to provide a solid source of information on the diverse and complex field of sound synthesis. As well as the details of the techniques of synthesis, some practical applications are described to show how synthesis can be used to make sounds.

It is designed to meet the requirements of a wide range of readers, from enthusiasts to undergraduate level students. Wherever possible, a non-mathematical approach has been taken, and the book is intended to be accessible to readers without a strong scientific background.

This book brings together information from a wealth of material which I have been collecting and compiling for many years. Since the early 1970s I have been involved in the design, construction and use of synthesizers. More recently this has included the reviewing of electronic musical instruments for *Sound On Sound*, the leading hi-tech music recording magazine in the United Kingdom.

The initial prompting for this book came from Francis Rumsey of the University of Surrey's Music Department, with support from Margaret Riley at Focal Press. I would like to thank them for their enthusiasm, time and encouragement throughout the project. I would also like to thank my wife and children for putting up with my disappearance for long periods over the last year.

Martin Russ

About this book

This book is divided into eight chapters, followed by References, a Glossary, a Jargon guide, and finally, an Index. The jargon section is designed to try and prevent the confusion that often results from the wide variation in the terminology which is used in the field of synthesizers. Each entry consists of the term which is used in this book, followed by the alternative names which can be used for that term.

Book guide

The chapters can be divided into five major divisions:

- Background
- Techniques
- Applications
- Miscellaneous
- Reference

Background: Chapter 1 sets the background and places synthesis in a historical perspective.

Techniques: Chapters 2, 3, 4 and 5 describe the main methods of producing and manipulating sound.

Applications: Chapters 6 and 7 show how the techniques described can be used to synthesize sound and music.

Miscellaneous: Chapter 8 has some speculation on future developments.

Reference: References, Glossary, Jargon guide, and Index.

Chapter section guide

Within each chapter, there are sections which deal with specific topics. The format and intention of some of these may be unfamiliar to the reader, and thus they deserve special mention.

Example Instruments

These sections are illustrated with block diagrams of the internal function and front panel controls of some representative example instruments, together with some notes on their main features. This should provide a more useful idea of their operation than just black and white photographs. Further information and photographs of a wide range of synthesizers and other electronic musical instruments can be found in Julian Colbeck's comprehensive *Keyfax* books (Colbeck, 1985–). Details on some specific instruments can be found in Mark Vail's *Vintage Synthesizers*, which is a collection of articles from the American magazine *Keyboard* (Vail, 1993).

Time lines

The time lines are intended to show the development of a topic in a historical context. Reading text which contains lots of references to dates and people can be confusing. By presenting the major events in time order, the developments and relationships can be seen more clearly. The time lines are deliberately split up so that only entries relevant to each chapter are shown. This keeps the material in each individual time line succinct.

Questions

Each chapter ends with a few questions, which can be used as either a quick comprehension test, or as a guide to the major topics covered in that chapter.

1 Background

1.1 What is synthesis?

'Synthesis' is defined in the 1994 edition of the *Chambers Dictionary* as 'building up; putting together; making a whole out of parts'. The process of synthesis is thus a creative bringing together, and it is this artistic aspect which is often overlooked in favour of the more technical aspects of the subject. Although a synthesizer may be capable of producing almost infinite varieties of output, controlling and choosing them requires human intervention and skill.

The word 'synthesis' is frequently used in just two major contexts: the creation of chemical compounds and the production of electronic sounds. But there are a large number of other types of synthesis.

1.1.1 Types

All synthesizers are very similar in their concept – the major differences are in their output formats and the way that they produce that output. Just some of the types of synthesizer are:

- texture synthesizers, used in the graphics industry
- video synthesizers, used to process video signals
- colour synthesizers, used as part of 'son et lumiere' presentations
- speech synthesizers, used in computer and telecommunications applications

- sound synthesizers, used to create and process sounds and music
- word synthesizers, more commonly known as authors!

Synthesizers have two basic functional blocks: a control interface, which is how the parameters which define the end product are set; and a 'synthesis engine' which interprets the parameter values and produces the output. In most cases there is a degree of abstraction involved between the control interface and the synthesis engine itself. This is because the complexity of the synthesis process is often very high, and it is necessary to reduce the apparent complexity of the control by using some sort of simpler conceptual model. This enables the user of the synthesizer to use it without requiring a detailed knowledge of the inner workings. This idea of models and abstraction of interface is a common theme which will be mentioned many times.

1.1.2 Sound synthesis

Sound synthesis is the process of producing sound. It can re-use existing sounds by processing them, or it can generate sound electronically or mechanically. It may use mathematics, physics, or even biology – and it brings together art and science in a mix of musical skill and technical expertise. Used carefully, it can produce emotional performances which paint sonic landscapes with a rich and huge set of timbres, limited only by the imagination of the creator.

Sounds can be simple or complex, and the methods used to create them are diverse. Sound synthesis is not solely concerned with sophisticated computer-generated timbres, although this is often the most publicized aspect. The wide availability of high quality recording and synthesis technology has made the generation of sounds much easier for musicians and technicians, and future developments promise even easier access to ever more powerful techniques. But the technology is nothing more than a set of tools which can be used to make sounds: the creative skills of the performer, musician or technician are still essential.

1.1.3 Synthesizers

Sounds are synthesized using a sound synthesizer. Although synthesizer can be spelt with a 'zer' or 'ser' ending, the 'zer' ending will be used in this book. Also, the single word 'synthesizer' is used here to imply 'sound synthesizer', rather than a generic synthesizer.

The synthesis of sounds has a long history. The first synthesizer might have been an early ancestor of homo sapiens hitting a hollow log, or perhaps learning to whistle. Singing uses a sophisticated synthesizer whose capabilities are often forgotten: the human vocal tract. All musical instruments can be thought of as being 'synthesizers', although few people would think of them in this context. A violin or clarinet is viewed as being 'natural', whereas a synthesizer is seen as 'artificial' – even though all of these instruments produce a sound by essentially synthetic methods.

More recently, the word 'synthesizer' has come to only mean an electronic instrument which is capable of producing a wide range of different sounds. The actual categories of sounds which qualify for this label of synthesizer are also very specific: purely imitative sounds are frequently regarded as nothing other than recordings of the actual instrument, in which case the synthesizer is seen as little more than a replay device. In other words, the general public seems to expect synthesizers to produce 'synthetic' sounds. As synthesizers become better at fusing elements of real and synthetic sounds, the boundaries of what is regarded as 'synthetic' will have to become more diffuse.

Forms

Synthesizers come in several different varieties, although many of the constituent parts are common to all of the types. Most synthesizers have one or more audio outputs; one or more control inputs; some sort of display; and buttons or dials to select and control the operation of the unit. The major split is into performance and modular forms.

- **Performance synthesizers** have a standard interconnection of their internal synthesis modules already built in. It is usually not possible to change this significantly, and so the signal flow always follows a set path through the synthesizer. This enables the rapid patching of commonly used configurations, but does limit the flexibility. Performance synthesizers form the vast majority of commercial synthesizer products.
- Conversely, **modular synthesizers** have no fixed interconnections, and the synthesis modules can be connected together in any way. Changes can be made to the connections whilst the synthesizer is making a sound, although the

usual practice is to set up and test the interconnections in advance. Because more connections need to be made, modular synthesizers are harder and more time-consuming to set up, but they do have much greater flexibility. Modular synthesizers are much rarer than performance synthesizers, and are often used for academic or research purposes.

Both performance and modular synthesizers can come with or without a music keyboard. The keyboard has become the most dominant method of controlling the performance aspect of a synthesizer, although it is not necessarily the ideal controller. Synthesizers which do not have a keyboard are often referred to as expanders or modules, and these can be controlled either by a synthesizer which does have a keyboard, or from a variety of other controllers. It has been said that the choice of a keyboard as the controller was probably the biggest set-back to the wide acceptance of synthesizers as a musical instrument. Chapter 7 describes some of the alternatives to a keyboard.

1.1.4 Sounds

Synthesized sounds can be split into simple categories like 'imitative' or 'synthetic'. Some sounds will not be easy to place in a definite category, and this is especially true for sounds which contain elements of both real and synthetic sounds.

Imitative sounds often sound like real instruments, and they might be familiar orchestral or band instruments. In addition, imitative sounds may be more literal in nature: sound effects. In contrast, synthetic sounds will often be unfamiliar to anyone who is used to hearing only real instruments, but over time a number of clichés have been developed: the 'string synth' and 'synth brass' being just two examples.

Synthetic sounds can be divided into various types, depending on their purpose. 'Imitations' and 'emulations' are intended to provide many of the characteristics of real instruments, but in a synthetic way. The many 'electronic' piano sounds are examples of this sort of emulated sound. 'Suggestions' and 'hints' are sounds where the resulting sound has only a slight connection with any real instrument. 'Alien' and 'off the wall' sounds are usually entirely synthetic in nature. Of course, most synthesizers can also produce 'noise', which is perhaps the most unpitched and unusual sound of all. All of these types of synthetic sound can be used to make real sounds more interesting (see Section 6.8).

1.1.5 Synthesis methods

There are many techniques which can be used to synthesize sound. Many of them use a 'source and modifier' model as a metaphor for the process which produces the sound: a raw sound source produces the basic tone, which is then modified by filters and envelopes to create the final sound. Another name for this model is the 'excitation and filter' model. The use of this model can be seen most clearly in analogue subtractive synthesizers, but it can also be applied to other methods of synthesis – S&S (sample and synthesis) or physical modelling, for example.

Some methods of synthesis are more complex: FM, harmonic synthesis, FOF (see Section 5.5), and granular synthesis. For these methods, the metaphors of a model can be harder to grasp, and this may be one of the reasons why the 'easier to understand' methods like subtractive and S&S synthesis have been so commercially successful.

1.1.6 Analogue synthesis

'Analogue' refers to the use of audio signals which can be processed using filters and amplifiers. Analogue synthesis methods split into three basic areas, although there are crossovers between them. The basic types are:

Subtractive

Subtractive synthesis takes a 'raw' sound which is usually rich in harmonics, and filters it to remove some of the harmonic content. The raw sounds are traditionally simple mathematical waveshapes: square, sawtooth, triangle and sine, although modern subtractive synthesizers tend to provide longer 'samples' instead of single cycles of waveforms. The filtering tends to be a resonant low-pass filter, and changing the cut-off frequency of this filter produces the characteristic (and clichéd) 'filter sweep' sound which is strongly associated with subtractive synthesis.

Additive

Additive synthesis adds together lots of sine waves with different frequencies to produce the final timbre. The main problem with this method is the complexity of controlling large numbers of sine waves – but see Section 1.1.7 on digital synthesis too.

Wavetable

Wavetable synthesis extends the ideas of subtractive synthesis by providing much more sophisticated waveshapes as the raw starting point for subsequent filtering and shaping. More than one cycle of a waveform can be stored, or many waveforms can be arranged so that they can be dynamically selected in real-time – this produces a characteristic 'swept' sound which can be subtle, rough, metallic or even glassy in timbre.

1.1.7 Digital synthesis

Digital technology replaces signals with numerical representations, and then uses computers to process those numbers. Digital methods of synthesizing sounds are more varied than analogue methods, and research is still continuing to find new ways of making sounds. Some of the types that you may encounter include:

FM

Frequency modulation is the technical term for the way that FM radio works, where the audio signal of music or speech is used to modulate a high-frequency carrier signal which then broadcasts the audio as part of a radio signal. In audio FM, both signals are at audio frequencies, and all sorts of complex aliasing, frequency mirroring and phase inversions happen – which can produce a wide range of timbres. (In fact, most of the effects that audio FM uses are exactly the sort of distortions and problems that you try to avoid in radio FM!) The main problem with FM is that it is not possible to programme it 'intuitively' without a lot of practice, but its major advantage is that it requires very little memory to store a large number of sounds. FM is used in some sound cards and portable keyboards, although it is currently out of fashion.

Wavetable

Wavetable synthesis uses the same idea as the analogue version, but extends the basic idea into more complex areas. The waveshapes are usually complete but short segments of real samples, and these can be looped to provide sustained sections of sound, or several segments of sound can be joined together to produce a composite 'sample'. Often this is used to 'splice' the start of one sound onto the sustain part of another. Because complete samples are not used, this method makes very efficient use of available memory space, but this results in a loss of quality. Wavetable synthesis is used in low-cost, mass-market sound cards and MIDI instruments.

Sample replay

Sample replay is the ultimate version of wavetable. Instead of looping short samples and splicing them together, sample replay does just what its name states: it replays complete samples of sounds, with perhaps one loop for the sustained section of the sound. Sample replay uses lots of memory, and is thus only used in more expensive sound cards and MIDI instruments.

Additive

Digital techniques make the task of coping with lots of sine waves much easier, and digital additive synthesizers have been more successful than analogue versions, but they are still a very specialized field. You are unlikely to find a sound card which uses just additive synthesis.

S&S

S&S is an acronym for samples and synthesis, and uses the techniques of wavetable and sample replay, but adds in the filtering and shaping of subtractive synthesis, and all in a digital form. This method is widely used in MIDI instruments, sound cards and professional electronic musical instruments.

Physical Modelling

Physical modelling uses mathematical equations which attempt to describe how the physics of an instrument works. The results can be stunningly realistic, very synthetic, or a mixture of both. The most important feature is the way that the model responds in much the same way as a real instrument. The high processing demands of physical modelling mean that it is currently only found in professional equipment.

1.2 Beginnings

The beginnings of sound synthesis lie with the origins of our species. Many mammals have an excellent hearing sense, and this serves a diverse variety of purposes: advance warning of danger; tracking prey; and communication. In order to be effective, hearing needs to monitor the essential parts of the audio spectrum. This can involve very low frequencies in some underwater animals, or ultrasonic frequencies for echo location purposes in bats, and the dynamic range required can be very large.

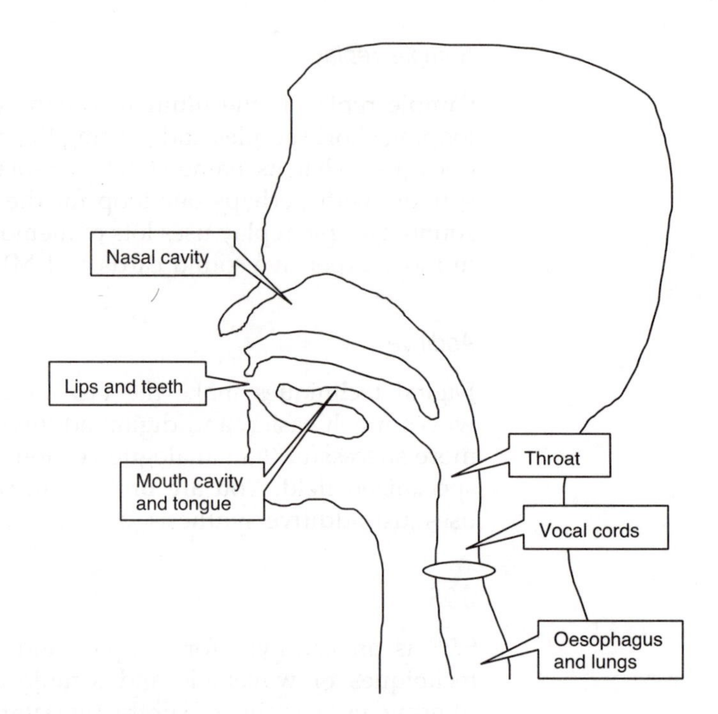

Figure 1.2.1 The human voice is a complex and sophisticated synthesizer capable of producing speech and singing sounds. The main sound source is the vocal cords, although some sounds are produced by the interactions between the lips, tongue and teeth with air currents. The throat, nose, mouth, oesophagus and lungs form a set of resonant cavities that filter the sounds and the mouth shape is dynamically variable.

Human hearing is more limited. Sounds from about 15 Hz to about 18 kHz can be heard, although the upper limit reduces with age. The dynamic range is more than 120 dB: which is a ratio of 10^{12}:1. With two ears and the complex processing achieved by the brain, the average human being can accurately locate sounds and resolve frequency to fractions of a hertz, although this performance is dependent on the frequency, level and other factors.

Human beings also have a sophisticated means of producing sound: the vocal tract. The combination of vocal cords, throat, tongue, teeth, mouth cavity and lips provides a versatile way of making a wide variety of sounds: a biological sound synthesizer. The development of this particular instrument is long and still ongoing – it is probably the oldest and most important musical instrument.

The mixture of sophisticated hearing and an in-built sound synthesizer, plus the everyday usage via speech, singing or whistling, makes the human being a perceptive listener. The combination of sound production and analysis forms a powerful feedback mechanism. Making sounds and listening to sounds is a fundamental part of human interactivity. Using the example of a baby, it is possible to see the range of possibilities:

- **Listening**. Pregnant mothers are often aware that sudden noises can startle a baby in the womb.
- **Mouth**. Most parents will confirm that babies are capable of making a vocal noise from just after birth!
- **Shaking**. Once control over hands and feet is possible, a baby will investigate objects by interacting with them. The rattle is specifically designed to provide audible feedback when it is shaken.
- **Singing**. Part of the process of learning to speak involves long periods of experimentation by the infant, where the range of possible sounds is explored.
- **Speech**. The singing sounds are then reduced down to the set of sounds which are heard from the parent's own speech.
- **Blowing**. Blowing (and spitting!) are part of the learning process for making speech sounds. Blowing into tubes and whistles may lead to playing real musical instruments.
- **Percussive pitching**. 'Open mouth' techniques for making sounds include slapping the cheek or top of the head, or tapping the teeth. The throat and mouth cavity are then altered to provide the pitching of the resulting sound.
- **Whistling**. Whistling requires the mastering of a musical instrument which is created by the lips.

I have arranged these in approximate chronological order, although the development of every human being is different. The important information here is the wide range of possible ways that sounds can be made, and the degree of control which is possible. Singing and whistling are both highly expressive musical instruments, and it is no accident that many musical instruments also use the mouth as part of their control mechanism. With such a broad collection of sounds, humankind has developed a rich and diverse set of musical and spoken sounds.

Beyond this human-oriented synthesis, there are many possibilities for making other sounds. Striking a log or other resonant object will produce a musical tone, and blowing across and through tubes can make a variety of sounds. A bow and arrow may be useful for hunting, but it can also produce an interesting twanging sound as well. From these, and a large number of other ordinary objects, human beings have produced a number of different families of musical instruments. And this process is still continuing.

In the twentieth century, a number of new instruments have been developed: one example is the electric piano. The word 'electric' is almost an misnomer for this particular instrument, because the actual sound is produced by metal rods vibrating in front of coils and thus inducing a voltage in the coils. No electricity is used to produce the sound – it is merely that the output of an electric piano is an electrical signal rather than an acoustic one, and so it needs to be amplified in order to be heard. Naturally, such an instrument could not have been produced before electricity came into common usage, since it depends on amplifiers and loudspeakers.

The synthesizer is even more dependent on technology. Advances in electronics have accelerated its development, and so the transition from simple valve-based oscillators to sophisticated digital tone generators using custom silicon chips has taken less than a hundred years. If semiconductors are taken as the starting-point of electronics, then the major developments in the electronic music synthesizer have only really occurred in the last half of the twentieth century.

1.3 Telecoms research

Much of the research effort expended by the telecommunications industry in the last century has been focused on sound, since the transmission of the human voice has been the major source of revenue. With the advent of reliable digital transmission techniques, communications are becoming increasingly computer-oriented. But the human voice is likely to remain one of the major sources of traffic for the foreseeable future.

Although the invention of the telephone showed that it was possible to transmit the human voice from one location to another by electrical means, this was not the reason for the commercialization of telephony. Familiarity now with the telephone makes it difficult to appreciate how strange the concept of talking to someone at a distance was at the time: why not go and talk to them face-to-face instead? The driving force behind the adoption of the telephone was actually musical – the telephone made it possible to broadcast a musical performance to many people. Again, long usage of radio and television have removed any sense of wonder about being able to hear a concert without actually being there. But at the turn of the century this was amazing!

Thaddeus Cahill's teleharmonium is one example of how telecommunications was used to provide musical entertainment. Developed from prototypes in the 1890s, the 1906 commercial version in New York was essentially a large set of power generators which produced electrical signals at various frequencies, and these could be distributed along telephone lines for the subscribers to listen to. The teleharmonium can be thought of as a 200 ton organ connected to lots of telephones rather than just one loudspeaker. As microphone technology developed, live performances by musicians could also be distributed in the same way. The ability to be able to talk to someone by telephone was a curious side-effect of this musical distribution system.

Telecommunications approaches sound from a technical viewpoint, and so a great deal of research was put into developing improved performance microphones and loudspeakers, as well as increasing the distance over which the sound could be carried. Speech is intelligible with levels of distortion that render music almost unlistenable, and so as the telephone began to be used more and more for speech communications, the research tended to concentrate on the speech transmission. This is one of the reasons that the telephone of today has a restricted bandwidth and dynamic range: it is designed to produce an acceptable level of speech intelligibility, but in as small a bandwidth as possible. The bandwidth of 300 Hz to 3.4 kHz is still the underlying standard for telephony, and is unlikely to change in the foreseeable future.

One example of the way that telecommunications research can be used for electronic musical purposes is the invention of the vocoder. Bell Telephone Laboratories invented the vocoder in the 1930s as a way of trying to process audio signals. The word comes from VOice enCODER, and the idea was to try and split the sound into separate frequency bands and then transmit these more efficiently. It was not successful at the time, although many modern military communications systems use digital descendants of vocoder technology. But the vocoder was rediscovered and adopted by electronic music composers in the 1950s.

By the 1950s, telephones were in wide use for speech, and the researchers turned back to the musical opportunities offered by telephony. Lord Rayleigh's influential work *The Theory of Sound* had laid the foundations for the science of acoustics back in 1878, and Lee de Forest's triode amplifier of 1906 provided the electronics basis for controlling sound. E.C. Wente's condenser microphone in 1915 provided the first high quality audio microphone, and the tape recorder gave a way to store sounds.

At the German radio station NWDR in Cologne in 1951, Herbert Eimert began to use studio oscillators and tape recorders to produce electronically generated sounds. Rather than assemble test gear together, researchers at the RCA Laboratories in the United States produced a dedicated modular synthesizer in 1955 which was designed to simplify the tedious production process of creating sounds by using automation. A mark II model followed in 1957, and this was used at the Columbia-Princeton Electronic Music Center for some years. Although the use of punched holes in paper tape to control the functions now appears primitive, the RCA synthesizer was one of the first integrated systems for producing electronic music.

Work at Bell Telephone Laboratories in the 1950s and 1960s led to the development of pulse code modulation, a technique for digitizing or sampling sound and thus converting it into digital form. As is usual in telecommunications technology, the description and its acronym: pulse code modulation and PCM, are in a technical language which conveys little to the non-engineer. What PCM does is actually very straightforward. An audio signal is 'sampled' at regular intervals, and this gives a series of voltage values: one for each interval. The voltages represent the value of the audio signal at the instant that the sample was made. These voltages are converted into numbers, and these numbers are then converted into a series of electrical pulses, where the number and organization of the pulses represent the size of the voltage. PCM thus refers to the coding scheme used to represent the numbers as pulses. (You may like to compare PCM with PWM (pulse width modulation) as described in Chapter 2.)

PCM forms the basis of sampling. A great deal of work was done to formalize the theory and practice of converting audio signals into digital numbers. The concept of sampling at twice the highest wanted frequency is called the Nyquist criterion after work published by Nyquist and others. The filtering required to prevent unwanted frequencies being heard was also developed as a result of telecommunications research. Further work in the 1970s led to the invention of the digital signal processor (DSP), which was required in order to carry out the complex numerical calculations which were needed to enable audio coding algorithms to be developed. DSPs have since been used to produce the many types of digital synthesizer.

Current telecommunications research continues to explore the outermost limits of acoustics, physics and electronics, although since telecommunications is now almost solely concerned with computers, the emphasis is more and more on intercommunications between computers, not between people.

1.4 Tape techniques

1.4.1 The tape recorder

The tape recorder has been a major part of electronic music synthesis almost from the very beginning. It enables the user to splice together small sections of magnetic tape which represent audio, and then replay the results. This has the important elements of 'building up from small parts' that is the basis of the definition of synthesis.

The principle of the tape recorder is not new. The audio signal is converted into a changing magnetic field, which is then stored onto iron. Early examples recorded onto iron wire, then onto steel ribbon. There were also experiments with the use of paper-backed tape, but the most significant breakthrough was the use of plastic coated with magnetic material, which was developed in Germany in 1935. But it was not until after the end of World War II in 1945 that tape recording started to become widely available as a way of storing and replaying audio material. The tape that was used consisted of a thin acetate plastic tape coated with iron oxide, and polyester film still forms the backing of magnetic tape. More details of the technique of magnetic recording are given in Chapter 4.

Tape recorders are very useful for synthesizing sounds because they allow permanent records to be kept of a performance, or they allow a performance to be 'time-shifted': recorded for subsequent playback later. This may appear to be obvious to the modern reader, but before tape recording, the only way to record sound for later playback was literally to make a record! It is not feasible to break up and re-assemble records, and so when it was first introduced, the tape recorder was a genuinely new and exciting musical tool.

Pitch and speed

A tape recording of a sound ties together two aspects of that sound: the pitch and the duration. If you record the sound at 15 inches per second (ips), then playing back at twice the speed, 30 ips, will double the pitch and so it will be transposed up by one octave. But the duration will be halved, and so a one second sound will only last for half a second when played back at twice the recording speed. Breaking this interdependence is not at all easy using tape recorder technology, although it is relatively simple using digital techniques.

Being able to change the pitch of sounds once they are recorded can simplify the process of producing electronic sounds using oscillators. By changing the speed at which the sound is recorded, the same oscillators can be used to produce tape segments which contain the same sound, but shifted in pitch by one or more octaves. This avoids some of the problems of continuously retuning oscillators – although it does depend on the tape speeds of the tape recorders being accurate.

Unfortunately, the tape speed of early tape recorders was not very accurate. Long-term drift of the speed affects the pitch of anything recorded, and so required careful monitoring. Short-term variations in tape speed are called wow and flutter. Wow implies a slow cyclic variation in pitch, whilst flutter implies a faster and more irregular variation in pitch. Depending on the type of sound, the ear can be very sensitive to pitch changes. Wow and flutter can be very obvious on solo piano playing, whilst some orchestral music can actually sound better!

Of course, changing the tape speed can be used as a creative tool: adjusting the tape speed whilst recording will permanently store the pitch changes in the recording, whilst changing the replay tape speed will only affect playback (but the speed changes are probably not as easy to reproduce on demand). Deliberately introducing wow and flutter can also be used to introduce vibrato and other pitch shifting effects.

Splicing

Once a sound has been recorded onto tape, it is then in a physical form which can be manipulated in ways which would be difficult or impossible for the actual sound itself. Cutting the tape into sections and then splicing them together allows the joining and juxtaposition of sounds. The main limits to this technique are the accuracy of finding the right place on the tape, and the length of the shortest section of tape which can be spliced together. Each join in a piece of tape produces a potential weak spot, and so an edited tape may need to be recorded onto another tape recorder. Each time that a tape is copied, the quality is degraded slightly, and so there is a need to compromise between the complexity of the editing and the fidelity of the final sound.

Reversing

Reversing the direction of playback of tape makes the sound play backwards. Unfortunately, because domestic tape recorders are designed to record in stereo on two sides (known as quarter-track format) merely turning the tape around does not work. Playing the back of the tape (the side which has the backing, rather than the oxide visible) does allow reversing of quarter track tapes, but there is significant loss of audio quality. Professional tape recorders use the mono full-track and stereo half-track formats, where the tape direction is unidirectional, and these can be used to produce reversed audio (although the channels are swapped on a half-track tape!).

Playing sounds backwards has two main audible effects:

- Most naturally occurring musical instrument sounds have a sharp attack and a slow release or decay time, and this is reversed. This produces a characteristic 'rushing' sound or 'squashed' feel, since the main rhythmic information is on the beats and these are now at the end of the notes.
- Any reverberation becomes part of the start of the sound, whilst the end of the sound is 'dry'. Echoes precede the notes which produced them. Both of these serve to reinforce the crescendo effect.

Tape loops

Splicing the end of a section of tape back onto the start produces a loop of tape, and the sound will thus play back continuously. This can be used to produce repeated phrases, patterns and rhythms. Several tape loops of different lengths played back simultaneously can produce complex polyrhythmic sequences of sounds.

Sound on sound

Normally, a tape recorder will erase any pre-existing magnetic fields on the tape before recording onto it using its erase head. By turning this off, any new audio which is recorded will be mixed with anything already there. This is called 'sound on sound' because it literally allows sounds to be layered on top of each other. As with many tape manipulation techniques, there is a loss in the quality each time that this technique is used – specifically for the pre-existing audio in this case.

Delays, echoes and reverberation

By using two tape recorders, where one records audio onto the tape, and the second plays back the same tape, it is possible to produce time delays. The time delay can be controlled by altering the tape speed (which should be the same on both tape recorders) and the physical separation of the two recorders. Some tape recorders have additional playback heads, and these can be used to provide short time delays. Dedicated machines with one record head and several playback heads have been used to produce artificial echoes, the Watkins CopyCat being one example. The use of echo and time delays has become part of the performance technique of many performers, from guitarists to synthesists.

By taking the time delayed signal and mixing it into the recorded signal, it is possible to produce multiple echoes from only one playback head (or the playback tape recorder). With multiple playback heads spaced irregularly, it is possible to use this feedback to simulate reverberation. With too much feedback, the system may break into oscillation, and this can be used as an additional method of synthesizing sounds, where a stimulus signal is used to initiate the oscillation.

Multi-tracking

Although early tape recorders had only one or two tracks of audio, experiments were carried out on producing tape recorders with more tracks. Linking two tape recorders together to give additional tracks was very awkward. Quarter-track tape recorders used four tracks, although usually only two of these could be played at once. Modified heads produced tape recorders with four separate tracks, and these 'multi-track' tape recorders were used to produce recordings where each of the tracks was recorded at a different time, with the complete performance only being heard when all four tracks were replayed simultaneously. This allowed the production of complete pieces of complex music using just one performer. Eight track recorders followed, then 16 track machines, then 24, and additional tracks can now be added by synchronizing two or more machines together to produce 48 and even 96 track tape recorders.

1.4.2 Found sounds

Found sounds are ones which are not pre-prepared. They are literally recorded as they are 'found' – in situ. Trains, cars, animals, factories and many other locations can be used as sources of found sounds. The term 'prepared sounds' is used for sounds which are specially set up, initiated and then recorded, rather than spontaneously occurring.

1.4.3 Collages

Just as with paper collages, multiple sounds can be combined to produce a composite sound. Loops can be very useful in providing a rhythmic basis, whilst found sounds, transposed sounds and reversed sounds can be used to add additional timbres and interest.

1.4.4 Musique concrète

'Musique concrète' is a French word which has come to be used as a description of music produced from ordinary sounds which are modified using the tape techniques described above. Pierre Schaeffer coined the term in 1948 as music made ' from . . . existing sonic fragments'.

1.4.5 Other methods

Although magnetic tape provides a versatile method of recording and reproducing sounds using a physical medium, it is not the only way. The optical technique used for the sound track on film projectors has also been used. The sound is produced by controlling the amount of light which passes through the film to a detector. Conventional film uses a 'slot' which varies in width, although it is also possible to vary the transparency or opacity of the film. Optical systems suffer from problems of dynamic range, and physical degradation due to scratches, dust and other foreign objects.

Chapters 2 and 4 deal with optical techniques in more depth.

Finally, disc manipulation is often overlooked because vinyl discs are often perceived as playback-only devices. It is quite possible to produce many of the effects of tape using a turntable or disc-cutting lathe. For example:

- Loops can be produced by closing a groove so that it joins back onto itself, although this requires care with positioning of the groove on the disc and the rotation speed, since these both set the length of the loop.
- Large pitch changes can be produced by using turntables with large speed ranges.
- Some pickups can be used in reverse play to reverse playback of sounds.
- Multiple pickups can be arranged on a disc so that echo effects can be produced.
- 'Scratching' involves using a turntable with a slip-mat under the disc, a bidirectional pickup cartridge, and considerable improvisational skill from the operator, who controls the playing of fragments of music from standard LP discs by playing them forwards and backwards, repeating and mixing together two (or more) discs at once.

1.5 Experimental versus popular musical uses of synthesis

There is a broad spectrum of possible applications for synthesis. At one extreme is the experimental research into the nature of sound, timbre and synthesis itself, whilst at the opposite extreme is the use of synthesizers in making popular music. In between these two, there is huge scope for using synthesis as a useful and creative tool.

1.5.1 Research

Research into music, sound and acoustics is a huge field. Ongoing research is being carried out into a wide range of topics. Just some of these include:

- alternative scalings
- alternative timbres
- processing of sounds
- rhythm, beats, timbre, scales etc.
- understanding of how instruments work.

Much of this work involves multi-disciplinary research. For example, trying to work out how instruments work can require knowledge of physics, music, acoustics, electronics, computing and more. Some of the results of this research work can find application in commercial products: Yamaha's DX series of FM synthesizers are just one of the many examples of the conversion of academic theory into practical reality. This is covered in more detail in Section 7 of this chapter.

1.5.2 Music

Music encompasses a huge variety of styles, sounds, rhythms and techniques. Some of the types of music in which a strong synthesized content may be found are:

- **Pop music**: Popular music has some marked preferences: it frequently uses 4/4 bars, and preferentially uses a strongly clichéd set of timbres and song forms – especially key changes to mark the end of a song. It often has a strong rhythmic element, which reflects one of its purposes: music to dance to.
- **New Age music**: New Age music mixes both natural and synthetic instruments into a form which concentrates on slower tempos than most popular music, and is more concerned with atmosphere.

- **Classical music**: Although much of classical music uses a standard palette of timbres which can be readily produced by an orchestra, the augmentation by synthesizers is not unknown in some genres – particularly music intended for film, TV and other media purposes.
- **Musique concrète**: Although musique concrète uses natural sounds as the source of its raw material, the techniques that it uses to modify those sounds are often the same as those used by synthesizers.
- **Electronic music**: Electronic music need not be produced by synthesizers, although this is often assumed to be the case. As with popular music, a number of clichés are commonly found: the 8 or 16 beat sequence and the resonant filter sweep are two examples.

Crossovers

There are some occasions when the boundaries between experimental uses of synthesizers cross over into more popular music areas, and vice-versa. The use of synthesizers in orchestras occurs when conventional instrumentation is not suitable, or when a specific rare instrument cannot be hired. Music which is produced for use in many areas often needs to have elements of orchestral and non-orchestral instrumentation – adding synthesizer parts can enhance and extend the timbres available to the composer or arranger, and it avoids any need for the synthesizer to attempt to emulate a real orchestra. Conversely, the use of orchestral instruments in experimental works also happens.

1.6 Electro-acoustic music

The study of the conversions between electrical energy and acoustic energy is called electro-acoustics. Unlike previous centuries, where the development of mechanically-based musical instruments has dominated the study of musical acoustics, in the twentieth century, innovation has been concentrated on instruments that are electronic in nature. It is thus logical that the term electro-acoustics should also be used by musicians to describe music which is made using electronic musical instruments and other electronic techniques.

Unfortunately, the term 'electro-acoustic music' is not always used consistently, and it can also apply to music where acoustic instruments are amplified electronically. The term 'electronic music' implies a completely electronic method of generating the sound, and so represents a very different way of making music.

In practice, both terms are now widely used to mean music which utilizes electronics as an integral part of the creative process, and so it covers such diverse areas as amplified acoustic instruments (where the instruments are not merely made louder), music created by synthesizers and computers, and popular music from a wide range of genres (pop, dance, techno, etc.). Even classical music performed by an orchestra, but with additional electronic instrumentation, or even post-processing of the recorded orchestral sound, could be considered to be 'electronic'.

1.6.1 Electro-acoustics

Electro-acoustics is science tempered by human interaction and art. In fact, the close linking between the human being and most musical instruments, as well as the space in which they operate, can be a very emotional one. The electronic nature of many synthesizers does not fundamentally alter this relationship between human and instrument, although the details of the interface are still very clumsy. As synthesizers develop, they should gradually become performer-oriented, and less technological, which should make their electro-acoustic nature less and less important. Many conventional instruments have histories of many hundreds of years, whereas electro-acoustic music is less than a century old, and synthesizers are less than 50 years old. Electro-acoustics is comparatively new.

1.7 From academic research to commercial production. . .

Synthesizers can be thought of as coming in two forms: academic and commercial. Academic research produces prototypes which are typically innovative, fragile and relatively extravagant in their use of resources. Commercial synthesizers are often cynically viewed as being almost the exact opposite: minor variations on existing technology; robust, perhaps even over-engineered; and very careful to maximize the use of available resources. Production development of research prototypes is often required to enable successful exploitation in the marketplace – although this is a difficult and exacting process, and there have been both successes and failures.

In order to be a success, there are a number of criteria that need to be met. Moving from a prototype to a product can require a complex exchange of information from the inventor to the manufacturer, and may often need additional development work. Custom chips or software may need to be produced, and this can introduce long delays into the timescales, as well as a difficult testing requirement. Management tasks like organizing contracts, temporary secondment of personnel, patents and licensing issues, all need to be monitored and controlled.

Even when the product has been produced, it needs to be promoted and marketed. This requires a different set of skills, and in fact, many successful companies split their operations into 'research and development' and 'sales and marketing' parts. The synthesizer business has seen many companies with ability in one of these fields, but the failures have often been a result of a weakness in the other field. Success depends on talent in both areas, and the interchange of information between them. Very few of the companies who started out in the 1960s are still active – however, the creative driving force behind these companies, which is frequently only one person, is often still working in the field, albeit in a different company.

Apart from the development issues, the other main difference between academic research and commercial synthesizer products is the motivation behind them. Academic research is aimed at exploring and expanding of knowledge, whilst commercial manufacturers are more concerned with selling products. Unfortunately, this often means that products need to have wide appeal, simple user interfaces, and easy application in the popular music industry. The main end market for electronic musical instruments is where they are used to make the music that is heard on TV, radio, films and CDs, and the development process is aimed at just this area.

What follows are some brief notes on some of these 'developed' products.

1.7.1 Analogue modular

Analogue synthesizers were initially modular, and were probably aimed at academic and educational users. The market for 'popular' music users literally did not exist at the time. The design and approach used by early modular synthesizers was similar to those of analogue computers. Analogue computers were used in academic, military and commercial research institutions for much the same types of calculations that are now carried out by digital computers.

Pioneering work by electronic music composers using early modular synthesizers was more or less ignored by the media until Walter Carlos released some recordings of classical Bach produced using a Moog modular synthesizer. The subsequent release of this material as the 'Switched on Bach' album quickly became a major success with the public, and the album became one of the best selling classical music records ever. This success led to enquiries from the popular music business, with the Beatles and the Rolling Stones being early purchasers of Moog modular synthesizers.

1.7.2 FM

Frequency modulation as a means of producing audio sounds was first comprehensively described by John Chowning, in a paper entitled: 'The Synthesis of Complex Audio Spectra by Means of Frequency Modulation', published in 1973. At the time, the only way that this type of FM could be realized was by using digital computers, which were expensive and not widely available to the general public.

As digital technology advanced, some synthesizer manufacturers began to look into ways of producing sounds digitally, and Yamaha bought the rights to use Chowning's 1977-patented FM ideas. Early prototypes used large numbers of simple TTL logic chips, but these were quickly replaced by custom-designed chips which compressed these onto just a few more complex chips. The first functional all-digital FM synthesizer designed for consumer use was the Yamaha GS1, which was a pathfinder product designed to show expertise and competence, as well as to test the market. Simple preset machines designed for the home market followed. Although the implementation of FM was very simple, the response from musicians and players was very favourable.

The DX1, DX7 and DX9 were released in late 1982, with the DX1 intended as the professional player's instrument, the DX7 a mid-range, cut-down DX1, and the DX9 as the low-cost, large-volume 'best seller'. What actually happened is very interesting. The DX9 was so restricted in terms of functionality and sound that it did not sell at all well, whilst the DX7 was hugely in demand, and the DX1 was interpreted as being a 'super' DX7 for a huge increase in price. Inevitably, it took Yamaha some time to increase the production of the DX7 to meet the demand, and this scarcity only served to make it all the more sought-after! By the time that the mark II DX7 was released, about a quarter of a million DX7s had been sold, which at the time was a record for a synthesizer.

The popularity of the DX7 was responsible for the mark II instrument, which was a major redesign, not a new instrument – a very rare approach, and one which shows how important the DX7, and FM had become. For several years, between 1983 and 1986, Yamaha and FM enjoyed a popularity which ushered in the transition from analogue to digital technology. It also began the trend away from user programming, and towards the selling of pre-prepared sounds or patches. The complexity of programming FM meant that many users did not want to learn, and so purchased sounds from specialist companies which marketed the results of a small number of 'expert' FM programmers.

1.7.3 Sampling

Sampling is a musical re-use of technology which was originally developed for telephony applications. The principles behind the technique were worked out early in the twentieth century, but it was not until the invention of the transistor in the 1950s that it became practical to convert continuous audio signals into discrete digital samples using pulse code modulation (PCM).

Commercial exploitation of sampling began with the Fairlight CMI (computer musical instrument) in 1979. Although this began as a wavetable synthesizer, as the size of the wavetables increased, it rapidly evolved into an expensive and fashionable professional sampling instrument, initially with only 8-bit sample resolution. Another 8-bit instrument, the Ensoniq Mirage, was the first instrument to make sampling affordable. E-mu released the Emulator in 1979, drum machines like the LinnDrum were released in 1979, and sampling even began to appear on low-cost 'fun' keyboards designed for consumer use in the home, during the mid 1980s.

Eight-bit resolution was replaced by 12 bits in the late 1980s, and 16 bits became widely adopted in the early 1990s. CD-ROMs of pre-prepared samples have replaced DIY sampling for the vast majority of users. Samplers have effectively become replay-only devices, with only a few creative individuals and companies producing samples – and lots of musicians using them.

1.7.4 Physical modelling

Physical modelling seems to have made the transition from research to product in a number of parallel paths. There have been several speech coding schemes based on modelling the way that the human voice works, but these have been restricted to mainly telecommunications and military applications – only a few of these have found musical uses (see Chapter 5). Research results which have been reporting the gradual refinement of modelling techniques for musical instruments have been released, notably by Julius Smith.

The development of electronic musical instruments is still continuing. The role of research is as strong as ever, although the pace of development is accelerating. Digital technology is driving synthesis towards general purpose computing engines with customized audio output chips, and this means that the software is increasingly responsible for the operation and facilities that are offered – not the hardware. By the middle of 1995, several companies had products which used general purpose DSPs to synthesize their sounds, and which used mixtures of synthesis technologies to produce those sounds: FM, additive, emulations of analogue synthesis, physical modelling, and more.

1.8 Synthesized classics

One of the major forces which popularized the use of synthesizers in popular music was the use of synthesizers to produce recorded performances of classical music. Because these could be assembled onto tape with great precision, the timing control and pitch accuracy which were used were on a par with the best of human realizations, and so the results could be described as 'virtuoso' performances. In the 1950s, this suited the mood of the time, and so a large number of electronically produced versions of popular classical music were produced. This has continued to the present day. Some of the large number of recordings are listed in table 1.8.1.

There is no 'correct' way to use synthesizers to create music, although there are a number of distinct 'styles'. Individual synthesists have their own preferences, although some have less fixed boundaries than others, and can move from one style to another within a piece of music. I do not know of any formal means of categorizing such styles, and so propose the following divisions:

- **Imitative**: Imitative synthesis attempts to use electronic means to realize a performance which is as close as possible to a recording of a conventional orchestra, band or group of musicians. The timbres and control techniques which are used are intended to mimic the real-world sounds and limitations of the instrumentation. Many film sound-tracks fall into this category.
- **Suggestive**: This style does not necessarily use imitative instrumental sounds, but rather, aims to produce an overall end result which is still suggestive of a conventional performance.

Table 1.8.1 Examples of synthesized classics

Artist(s)	*Title*	*Record company*	*Date*	*Catalogue number*
Walter Carlos	Switched on Bach	CBS	1973	MK 7194
Wendy Carlos	Switched on Bach 2000	Telarc Records	1992	CD 80323
Suzanne Ciani	Seven Waves	Private Music	1988	2046-2-P
Julian Colbeck and Jonathan Cohen	Back to Bach	Editions EG Records	1992	EEG 2104-2
Electrophon (Brian Hodgson and Dudley Simpson)	In a Covent Garden	Polydor Records	1973	2383 210
Metlay Team	Band of Fire	Atomic City	1991	ATOM CD-07
The Walter Murphy Band	A Fifth of Beethoven	Private Stock Records	1976	PVLP 1009
Karlheinz Stockhausen	Kontakte	Wergo Records	1960	WER 60009
Isao Tomita	Snowflakes are Dancing	RCA	1974	ARL1-0488
Kit Watkins	A Different View	IC Records	1991	IC 720.142
Charles Wuorinen	Time's Encomium	Nonesuch Records	1969	H-71225

- **Sympathetic**: Although using instrumental sounds and timbres which may be well removed from those used in a normal performance, a 'sympathetic' realization of a piece of music aims to choose sounds which are in keeping with some element of the conventional performance.
- **Synthetic**: Electronic music aims to free the performer from the constraints of conventional instrumentation, and so this category includes music where there is little that would be familiar in tone to a casual listener.

1.8.1 Recordings

There are a large number of recordings of popular classical music available which have been performed using electronic music equipment. These vary widely in style, sound quality, competence and 'collectability'. The examples listed in Table 1.8.1 from the author's collection have been chosen for their breadth of subjects and treatments.

1.9 Synthesis in context

Sound synthesis does not exist in isolation. It is but one method of producing sound and music. There are a large number of non-synthetic, non-electronic methods of producing musical sounds.

All musical instruments synthesize sounds, although most people would probably use the word 'make' rather than 'synthesize' in this context. In fact, the word synthesize has come to mean something which is un-natural – synthetic implies something which is similar to, but inferior to 'the real thing'. The ultimate example of this view is the sound synthesizer, which is often described as being capable of emulating any type of sound – but with the proviso that the emulation is usually not perfect.

1.9.1 'Synthetic' versus 'real'

As with anything new and/or different, there is a certain amount of prejudice against the use of electronic musical instruments in some contexts. This is often expressed in words like: 'What is wrong with *real* instruments?' and is frequently used to advocate the use of orchestral instruments rather than an electronic realization using synthesizers.

There are two major elements to this prejudice: unfamiliarity and fear of technology. Many people are very used to the sounds and timbres of conventional instrumentation, especially the orchestra. In contrast, the wider palette of synthetic sound is probably very unfamiliar to many casual listeners. So an unsympathetic rendering of a pseudo-classical piece of music, produced using sounds which are harsh, unsubtle, and obviously synthetic in origin, is almost certain to elicit an unfavourable response in many listeners. In contrast, careful use of synthesis can result in musical performances which are acceptable even to a critical listener – especially if no clues are given as to the synthetic origins.

The technological aspect is more complex. Although the pianoforté was once considered too new and innovative to be considered for serious musical uses, it has now become accepted by familiarity. The concept of assembling together a large number of musicians into an orchestra was a more gradual process, but the same transition from 'new' to 'accepted' still occurred. It seems that there may be an inbuilt 'fear' of anything new or which is not understood. This extends far beyond the synthesizer: computers and most other technological inventions can suffer from the same aversion.

Attempting to draw the line between what is acceptable technologically and what is not can be very difficult. It also changes with time. Arguably, the only 'natural' musical instrument is the human voice, and anything which produces sounds by any other method is inherently 'synthetic'. This includes all musical instruments from simple tubes through to complex computer-based synthesizers. There seems to be a gradual acceptance of technological innovation over time, which results in the current wide acceptability of musical instruments which may well have been invented more than 500 years ago.

1.9.2 Film scoring

Film scoring is an excellent example of the way that sound synthesis has become integrated into conventional music production. The soundtracks of many films are a complex mixture of conventional orchestration combined with synthesis, but a large number of films have soundtracks which have been produced entirely electronically, with no actual orchestral content at all, although the result sounds like a performance by real performers. In some cases, although the music may sound 'realistic' to a casual listener, the performance techniques may well be beyond the capability of human performers, and some of the timbres used can be outside of the repertoire of an orchestra.

There are advantages and disadvantages to the 'all synthetic' approach. The performer who creates the music synthetically has complete control over all aspects of the final music, which means that changes to the score can be made very rapidly, and this flexibility suits the restraints and demands which can result from film production schedules. But giving the music a human 'feel' can be more difficult, and arguably more time consuming, than asking an orchestra to interpret the music in a slightly different way. The electronic equivalent of the conductor is still some way in the future, although there is considerable academic research into this aspect of controlling music synthesizers.

Mixing conventional instrumentation with synthesizers is also used for a great deal of recorded music. This has the advantage that the orchestral instruments can be used to provide a basic sound, and additional timbres can be added into this to add atmosphere or evoke a specific feel. Many of the sounds used in this context are clichés of the particular time when the music was recorded. For example, soundtracks from the late 1970s often have a characteristic 'drum synthesizer' sound with a marked pitch sweep downwards – the height of fashion at the time, but quaint and hackneyed to modern ears.

1.9.3 Sound effects

The real world is noisy, but the noises are often unwanted. For film and television work, background noise, wind noise and other extraneous sounds often mean that it is impossible to record the actual sound whilst recording the pictures, and so sound effects need to be added later. Everyday sounds like doors opening; shoes crunching on gravel paths; switches being turned on or off; cans of carbonated drink being opened; and more are often required. Producing these sounds can be a complex and difficult process, especially since many sounds are very difficult to produce convincingly.

Years of exposure to film and television have produced a set of clichéd sounds which are often very different from reality. For example, does a real computer produce the typical busy whirring and bleeping sounds that are often used for anything in the context of computers or electronics? Sliding doors on spaceships always seem to open and close with a whoosh of air, which would seem to suggest a serious design fault. The guns in Western films suffer from a large number of ricochets – and fight scenes often contain the noises of large numbers of bones being broken, whilst the combatants seem relatively uninjured.

Many of the sounds that you hear on film or television are dubbed on afterwards. Many are 'synthesized' live using props, although often the prop is not what you might expect: rain can be emulated by dropping rice onto a piece of cardboard, for example. But many sounds are produced synthetically, especially when the real-world sound does not match expectations. One example is the noise made when a piece of electronic equipment fails catastrophically: often nothing is heard apart from a slight clunk or hum – which is completely unsuitable for dramatic use. A loud and spectacular sound is needed to accompany the unrealistic shower of sparks and smoke which billow from the equipment.

This use of sounds to enhance the real world can also be used to extremes, especially in comedy. A commonly used set of 'cod' or comic sounds has become as much a part of the film or television medium as the 'fade to black'. Laboratory equipment blips and bloops, and elastic bands twang in an unrealistic but amusing way – the exaggeration is the key to making the sound effect funny. Many of these sounds are produced using synthesizers, or a combination of prop and subsequent processing in a synthesizer.

1.10 Electronics and acoustics – fundamental principles

A knowledge of the electronic aspects of acoustics can be very useful when working with synthesizers, because synthesizers are just one of many tools which can be used to assist in the creation of music. So this section provides some background information on acoustics and electronics. Because some of the terminology used in this section uses scientific unit symbols, some additional information on the use of units is also provided.

1.10.1 Acoustics

Acoustics is the science of sound. Sound is concerned with what happens when something vibrates. The vibration can be produced by vocal cords vibrating, wind whistling through a hole, a guitar string being plucked, a gong being struck, a loudspeaker being driven back and forth by an amplified signal, and more. Although most people think of sound as being carried only through the air, sound can also be transmitted through water, metals, wood, plastics and many other materials.

Whilst it is often easy to observe an object vibrating, sound waves are less tangible. The vibrations pass through the air, but the actual process of pressure changes is hard to visualize. One effective analogy is the stretched spring which is vibrated at one end – the actual compression and rarefactions (the opposite of compression) of the 'waves' can then be seen travelling along the spring. Trying to then amalgamate this idea of pressure waves on springs with the ripples spreading out on a pond is more difficult. As with light, the idea of spreading out from a source is hard to reconcile with what happens – people see and hear things, and waves and beams seem like very abstract notions. In real life, the only way to interact with sound waves is with your ears, or for very low frequencies, your body.

When an object vibrates, it moves between two limits – a vibrating string provides a good example, where the eye tends to see the limits (where the string is momentarily stopped whilst it changes direction) rather than where it is moving. This movement is coupled to the air (or another transmission medium) as pressure changes. The rate at which these pressure changes happen is called the frequency. The number of cycles of pressure change which happen in one second is measured in a unit called hertz (cycles per second is an alternative unit). The

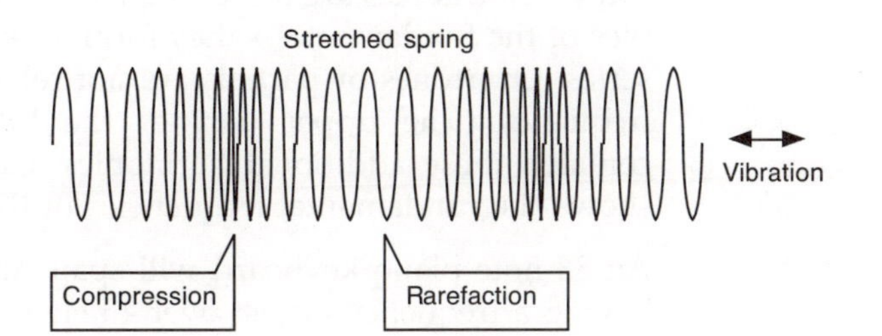

Figure 1.10.1 Pressure changes in the air can be thought of as being similar to a stretched spring which is vibrated at one end. The resulting pressure 'waves' can be seen travelling along the spring.

time for one complete cycle of pressure change is called the period, and is measured in seconds.

Pitch and frequency

Frequency is also related to musical pitch. In many cases, the two are synonymous, but there are some circumstances in which they are different. In this book frequency will be used in a technical context, whilst pitch will be used when the subject is musical in nature.

The frequency usually used for the note A just above middle C (so called because it is written in the middle: between the bass and treble staves) is 440 Hz. There are local variations on this 'A-440 standard', but most electronic musical equipment can be tuned to compensate. Human hearing starts at about 20 Hz, although this depends on the loudness and listening conditions. Frequencies lower than this are called sub-sonic or, more rarely, infrasonic. For high frequencies human hearing varies with age and other physiological effects (like damage caused by over-exposure to very loud sounds, or ear infections). For a normal teenager, frequencies of up to 18 kHz (18 000 Hz) can be heard – the 15 625 Hz line whistle from a 625-line PAL TV is one useful indicator. The aging process means that an average middle-aged person will probably only be able to hear frequencies of perhaps 12 or 13 kHz. The 'hi-fi' range of 20 Hz to 20 kHz is thus well in excess of most listeners' ability to hear, although there is some debate about the ability of the ear to discern higher frequencies in the presence of other sounds (most hearing tests are made with isolated tones in quiet conditions).

Notes

The fundamentals of most musical notes are in the lower part of this range. The fundamental is the name given to the lowest major frequency which is present in a sound. The fundamental is the pitch which most people would whistle when attempting to reproduce a given note. Harmonics, overtones or partials are the names for any additional frequencies which are present in a sound. Harmonics are those frequencies which are integer multiples of the fundamental – they form a series called the harmonic series. Overtones or partials are not related to the fundamental frequency. The upper part of the human frequency range contains these additional harmonics and partials. Table 1.10.1 shows the fundamental frequencies of the musical notes.

An 88-note piano keyboard will span A0 to C8, with the top C having a frequency of just over 4 kHz.

Table 1.10.1 Note frequencies in Hz

C	C#	D	D#	E	F	F#	G	G#	A	A#	B	Octave
16.35	17.32	18.35	19.45	20.60	21.83	23.12	24.450	25.96	27.50	29.14	30.87	0
32.70	34.65	36.71	38.89	41.20	43.65	46.25	49.00	51.91	55.00	58.27	61.74	1
65.41	69.30	73.42	77.78	82.41	87.31	92.50	98.00	103.83	110.00	116.54	123.47	2
130.81	138.59	146.83	155.56	164.81	174.61	185.00	196.00	207.65	220.00	233.08	246.94	3
261.63	277.18	293.66	311.13	329.63	349.23	369.99	392.00	415.30	440.00	466.16	493.88	4
523.25	554.37	587.33	622.25	659.26	698.46	739.99	783.99	830.61	880.00	932.33	987.77	5
1046.50	1108.73	1174.66	1244.51	1318.51	1396.91	1479.98	1567.98	1661.22	1760.00	1864.66	1975.53	6
2093.00	2217.46	2349.32	2489.02	2637.02	2793.83	2959.96	3135.96	3322.44	3520.00	3729.31	3951.07	7
4186.01	4434.92	4698.64	4978.03	5274.04	5587.65	5919.91	6721.93	6644.88	7040.00	7458.62	7902.13	8
8372.02	8869.84	9397.27	9956.06	10548.08	11175.30	11839.82	12543.85	13289.75	14080.00	14917.24	15804.27	9
16744.04	17739.69	18794.55	19912.13	21096.16	22350.16	23679.64	25087.71	26579.50	28160.00	29834.48	31608.53	10
C	C#	D	D#	E	F	F#	G	G#	A	A#	B	Octave

Musical pitch is divided into octaves, and each octave represents a doubling of frequency. So A4 has a frequency of 440 Hz, whilst A5 has twice this frequency, 880 Hz. A3 has half the frequency, 220 Hz. Octaves are normally split into 12 parts, and the intervals are called semitones. The relationship between the individual semitones in an octave is called the scale. Table 1.10.1 shows the equal tempered scale, where the intervals between the semitones are all the same: many other scalings are possible. Since there are 12 tones, and the frequency doubles in an octave interval, then semitone intervals in an equal tempered scale are each related by the twelfth root of two, which is approximately 1.059463. The study of musical scales is a complex subject. For further information see Pierce, 1992. Semitones are split up into 100 cents, but most human beings can only detect changes in pitch of 5 cents or more. Cent intervals are related by the twelve-hundredth root of 2, which is approximately 1.00057779. As an example of what this represents in terms of frequency, for a note of 880 Hz, a cent is just under 0.51 Hz, and so 5 cents represents only 2.5 Hz!

Phase

When an object is vibrating, it repeatedly passes through the same position as it moves. A complete movement back and forth is called a cycle or an oscillation, which is why anything that produces a continuous vibration is called an oscillator. The particular point in a cycle at any instant is called the phase: the cycle is divided up into 360°, rather like a circle in geometry. Phase is thus measured in degrees, and zero is normally associated with either the start of the cycle, or where it crosses the resting position. The word 'zero crossing' is used to indicate when the position of the object passes through the rest position. A complete cycle conventionally starts at a zero-crossing,

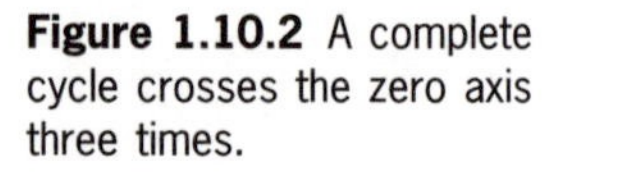

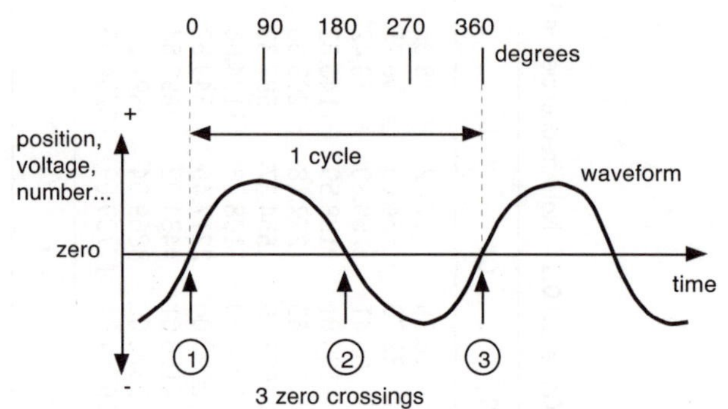

Figure 1.10.2 A complete cycle crosses the zero axis three times.

passes through a second one , and then ends at the third zero-crossing.

The change of position with time as an object vibrates is called the waveform. A simple oscillation will produce a sine wave, which looks like a smooth curve. More complex vibrations will produce more complex waveforms – a guitar string has a complex waveform because it produces a number of harmonics at once. If two identical waveforms are mixed together, the phase can determine what happens to the resulting waveform. If the two are in phase, i.e. they both have the same position in the cycle at the same instant, then they will be added together – this is called 'constructive interference', because the two waveforms add together as they 'interfere' with each other. Conversely, if the two waveforms are 180° out of phase, then the phases will be equal and opposite, and the two waveforms will tend to cancel each other out – this is called 'destructive interference'.

Beats

Slight differences of frequency between two waveforms can produce a different effect. Assuming that the two waveforms start at the same zero crossing, and with the same phase, then the waveform with the higher frequency will gradually move ahead of the slower waveform, and its phase will be ahead. This means that from an initial state of constructive interference, the waveforms will pass through destructive interference and then back to constructive interference repeatedly. The rate of passing through these adding and cancellation stages is determined by the difference in frequency. For a difference of one tenth of a hertz, it will take ten seconds for the cycle of constructive, destructive and constructive interference to occur. This cyclic variation in level of the mixed waveforms is called 'beating', and sounds like one sound which 'wobbles' in level. This beating is often used in analogue synthesizers to provide a 'lively' or 'interesting' sound.

If the difference in frequency between the two waveforms is increased, then the speed of the beats will increase. When the frequency of the beats is above 20 Hz, then the mixed sound begins to sound like two separate frequencies. As the difference increases, the two frequencies will pass through a series of ratios of frequency which sound pleasant, and others which sound unpleasant. The ratios between the two frequencies is called an interval – the easiest and most 'pleasant' interval is a ratio of 2:1 – an octave.

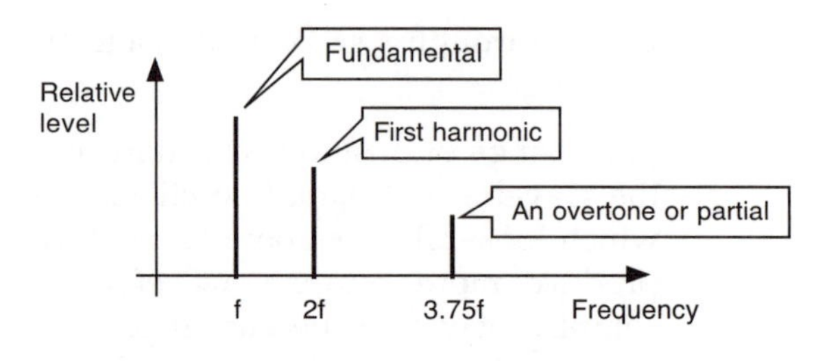

Figure 1.10.3 Timbre is set by the frequency content of a sound. In this example, the fundamental frequency of the sound is at frequency f, whilst there is a harmonic at twice this frequency, 2f. There is also an overtone or partial frequency at 3.75f.

Timbre

Timbre is a description of the contents of a sound. The timbre of a sound is determined by the harmonic content: the relationship between the level of the fundamental, the levels of the harmonics or overtones and their evolution in time (their envelopes – see below). Pure sounds tend to have only a few harmonics at low levels, whilst bright sounds tend to have many harmonics at high levels. Missing harmonics can also be important, and can produce 'hollow' sounding timbres. If the ratios of the frequencies between the fundamental and the other frequencies are not integers, then the timbre can sound bell-like or even like noise. The ability of the human ear to perceive timbre is related to the frequency. At low frequencies, the ear can detect phase differences and can follow changes in a large number of harmonics. As the frequency increases, then the phase discrimination ability of the ear diminishes above A-440, and the number of harmonics which can be heard decreases because of the response of the ear.

For example, a sound which has a fundamental of 100 Hz has harmonics at 100 Hertz intervals, and so the 150th harmonic is at 15 kHz. But a sound with a fundamental of 1 kHz has a 15th harmonic at 15 kHz. The number of audible harmonics are thus restricted as the fundamental frequency rises. Synthesizers provide comprehensive control over the frequency, phase and level of harmonics, and so give the user control of the timbre.

Timbre is derived from a French word. 'Tone colour' and 'tonal quality' are commonly used as synonyms for timbre.

Loudness

When a string vibrates, the size of the string and the amount of movement determine how much energy is transferred to the surrounding medium (usually air). The larger the amount of energy that is turned into changes in air pressure, the louder the sound will be. This can be demonstrated by using a tuning fork: it becomes much louder when it is placed on a table top, because

Table 1.10.2 Decibels

Sound pressure level (dB)	*Sound pressure (microbars)*	*Power (watts per square metre)*	*Power (watts per square metre)*	*Equivalent*	*Musical dynamic*
130	632	10 W	10	Threshold of pain	
120	200	1 W	1	Aircraft taking off	
110	63	100 mW	0.1	Loud amplified music	
100	20	10 mW	0.01	Circular saw	fff
90	6	1 mW	0.001	Train	ff
80	2	100 μW	0.0001	Motorway	f
70	0.6	10 μW	0.00001	Factory workshop	mf/mp
60	0.2	1 μW	0.000001	Street noise	p
50	0.06	100 nW	0.0000001	Noisy office	pp
40	0.02	10 nW	0.00000001	Conversation	ppp
30	0.06	1 nW	0.000000001	Quiet room	
20	0.002	100 pW	0.0000000001	Library	
10	0.006	10 pW	0.00000000001	Leaves rustling	
0	0.0002	1 pW	0.000000000001	Threshold of hearing	

then it is moving a much larger amount of air. The amount of movement of a vibrating object is called the amplitude of the vibration, whilst the amount of energy in the sound which is produced by the vibrating object is called the power or intensity of the sound. Power is measured in watts, but a relative logarithmic scale is commonly used to avoid large changes in units: dB or decibels. Named after Alexander Graham Bell, the pioneer of telephony, decibels are used to indicate the relative difference between sound intensities or sound pressure levels. The perception of sound power or level by humans is subjective: a change in sound power of 1 dB is 'just audible', whilst for something to sound 'twice as loud', the change is about 10 dB.

Musicians use an alternative relative measure for sound level. The 'dynamics marks' used on musical scores provide guidance about the loudness of a specific note. These range from ppp (pianississimo = very softly) to fff (fortississimo = very loudly), although this tends to be a subjective measure, and is also dependent on the instrument producing the sound. On average, the range covered by dynamics marks is about 50 or 60 dB, which represents a ratio of about a million to one in sound intensity.

'Loudness' is a specific term which means the subjective intensity of a sound, as opposed to intensity, which can be objectively measured by a sound intensity meter. The human hearing response to different frequencies is not flat: sounds between 3 and 5 kHz will sound louder than lower or higher pitched

Table 1.10.3 Dynamics

Musical dynamic	*Name*	*Description*	*dB (approx)*
fff	Fortississimo	Loudest	100
ff	Fortissimo	Very loud	93
f	Forte	Loudly	85
mf	Mezzo-forte	Moderately loud	78
mv	Mezza-voce	Medium tone	70
mp	Mezzo-piano	Moderately soft	62
p	Piano	Softly	55
pp	Pianissimo	Very softly	47
ppp	Pianississimo	Softest	40

sounds, and a graph can be plotted showing this response – called an equal loudness contour. This topic is covered by the science of psycho-acoustics, which is the study of the inter-relationship between sound and its perception. Loudness is commonly used (incorrectly from a technical viewpoint) as a synonym for sound intensity.

Since sound is just pressure waves moving through a transmission medium like air, it can be measured in terms of the pressure changes which are caused. The unit for such pressure changes is the bar, although in common with many scientific units, smaller sub-divisions like millibars or microbars are more likely to be encountered in normal acoustics measurements. Since sound loudness is dependent on the response of the ear, it is measured in phons, where the phon is based on a subjective measure of the apparent loudness of sounds at different frequencies and intensities.

Envelopes

Sounds do not start and stop instantaneously. It takes a finite time for a string to start vibrating, and time for it to reduce to a stationary state. The time from when an object is initiated into a vibrating state is called the attack time, whilst the time for the vibration to decay to a stationary state again is called the decay time. For instruments that can produce a continuous sound, like an organ, the decay time is defined as the time for the sound to decay to the steady-state 'sustain' level, whilst the end of the vibration is called the release time.

Some instruments have long attack, decay and release times: bowed stringed instruments for example. Plucked stringed instruments have shorter attack times. Some instruments have very fast attack times: pianos, percussion. Very short times are

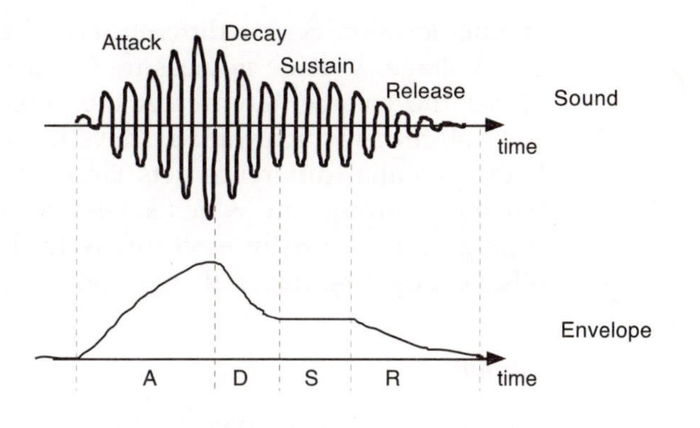

Figure 1.10.4 An envelope is the change in volume with time.

often called transients. The combination of all the stages of a sound is called an envelope. It shows the change in volume of the sound plotted against time.

The word envelope can also be used in a more generic sense: it then refers to any complex time function. A typical example might be the envelope of one harmonic within a sound, which you would find in an additive synthesizer (see Chapter 2).

Gain and attenuation

The amplitude of a sound is a measurement of the extremes of its waveform: the most positive and most negative voltages. If the amplitude changes, then the ratio between the original and the changed amplitudes is called the gain. Gains can be positive or negative, and can refer to amplitude or power – and are usually measured in dB. Gains of less than one are called attenuation, so large attenuations mean that the audio signal can become very small, whilst large gains mean that the signal can become very large.

1.10.2 Electronics

Electronics is concerned with the study and design of devices which use electricity. Specifically it is concerned with the movement of electrons – tiny particles which carry a minute electrical charge and so produce electric currents.

Voltage

Electrons flow through a conducting medium if there is a difference in the distribution of electrons – which means that there is an excess of electrons in one location, and too few electrons in another location. Such a difference is called a potential difference, or a voltage. Voltage is measured in a unit called the volt. The higher the voltage, the greater the potential difference, and the more electrons which want to move from one location to another. If the potential difference gets large enough, then the electrons will jump through air, which is what a spark is: electrons flowing through air. Normally electrons only flow through metals and other conducting materials in a more controlled manner.

Current

Current is the name given for the flow of electrons. Using water as an analogy, the current is the flow of water, whilst the potential difference is the height of the water tower above the tap. The higher the water tower, the greater the pressure, and the larger the flow when the tap is opened. To put things into some sort of perspective: a current of 1 ampere represents the movement of about 6 thousand million million electrons per second.

Resistors are materials which impede the progress of electrons. Most metals will allow electrons to pass with almost no resistance, although very few materials present no resistance to the flow of electrons. Materials which allow electrons to pass through with no resistance are called super-conductors: conductors because they 'conduct' electrons along, and super because they have no resistance to the flow of electrons. The word 'resistance' is actually used in electronics, but with a refined meaning: the resistance of a material is a measure of how hard it is for electrons to flow through it. Materials which do not allow electrons to flow are called insulators, whilst materials which do allow the flow of electrons are called conductors.

Resistors

Electronic components are made which have specific resistances, and these are called resistors. Resistance is measured in ohms, and is the voltage divided by the current. If a current of 1 amp (ampere is normally shortened to amp in common usage amongst electronics engineers) is flowing through a resistor, and there is a voltage of 1 volt across the resistor, then the resistance is 1 ohm.

$$R = V/I$$

where R = the resistance, V = the voltage, and I = the current.

Resistors can range in volume from very low resistances (fractions of ohms) for short lengths of metal wire, through to very high resistances (millions of ohms) for some materials

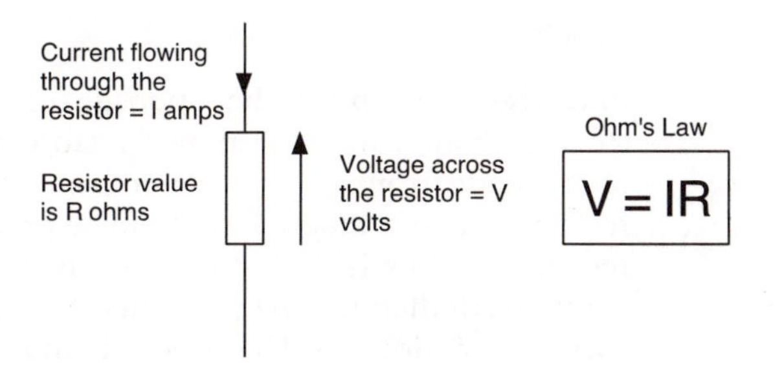

Figure 1.10.5 Another way of looking at the relationships between voltage, current and resistance is by considering the voltage across a resistor. If a current *I* is flowing through a resistance of *R* ohms, then the voltage which will be present across the resistor will be *V* volts, where $V = IR$.

which are on the borders of being insulators. For very high resistances, an alternative measurement is used: conductance. Conductance is the reciprocal of resistance, and it uses units called mhos, which shows that electronics engineers have a sense of humour!

When the current flows through the resistor, it produces heat. The amount of heat is determined by the product of the voltage across the resistor and the current. This is called the power which is given off by the resistor, and it is measured in watts.

$$\text{Power} = V \times I$$

If 1 amp flows through a resistor with a resistance of 1 ohm, then the voltage across that resistor will be 1 volt, and 1 watt of power will be dissipated as heat by the resistor. The small resistors that are found in most domestic electronic equipment, like radios and hi-fi, will be ¼ watt or ⅛ watt, and will be just under a centimetre long and a couple of millimetres across. A 'typical' value would be ten thousand ohms.

Capacitors

Having said that electrons carry a charge, and that the flow of charge is called a current, what happens if no current flows? Charge can be stored by having a device which stores electrons, and this is called a capacitor – since it has a 'capacity' for holding charge. You 'charge up' a capacitor by applying a voltage to it. Once it has stored a charge you can remove the voltage and the charge will stay in the capacitor (although it will gradually decay away in time). The size of a capacitor is measured in farads (named after Michael Faraday, a major pioneer in early electricity and magnetism experiments) and has the symbol F. This is a very large unit, so large that 1 F capacitors are very rare. Capacitors are normally measured in smaller sub-units of farads: μF, nF, or pF being the most common units. Large capacitors are often quoted in tens of thousands of μF, which represents a few hundredths of a farad.

Inductors

Inductors are almost the opposite of capacitors – instead of storing charge, they temporarily store current. An inductor is often made from a coil of wire, and the action of a current flowing causes a magnetic field to be produced. The energy from the current flow is thus stored as a magnetic field. If the current is removed, then the magnetic field will collapse, and produce a current as it does so. The energy is thus converted from current to magnetic field and back again. The 'size' of an inductor is measured in henrys, and again, this is such a large unit that hundredths and thousandths of henrys are much more likely to be found in common use.

Transistors

The transistor is a device which uses special materials called semi-conductors. Silicon and germanium are two examples. A semiconductor is a material whose resistance is normally very high, but to which the addition of tiny amounts of other elements can alter the resistance. By controlling exactly how these other elements are placed in the semi-conductor material it is possible to produce devices which can control the flow of currents. A transistor is one such device. It has three terminals: current flows between two of these only when the third has a small current flow too. The control current is much smaller than the main current flow, and so the device can be used as an amplifier. If the control current turns on and off, then the main current turns on and off too, so the transistor can also be used as a switching device. Transistors are the basis of almost all electronics.

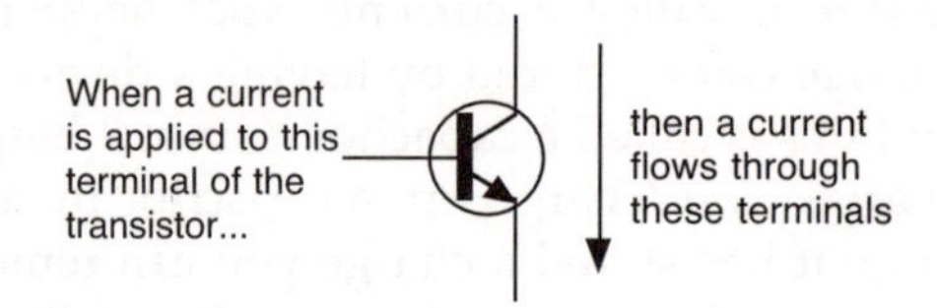

Figure 1.10.6 A transistor uses one current to control another. It can be used as an amplifer, a voltage-to-current converter, or a switch.

Transistors which use current as the control are called bipolar or junction transistors, whilst there are other types which use electric fields to control the main current, and these are called field effect transistors, abbreviated to FETs. FETs use very small currents indeed, and are widely used in electronics – particularly in making chips (see 'integrated circuits', below).

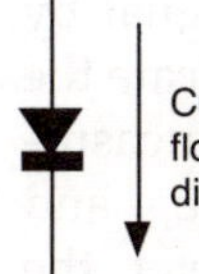

Figure 1.10.7 Current only flows through a diode in one direction.

Diodes

Diodes are simple semiconducting devices which allow a current to flow only one way. Inside there is a barrier which prevents the flow of electrons in one direction, but which breaks down and lets the electrons flow past in the other direction. When the current flows, the diode then behaves like a low-value resistor, and so some heat is produced. If the barrier is made from a special material, then the effective resistance is higher, but instead of heat, light is produced, and these are called light-emitting diodes, or LEDs.

The functions of both diodes and transistors used to be produced by using valves. Valves were small evacuated glass tubes which had a small heating element which was used to excite a special material so that it emitted electrons which travelled across the valve to a collecting plate. Current could only flow from the emitter (called a cathode) to the collecting plate (called the anode), and so a diode was produced. By putting a grid in between the cathode and anode, the flow of current could be controlled in much the same way as a transistor.

Integrated circuits

Integrated circuits, or ICs for short, are an extension of the process which is used to make transistors. Instead of putting just one transistor onto a piece of silicon, the first ICs 'integrated' a complete two transistor circuit onto one piece of silicon. As the technology developed, resistors and capacitors were added, and the number of transistors increased rapidly. By the mid 1990s, ICs made from hundreds of thousands of transistors had become common.

The development of sophisticated stand-alone computer ICs from very humble cash register origins has produced the microprocessor. Commonly known as a 'chip', microprocessors carry out a vast range of processing tasks: a typical item of consumer electronic equipment will contain several: a video cassette recorder (VCR) could have a 'chip' dedicated to dealing with the front panel and IR remote control commands. Another would keep track of the programming and time functions, whilst another might handle the tape transport mechanism. The use of microprocessor chips has had a major effect on the evolution of synthesizers: most notably in the change from analogue to digital methods of producing sounds.

Analogue electronics

Analogue electronics is concerned with signals: audio, video, instrumentation or control signals. Usually direct representations of the real-world value, but converted to an electrical signal by some sort of transducer or converter, analogue signals indicate the value by a voltage or current. For example, a device for measuring the level of a liquid in a tank might produce a voltage – and by connecting this voltage to a calibrated indicator or meter, the level can be monitored remotely.

Signals in analogue electronics are often shown as plots of the value against time. These waveforms are often interpreted and drawn as if they were centred on a value of zero. So the use of the term 'zero-crossing' does not necessarily mean that the waveform actually passes through a zero position, but merely that it passes an arbitrary line which is approximately mid-way between the highest and lowest points of the oscillation.

Op-amps

Operational amplifiers, or op-amps, are one of the basic building blocks of analogue electronics. Whilst transistors can be used as amplifiers, op-amps provide idealized, near-perfect gain blocks which are easy to control and use in circuits. Op-amps have a very large gain, and in normal use this is deliberately reduced by feeding some of the output back into the input – rather like the way that people tend to shout if they can't hear themselves.

The integrator is one example of an analogue processing element which can be created using an op-amp. By connecting the output of an op-amp to the input through a capacitor, the resulting circuit can only change its output slowly – at a rate set by the time that the capacitor takes to charge. An integrator can be used to convert a sudden change into a smooth transition, and is a simple filter circuit – it 'filters' out rapid changes.

The oscillator is a variation on the integrator. If a capacitor is arranged so that it acts as a timing element for an op-amp circuit,

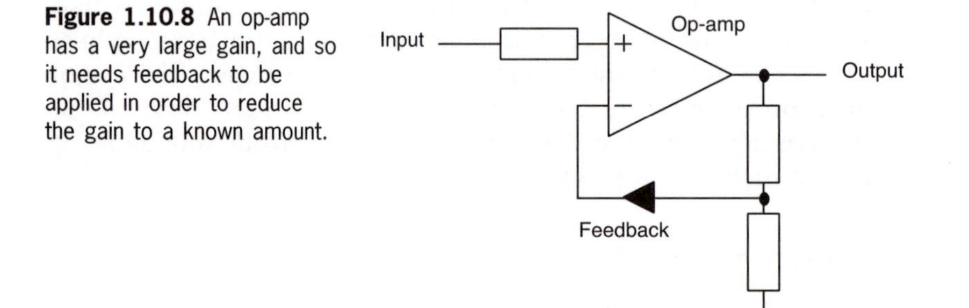

Figure 1.10.8 An op-amp has a very large gain, and so it needs feedback to be applied in order to reduce the gain to a known amount.

then the circuit will repeat the cycle of charging and discharging the capacitor continuously. This produces a repetitive output at a frequency set by the time it takes for the capacitor to charge and discharge.

Filters are sophisticated versions of integrators. They come in many forms, and most have a gain which is dependent on frequency – exactly the opposite of a hi-fi amplifier, which aims to produce a 'flat' or consistent gain for any input frequency. Filters have a wide range of uses in synthesizers.

Digital electronics

Digital electronics uses signals which represent real-world values as numbers. The numbers are held in binary form – voltages or current which can have only two values or states: on and off, or one and zero. By using groups of these two-valued voltages, any number can be stored in digital form. One familiar digital circuit is a light switch: assuming that there is no dimmer, then a light is either on or off.

Gates are simple electronic circuits which take one or more of these digital inputs and produce an output which is a logical function of them. For example, an output might only occur if both inputs are the same, or an output might be the opposite of the input. The rules for determining how these interactions take place is called Boolean algebra, and this is the branch of mathematics which is used to solve problems of the form: 'John does not cycle to work at the weekend. Bill travels to work, but only at the weekend. Simon has a car, and he gives a lift to a colleague on Saturday. Who does Simon give the lift to?' (It could be John or Bill – more information is needed to provide a more definitive answer).

Registers are simple circuits which can store a binary value. Sets of registers can be used to hold whole numbers, and are known as memory. Real-world values which can change are represented as sequences of numbers, and this can occupy large amounts of memory. Audio signals require high precision and the frequent measuring of the value, and so synthesizers, and especially samplers, may need to contain lots of memory chips in order to store audio signals.

Memory comes in two forms. Permanent storage is called read-only memory (ROM) and is used to store the instructions which control how a piece of equipment works – called the operating system. Digital information (data) is stored in ROM memory by physically breaking links inside the ROM with short bursts of high current. Temporary storage is called random access memory (RAM), since any of the data it contains can be directly accessed as it is required – in contrast to a serial memory like a tape, where you need to wind through the tape to get to the required data. Some variants of ROM memory can be erased and rewritten: instead of using a permament break in a wire, they store the data as charges on capacitors. These reprogrammable ROMs are called erasable programmable ROMs (EPROMs) or flash EPROMs – although this is commonly becoming abbreviated to just flash memory.

Microprocessors are stand-alone general purpose computers which are designed to carry out lots of logical operations very quickly and efficiently. They do this by having memory stores and registers for values, a special arithmetic section which can carry out logic functions on the values, and a way to control the movement and processing of the data: usually a sequence of instructions called a program.

Digital signal processors (DSPs) are microprocessors which have been optimized to deal with manipulating signals: often audio signals, although video and other types of signal are also possible. DSPs have a streamlined architecture and special circuitry to carry out functions rapidly and efficiently.

The preceding sections are not intended to be complete guides to either electronics or acoustics. Instead they aim to give an overview of some of the major concepts and terms which are used in these subjects. Further information can be found by following the references in the Bibliography.

1.10.3 Units

Technical literature is full of units, and these units are often prefixed with any of a number of symbols which show the relative size of the unit. A familiar example is the use of the metre for measuring the dimensions of a room, but kilometres are used for measuring the dimensions of a country. A kilometre is one thousand metres, and this is shown by the 'kilo' prefix to the basic unit: the metre.

Table 1.10.4 gives some conversions between units and prefixes.

So one microsecond (μs) is one millionth of a second, whilst one megahertz (MHz) is one million hertz.

When the size of computer memory is described, two prefixes are used which refer to powers of two instead of powers of ten. A Kbyte of memory does not mean one thousand bytes of

Table 1.10.4 Units

Name	*Symbol*	*Ratio*	*Ratio*
Tera	T	1 000 000 000 000	1 million million times
Giga	G	1 000 000 000	1 billion times (1 thousand million times)
Mega	M	1 048 576	1048576 times
Mega	M	1 000 000	1 million times
K	K	1 024	1 thousand and twenty-four times
Kilo	k	1 000	1 thousand times
Milli	m	1/1 000	1 thousandth
Micro	µ	1/1 000 000	1 millionth
Nano	n	1/1 000 000 000	1 billionth (1 thousand millionth)
Pico	p	1/1 000 000 000 000	1 million millionth

memory, instead it refers to 1 024 bytes of memory. Kbytes are sometimes mistakenly called Kilobytes. A similar confusion can arise over the use of the prefix M. One Mbyte of memory is 1 048 576 bytes of memory, and not a million! In this case, although the use of the word Megabyte in this context is technically wrong, it has entered into common usage and become widely accepted.

1.11. Digital and sampling

This section brings together the background principles behind the two major technologies used in digital musical instruments: digital and sampling.

1.11.1 Digital

The word 'digital' can be applied to any technology where sound is created and manipulated in a discrete or quantized way – as samples (numbers which represent the sounds) rather than continuous values. This tends to imply the use of computers and sophisticated electronics, although an emphasis on the technology is often a marketing ploy rather than a result of using digital methods to make sounds. Physical modelling synthesizers are an excellent counter example where much of the complexity of the digital processing is deliberately hidden from the user, and as a result, the synthesizer is perceived by the performer as merely a very flexible and responsive 'instrument'. Perhaps in the future we will see digital 'instruments' where the synthetic method of sound production is not apparent from the external appearance (although a power supply cable might be a useful clue!).

In order for digital techniques to work with sound, there needs to be a way to represent sounds and values as numbers. Digital systems use binary digits, or bits, as their basic way of storing and manipulating numbers. Bits tend to be organized into groups of 8, for various historical and mathematical reasons. A single bit can have one of two values: on or off, usually given the values 1 and 0 respectively. Eight bits can represent any of 256 values, from 0 to 255, or %0000 0000 to %1111 1111 in binary notation. The '%' is often used to indicate a binary number, and the binary digits (bits) are grouped into blocks of four to aid reading. A collection of eight bits is known as a byte – and the blocks of four are called nibbles(!). Sixteen bits can be used to represent numbers from 0 to 65 535, and more bits can provide larger ranges of numbers. The two bytes which make up a 16-bit 'word' are called the most significant and the least significant bytes, which is normally abbreviated to MSB and LSB. Notice that these simple numbers are integers – only whole numbers can be represented with this method.

Larger numbers, and especially decimal numbers, require a different method of representing them. Floating point numbers split the number into two parts: a decimal number part from 0 to 9.9, and a multiplier or exponent part, which is a power of ten. The value 2 312 could thus be regarded as being 2.312×1000, and this would be stored as 2.312×10^3 in floating point representation. For binary numbers, a power of two is used instead of a power of ten, but the principle of splitting the number into a decimal number and a multiplier is the same. Floating point numbers can be processed using either a microprocessor, or with special purpose arithmetic chips called digital signal processors, often abbreviated to DSPs. DSPs are optimized for the carrying out of complicated mathematical operations on numbers (some are designed for integers, whilst others are intended for floating point numbers) – and are typically used for filtering, equalization and 'effects' like echo, reverb, phasing and flanging.

1.11.2 Sampling

Sampling is the process of conversion from an analogue to a digital representation. Whereas an audio signal is a continuous series of values which can be displayed on an oscilloscope as a waveform, a digital 'signal' is a series of numbers. The numbers represent the value of the audio signal at specific points in time – and these are called samples. The sampling process has three stages:

1. The audio signal is 'sampled'.
2. The sample value is converted to a number.
3. The number is presented at an output port.

Samples are thus just numbers which represent the value of an audio waveform at a specific instant of time.

The opposite of sampling is the conversion from digital to analogue. This is called 'sample replay'. Replaying samples has three stages:

1. The number is presented to an input port.
2. The number is converted to an analogue value.
3. The analogue value forms part of an audio signal.

Sample replay is the basis of almost all digital synthesizers. Regardless of how the digital sample is produced, the conversion from digital to audio is what produces the sound that is heard. Chapter 3 shows how sample replay has progressed from single cycle waveforms to complex looped sample replay.

1.11.3 Conversion

The conversion process from analogue to digital and back again is at the heart of sampling technology. A complete digital audio conversion system, as used in a sampler, a direct-to-disc recording system, or a digital effect processor, typically consists of two sections. An 'analogue-to-digital' section converts the audio signal into digital form and temporarily stores it in the sample RAM memory:

- audio signal
- anti-aliasing filters
- sample and hold
- ADC chip
- sample RAM containing digital sample values.

whilst the 'digital-to-analogue' section reverses the process and converts the digital representation of the audio back into an analogue audio signal:

- sample RAM containing digital sample values
- DAC chip
- deglitcher
- reconstruction filter
- audio signal.

The majority of the actual conversion is achieved by two chips: analogue-to-digital conversion is carried out by an ADC chip, whilst the reverse process of digital-to-analogue conversion is done by a DAC chip. DACs are also commonly used inside ADCs.

Although the names are different, the circuits which make up the two parts on either side of the sample RAM memory have

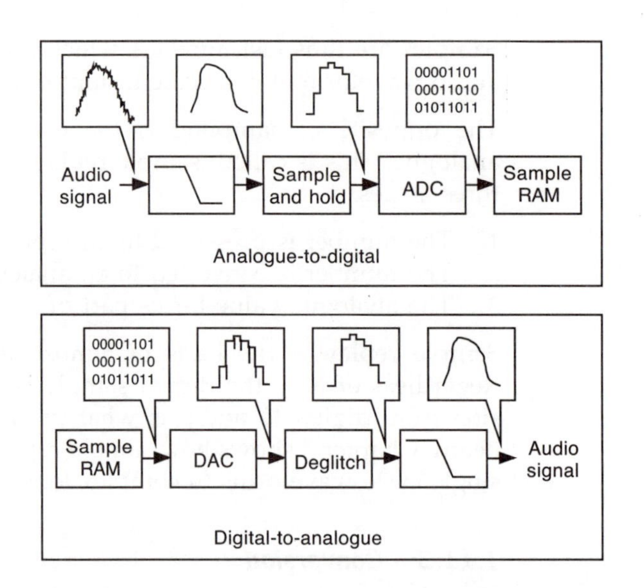

Figure 1.11.1 An overview of a complete sampling system,

very similar functions. The anti-aliasing filters prevent any unwanted audio frequencies from being converted by the ADC, whilst the reconstruction filter prevents any additional frequencies produced by the DAC's stepped waveform output from being heard at the audio output. The sample and hold circuit improves the quality of the conversion by presenting a constant level whilst the conversion is taking place, and the deglitcher prevents any momentary unwanted outputs from the DAC from being converted back into audio clicks.

Digital-to-analogue

A typical digital-to-analogue converter (DAC) has three parts:

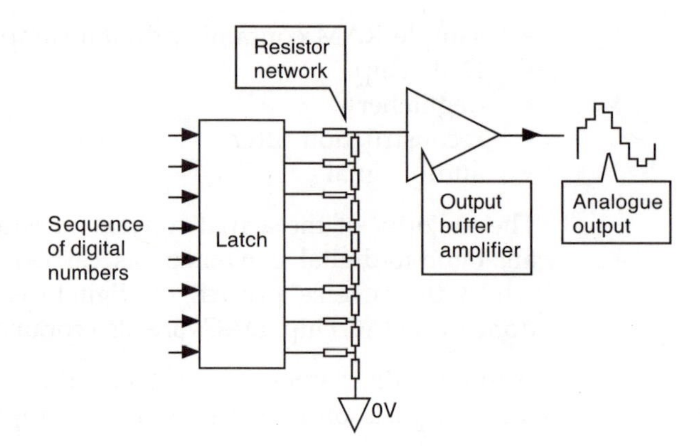

Figure 1.11.2 A digital-to-analogue converter, or DAC, converts digital numbers to analogue voltages by using a network of resistors. The network is arranged so that the bits in the latch change the output voltage depending on their value, so the most significant bits have the largest effect.

- a latch to hold the digital number
- a network of resistors to convert the number to a voltage
- an output buffer amplifier.

The latch holds the digital number which represents the sample value. The network of resistors is arranged so that the bits in the digital number produce voltages which are proportional to their position in the number. Large value bits produce big voltages, and small value bits produce small voltages. These voltages are added together by the output amplifier, whose output is thus an analogue voltage which represents the value of the digital number.

Analogue-to-digital

In a typical analogue-to-digital converter (ADC), the audio waveform is examined at regular intervals of time and the value is held in an analogue memory circuit called a sample and hold. The sample and hold circuit is the point in the conversion circuitry where the audio signal is actually 'sampled', and it is designed to capture the instantaneous value of the voltage at that point in the waveform and hold it whilst the conversion process proceeds. If the sample and hold circuit takes too long to capture the audio waveform, or if the held value changes whilst the conversion is taking place, then this can degrade the quality of the conversion.

Once the sample and hold circuit has captured the value of the audio waveform, the held value is compared with a value which is produced by a counter and a DAC – in much the same type of circuit as that found in a wavetable synthesizer producing a sawtooth waveform. The counter counts up from zero, and as it does so, the ascending count of numbers is converted into a rising voltage by the DAC. A comparator circuit looks at the held value from the sample and hold circuit and the output of the DAC, and when they are the same, the comparator indicates that the two are equal, and the counter output is conveyed to the output of the ADC and held in a latch. The output of the ADC now holds a number which represents the value of the sample. The counter then resets and the ADC can begin to process the next value from the sample and hold circuit.

The detailed operation of some ADCs may differ from this, but the principle is the same – the audio signal is sampled, the sample value is converted to a numerical representation of the value, and the number appears at the output port of the ADC. Some ADCs achieve the conversion process in different ways,

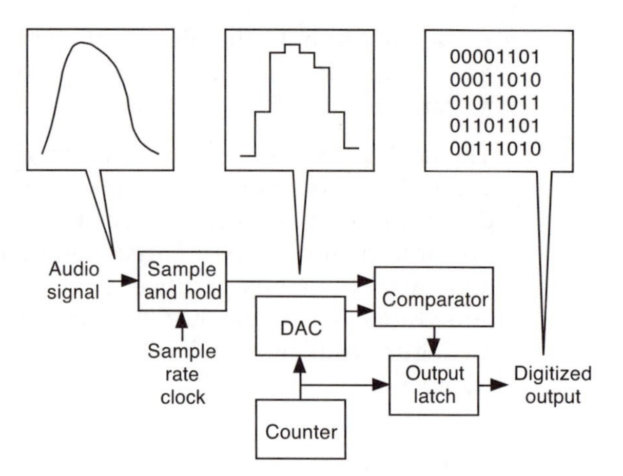

Figure 1.11.3 An analogue-to-digital converter takes an audio signal, samples it and then compares the value with the output of a DAC driven by a counter. When the two values are the same, the comparator latches the output of the counter into the ADC output latch. The analogue signal sample has then been converted into an equivalent digital value. This process repeats at the sampling frequency.

and the output format of the number can be either serial (one stream of bits) or parallel (several streams carrying a complete sample).

1.11.4 Sampling theory

In order for the process of taking a sample of the audio signal and transforming it into a number to work correctly, a number of criteria have to be met. Firstly, the rate at which the samples are taken must be at least twice the highest frequency which is required to be converted – which in practice means that the input is normally filtered so that the highest frequency which can be present is known. Secondly, the samples must be taken at regular intervals – any jitter or uncertainty in the timing can significantly degrade the conversion quality. Finally, the numbers used to represent the signal must have enough resolution to adequately represent its dynamic range.

1.11.5 Sample rate

The simplest representation of a waveform at a given frequency is two sample values – ideally the top and bottom peaks. The time between these two peaks represents half of the period of the waveform drawn through them, which assumes that it is a sine wave – and if this is the highest frequency, then it must be a sine wave. So because two points are needed, the sampling needs to be at least twice as fast as the frequency which is being sampled. This requirement that the sampling frequency is at least twice the highest frequency in the signal is called the Nyquist criterion. Note that if the sampling rate was exactly twice the frequency of the audio signal, then the two points

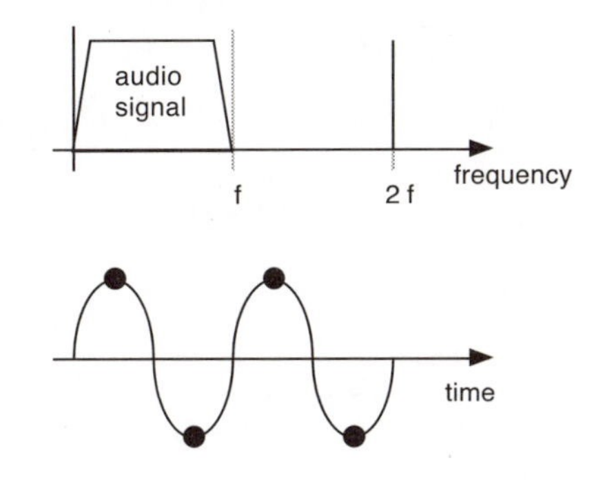

Figure 1.11.4 The Nyquist rate is twice the highest frequency which is present in the signal to be sampled. At least two samples are needed in order to provide a single cycle of a waveform.

would always be at exactly the same values on the waveform, which could include zero, and so there might be no output at all. Sampling is not normally done synchronously, and so a sampling rate which is at least twice that of the highest frequency in the audio signal will enable the same waveform to be reconstructed at the output of a subsequent digital-to-analogue converter section.

If the audio signal is sampled at a rate which is higher than twice the highest frequency which is present in the audio signal, then no additional information is provided by using a higher sampling rate. The Nyquist criteria thus represents the most efficient rate at which to sample a given audio signal with a specific highest frequency component. However, sampling at higher frequencies can simplify the design and implementation of the filtering and some other parts of the circuitry. In the limiting case, some ADCs sample at several hundred times the Nyquist rate and then process the resulting one-bit representation to produce the equivalent of more bits sampled at a lower rate. But the basic amount of information which is required in order to be able to reconstruct the audio signal still remains constant.

1.11.6 Filtering and aliasing

If any frequencies are present in the audio signal which are above the 'half sampling' frequency, then they will still be sampled, but the effect will be to make them appear to be lower in frequency – this process is called aliasing. It can be likened to a security camera which looks at a room for a few seconds every five minutes. If the room has someone inside the first time that the camera is active, but not the second time, then there are several scenarios for what has happened. The obvious case is that the person was present for the first five minutes and so was observed when the camera was active, but who then left before

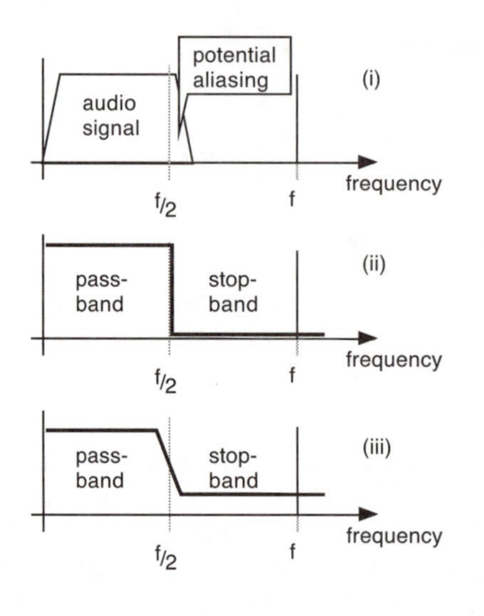

Figure 1.11.5 If an audio signal contains some frequencies which are higher than half of the sampling frequency, then aliasing can occur (i). An anti-aliasing filter prevents this by having a pass-band which is set so that frequencies above the half sampling frequency are in the stop-band of the filter. Theoretically this filter should pass everything below the half sampling frequency and stop everything above (ii). In practice, filters with a realizable cut-off slope and sufficient stop-band attenuation to prevent audible aliasing are used (iii).

the camera was active the second time. Alternatively, the person could have been in and out of the room several times, and just happened to be present the first time, but not the second. The important point is that the two cases appear the same from the viewpoint of the camera. Aliasing behaves in the same way – an aliased high frequency appears as a lower frequency in the digital representation, and it is not possible to reconstruct the original higher frequency. It is a 'one-way' process where information is lost or becomes ambiguous.

To prevent information being lost by the sampling process, anti-aliasing filters are used to constrain the audio signal to below the half sampling frequency. This ensures that only frequencies which can be reproduced are sampled, and so guarantees that the DAC will be able to output the same audio signal. The design of these filters affects the quality of the conversion, since they need to pass all frequencies below the half sampling rate, but completely reject all frequencies above the half sampling rate. 'Brick wall' filters with flat pass bands and high stop-band rejection are difficult to design and fabricate, and in practice, the cut-off frequency of the filter is set to slightly lower than the half sampling rate, and the stop-band rejection is chosen so that any frequencies which pass through to the conversion process will be so low that they will be lost in the inherent noise of the converter.

The half-sampling frequency sets the highest frequency which was present in the original audio signal before it was sampled, and this is the highest frequency which will be reproduced when the sample is replayed. So for a sample rate of 44 kHz, the highest frequency which can be reproduced by the replay circuitry will be just under 22 kHz. But reproducing sample values from a memory device also produces unwanted additional frequencies. Consider again the limiting case of two adjacent sample values, which represent a sine wave at just under the half-sampling frequency: when these are read out from a memory device, they will form the equivalent of a square-shaped waveform. Additional filtering is required to remove these extra frequencies, and a sharp low-pass filter with a cut-off frequency set to near the half-sampling rate is normally used. This filter is often called a reconstruction filter, and it limits the output spectrum of the sample replay to those frequencies which are below the half-sampling frequency. Any unwanted frequencies which are not removed by this filter are called aliasing frequencies.

Domestic CD players with a 44.1 kHz sample rate, and thus a half sampling rate of 22.05 kHz, normally quote the upper limit of the audio signal frequency response as being 20 kHz. Professional DAT recorders typically use 48 kHz sampling, and also quote a 20 kHz upper frequency response. This 44.1/48 kHz sample rate/20 kHz bandwidth has become a 'de facto' standard for samplers and digital synthesizers.

1.11.7 Resolution

The size of the numbers which are used to represent the sample values determine the fidelity with which the audio signal can be reproduced. In digital circuitry, the number of bits which are used to represent the sample value limits the range of available numbers. In the simplest case, a one-bit number can have only two values: 1 and 0. For each additional bit which is used, the number of available values doubles: so for two bits, four values are available. Three bits provide 8 values, four bits 16 values, and so on. In general, the number of available sample value numbers is given by:

$$D = 2^n$$

where D is the number of available values and n is the number of bits used. The number of available numbers to represent the sample values affects the precision of the digital version of the original audio signal. If only one bit is available, then only a very crude version of the audio signal is possible. As more bits are used to represent the sample values, then the ratio between the largest and the smallest number which can be represented increases – and it is the size of the smallest change which determines how good the resolution is. As the number of bits increases, then the detail which can be represented by the numbers improves. This reduces the distortion, and for a typical 16-bit conversion system the distortion will be more than 90 dB below the maximum output signal.

The number of bits which are used to represent a sample is important because it sets the limiting value on the output quality of the signal. The relationship can be approximated by the simple formula:

$$S = 6n \text{ dB}$$

where S is the signal-to-noise (and distortion) ratio (SNR): the ratio between the loudest audio signal and the inherent noise and distortion of the system, often called the dynamic range, measured in dB, and n is the number of bits. This is the performance of a perfect system, and represents the 'ideal' case: real-world digital audio systems will only approach these figures.

Table 1.11.1 Bits and signal-to-noise ratio

Number of bits	*Dynamic range*
8	48
10	60
12	72
14	84
16	96
18	108
20	120
22	132
24	144

Table 1.11.1 shows the number of bits versus the 'ideal' dynamic range. As the table shows, a 'CD quality' output should have a dynamic range of nearly 96 dB: 'better than 90 dB' is frequently quoted in manufacturer's specifications. Note that the entire audible range, from silence to painful, can be covered by 20 bits. It thus appears that using between 16 and 20 bits should be adequate for almost all purposes. Unfortunately, this is not the case, and the simple example of volume control illustrates the problem.

Suppose that a digital synthesizer design uses 16-bit numbers to represent the audio samples, and that the volume control is implemented by manipulating the digital audio signal. So for maximum output volume (0 dB), all of the 16 bits in the audio samples will be used in the replay of the signal. A crude method of reducing the volume could be achieved by using less bits: shifting the digital numbers to the right. Each bit which is removed reduces the volume by 6 dB, so a coarse volume control might work by shifting the digital words to the right so that less bits are used, with zeroes added from the left. So as bits are removed the volume decreases, but there is a corresponding decrease in the dynamic range of the signal. For an audio signal which is at –48 dB, only 8 of the original 16 bits are being used to produce the audio signal, which means that the output signal effectively has only an 8-bit resolution – the remainder of the signal has been filled with 8 zeroes.

Using only half of the available bits for an audio signal has two major effects on the audio. The reduction in dynamic range means that there is a corresponding increase in the background noise level, whilst the release of notes can become distorted, especially if reverb is used. This characteristic 'grainy' distortion is called 'quantization noise', and is caused by the transition between silence and audio represented by just one bit changing: effectively the audio waveform has been converted into a pulse wave.

Reducing the volume by having less bits in the output signal is thus very different from always using all the available bits and changing the volume with an analogue volume control. With the analogue control, the full bit resolution is always available, and so a –48 dB signal would still have the same dynamic range as the original sample – even if some of this is buried in the background electrical noise of the system. Some types of digital-to-analogue converter chips allow exactly this type of output. Multiplying DACs and floating point DACs can be used with two inputs: one of which represents the audio signal at the full bit resolution, whilst the other input represents the volume control bits. This type of 'fixed with scaling' conversion system is in widespread use. For example, telephones do not use linear coding but their basic performance is approximately 8 bit for signal-to-noise ratio, with about the equivalent of 12 bits for the dynamic performance – although the restricted bandwidth significantly affects the perceived quality. In synthesizers, the sample resolution is normally 16 bits, whilst the volume control can be 6 bits or more, which is sometimes translated as '24 bit DACs' in manufacturers' literature.

It should be noted that shifting digital numbers to the right is not a very useful way of making changes to the volume of a signal, since the 6 dB steps are very coarse. In practice, the numbers are reduced or increased by using a multiplication device: often a special purpose signal processing chip.

In audio signal terms, with 8-bit integer numbers at a rate of 8 kHz, the sound quality is comparable to a low quality cassette or telephone. Sixteen-bit numbers at a rate of 44.1 kHz are often quoted as being of 'CD audio quality', since this is the basic storage used by a CD player for audio.

1.12 MIDI

The musical instrument digital interface (MIDI) (Rumsey, 1994) has played a major role in the development of electronic music since 1983. Wherever possible, this book has deliberately avoided making explicit references to MIDI in order to prevent it becoming 'Yet another book on MIDI'.

For example, the envelopes described in Chapter 2 are mostly dealt with in terms of control voltages, gate pulses and trigger signals because these are likely to be the native interfacing for many analogue synthesizers, even though many users will also use a MIDI-to-CV converter box to enable the use of MIDI control.

Since some readers may not be familiar with MIDI, the remainder of this section provides some background information.

1.12.1 Overview

MIDI provides an interface for the exchange of information between electronic musical instruments and computers. It is based around musical events – except in rare circumstances, musical sounds are not conveyed via MIDI. Instead, MIDI carries information about what is happening, occurrences such as: when a note has been pressed; when a drum is hit; and when the sequencer has stopped. A MIDI-equipped keyboard will thus output information about what is happening on its own keyboard – so if some notes are played, it will output MIDI information as 'messages' which describe which notes are being played, as they are being played.

1.12.2 Ports

The MIDI interface is called a MIDI port. There are three types, although only one or two of the types may be present on a given piece of equipment.

- The in port accepts MIDI data.
- The out port transmits MIDI data.
- The 'thru' (American spelling: MIDI was originally specified in the USA) port merely transmits a copy of the MIDI data which arrives at the in port.

All MIDI ports look alike: they consist of 180° 5 pin DIN sockets – although each port will normally be marked with its function (in, out or thru).

1.12.3 Connections

Connecting MIDI ports together requires just one simple rule:

Always connect an out or a thru to an in.

In a MIDI 'network', information flows from a controller source to an information sink. A keyboard is often used as a source of control information, whilst a synthesizer module is usually an information sink. So the out port of the keyboard would be connected to the in port of the synthesizer module. MIDI messages would then flow from the keyboard to the synthesizer module, and the synthesizer could then be 'played' from the keyboard.

1.12.4 Channels

MIDI provides 16 separate channels – which can be thought of as TV channels. A piece of MIDI equipment can be 'tuned' so that it receives only one channel, and it will then only respond to MIDI messages which are on that channel. Alternatively, it is possible to set a piece of MIDI equipment so that it will respond to messages on any channel – called 'omni'. Some important MIDI messages are sent to all 16 channels.

1.12.5 Modes

MIDI has several modes of operation. The important ones are:

- Monophonic (one instrument: one note at once)
- Polyphonic (one instrument: several notes at once)
- Multi-timbral (several different instruments at once: several notes at once).

Modes are normally only important to users of guitar controllers or other specialized uses.

1.12.6 Program changes

Continuing with the TV analogy, MIDI calls sounds or patches 'programs'. The message which indicates that a program should change is called a 'program change' message. Any of 128 programs can be selected. If more programs are required, then a bank change message allows the selection of banks of 128 programs. For most applications, a program change number does not indicate a specific sound – but a specialized mapping called 'General MIDI' or GM does specify which program change number calls up what sort of sound from a sound module.

1.12.7 Notes

One of the commonest MIDI messages is the note on message. This indicates that a note has been played on a keyboard – although it could also mean that a sequencer is replaying a stored performance. The note on message contains information on the MIDI channel that is being used, what key has been played, and how quickly the key was pressed – called the 'velocity'. As a shorthand method of sending messages, a velocity of zero is taken to mean a note off message, even though a separate note off message exists. The MIDI note on message does not contain any timing information about when the key was pressed – the message itself is used to indicate that the key has just been pressed.

Other common note specific messages include:

- the Pitch Bend message, which transmits any changes in the position of the pitch bend wheel;
- the after-touch messages: polyphonic and monophonic – these transmit information about how hard the keys are being pressed once they have reached the end of their travel. This is intended as an additional control source for introducing vibrato or other modulation by increasing the finger pressure on a key which is being held down.

1.12.8 MIDI controllers

A MIDI controller is something which is used to control part of a performance – like the modulation wheel which is often found on the left hand side of the keyboard on many synthesizers, and which can be used to introduce vibrato or other modulation effects into the sound. Another example might be a foot volume pedal which plugs into a synthesizer – it controls the volume of the synthesizer directly, but it may also cause the synthesizer to transmit MIDI volume messages which indicate the position of the foot pedal.

There are a large number of possible controllers – with functions ranging from volume or portamento, through to one which can control the timbre of a sound or set an effect parameter. Only a few of the controllers are defined – many are deliberately left undefined so that manufacturers can allocate them for their own purposes.

1.12.9 System exclusive

Although there are lots of MIDI controller messages, there is an alternative way to provide control over a remote MIDI device. System exclusive (sysex) messages are designed to allow manufacturers to make their own MIDI messages. Sysex messages can be used to edit synthesizer parameters; to store sound data; and to transmit samples.

1.12.10 MIDI files

MIDI files are a way to move MIDI sequencer file information between different sequencers. The PC standard disk format allows MIDI files to be moved freely between any computers that can read 3.5 inch 720 Kbyte IBM PC-compatible floppy disks.

1.12.11 Reference

The book by Francis Rumsey (Rumsey, 1994) gives excellent detailed information on MIDI and is recommended reading. The official MIDI documentation (The MIDI Specification 1.0) is very formal, rather technical and not intended for the general reader.

1.13 Questions

This section is designed to act as a brief review of the subject covered in the preceeding chapter. The answers are in the text.

1. What is sound synthesis?
2. What is the difference between a modular and a performance synthesizer?
3. Outline the major methods of sound synthesis.
4. What is Acoustics?
5. What is Electronics?
6. Outline the processes that are required to take a product from laboratory prototype to commercial production.
7. Describe some ways in which synthesizers can be used to make music.
8. Categorize ten different sounds under the following categories: realistic, synthetic, imitative, suggestive or sympathetic.

Time line

Date	Name	Event	Notes
1582	Galileo	Galileo conceives the idea of using a pendulum as a means of keeping time	
1600	William Gilbert	Electricity is named after the Greek word for amber	William Gilbert was the court physician to Elizabeth I
1612	Francis Bacon	Publishes *New Atlantis*, which describes all sorts of now current sound 'wonders' in a passage starting: 'We also have sound houses...'	An essential quote in most books on electronic music
1657	Christian Huygens	Christian Huygens uses the pendulum to regulate the time-keeping of a clock	
1676	Thomas Mace	Thomas Mace uses a thread and a heavy round object to mark musical time	Also designed a lute with 50 strings in 1672
1696	Etienne Loulie	Etienne Loulie invents the 'chronometre', an improvement on Mace's idea, but with a variable length thread	
1700	J. C. Denner	Invented the clarinet	Single reed woodwind instrument
1752	Benjamin Franklin	Flies a kite in a thunderstorm to prove that lightning is electrical	
1756–1827	Ernst Chladni	Worked out the basis for the mathematics governing the transmission of sound	The 'father of acoustics'
1801	Jacquard	Jacquard punched cards invented	Basis of stored program control, as used in computers, pianolas etc.
1801	Valve trumpet	The modern valve trumpet is invented	Not all musical instruments are old!
1812	D. N. Winkel	Winkel invents a clockwork driven double pendulum timer – very much like a metronome	
1815	J. N. Maelzel, brother of Leonard Maelzel	Invents the metronome and patents it	Some dispute about Maelzel versus Winkel as to who actually invented the metronome
1818	Beethoven	Beethoven starts to use metronome marks in scores	
1820	Oersted	Discovery of electromagnetism	The basis of electronics
1821	Michael Faraday	Discovers the dynamo, and formalises link between magnetism, electricity, force and motion	Used in motors, microphones, solenoids
1837	Samuel Morse	Invents Morse code	
1844	Samuel Morse	Invents the electric telegraph	The first telegraph message was 'What hath God wrought?'
1846	Adolphe Sax	Invents the saxophone	
1849	Heinrich Steinweg	Steinway pianos founded by Heinrich Steinweg	
1862	Helmholtz	Published *On the Sensations of Tone*	Laid the foundations of musical acoustics

Date	Name	Event	Notes
1866–1941	Dayton Miller	Worked on photographing sound waves, and turned musicology into a science	
1868–1919	Wallace Sabine	Founded the science of architectural acoustics as the result of a study of reverberation in a lecture room at Harvard where he was professor of physics	
1876	Alexander Graham Bell	Invents the telephone	Start of the marriage between electronics and audio
1878	Lord Rayleigh	Publishes *The Theory of Sound*	Laid the foundations of acoustics
1887	Heinrich Hertz	Produces radio waves	
1895	Marconi	Radio telegraphy invented	
1896	Thomas Edison	Motion picture invented	
1897	Yamaha	Nippon Gakki (Yamaha) founded	
1901	Guglielmo Marconi	Marconi sends a radio signal across the Atlantic	
1901–	Harry Partch	Experimented with thirteenth tones and other microtonal scales	Mostly self-taught
1904–1915	Valve	Development of the valve	The first amplifying device – the beginnings of electronics
1906	Lee de Forest	Invented the triode amplifier	The beginnings of electronics
1906	Reginald Fessenden	First transmission of audio by the use of radio waves	
1908–	Oliver Messiaen	Serialism, Eastern rhythms and exotic sonorities	Some of his music uses up to 6 Ondes Martenot
1910–1920	Futurists	Futurists	Category of music
1912–	John Cage	Pinoeer in experimental and electronic music	Famous for 'prepared' pianos, and '4 minutes 33 seconds' – a silent work
1914	Hornbostel & Sachs	Published a classification of musical instruments based on their method of producing sound	Idiophones, membranophones, chordohones, aerophones etc
1916	Luigi Russolo	Categorizes sounds into 6 types of noise	Also invented the russolophone, which could make 7 different noises
1923	John Logie Baird	Begins experiments with light sources and disks with holes in them for scanning images	The beginnings of television and computer monitors
1925	John Logie Baird	First television transmission	...across an attic workshop!
1925–	Pierre Boulez	Pioneer of serialims and avante-garde music	
1929	Couplet & Givelet	Four voice, paper tape driven 'automatically operating oscillation type	Control was provided for pitch, amplitude, modulation, articulation and timbre
1930s	Bell Telephone Labs	Invented the vocoder – a device for splitting sound into frequency bands for processing	More musical uses than telephone uses!

Date	Name	Event	Notes
1933	Stelzhammer	Electrical instrument using electromagnets to produce a variety of timbres	
1934	John Compton	UK patent for rotating loudspeaker	
1940s	Arnold Schoenberg	12 tone technique and atonality	
1945	Metronome	First pocket metronome produced in Switzerland	
1945	Ronald Leslie	Patents rotating speaker system	
1947	Conn	Independent electromechanical generators used in organ	
1948	Baldwin	Blocking divider system used in organ	
1948	Pierre Schaeffer	'Concert of Noises' Futurist movement. Invented music concrete	Music concrete is made up of pre-existing elements
1949	Allen	Organs using independent oscillators	
1950	John Leslie	Re-introduction of Leslie speakers	This time they are a success
1950s	Charles Wuorinen	Quarter tones	
1955	E.L.Kent	Kent Music Box in Chicago. Inspired RCA mark II synthesizer	
1955	Louis & Bebe Barron	Soundtrack to 'Forbidden Planet' is a 'tour de force' of music concrete using synthetic sounds	
1955–1956	Stockhausen	'Gesang der Junglinge' mixes natural sounds with purely synthetic sounds	
1957	RCA	RCA Music Synthesizer mark II	Used punched paper tape to provide automation
1958	Edgard Varese	Produced some 'electronic poems' for the Brussels Expo	
1960s	Wurlitzer, Korg	Mechanical rhythm units built into home organs by Wurlitzer and Korg	
1962	Ligetti	Ligetti uses the metronome as a musical instrument	Actually he uses 100 of them in concert
1962	Telstar	The first telecommunications satellite to transmit telephone and television signals	
1965	Early Bird	First geo-stationary satellite	
1966	Rhythm machine	Rhythm machines appear on electronic organs	Non-programmable, and very simple rhythms
1968	Walter Carlos	'Switched On Bach', an album of 'electronic realisations' of classical music, becomes a best seller	Moog synthesizers suddenly change from obscurity to stardom
1970	Tom Oberheim	Oberheim Electronics founded	US company
1972	E-mu	E-mu founded by Dave Rossum. Initial products are custom modular synthesizers	
1972	Hot Butter	'Popcorn' becomes a hit single	
1972	Roland	Ikutaro Kakehashi founds Roland in Japan, designed for R&D into electronic musical instruments	First products are drum machines

Date	Name	Event	Notes
1974	George McRae	'Rock Your Baby' is first record to completely replace the drummer with a drum machine	
1974	Kraftwerk	'Autobahn' album is huge success. A mix of music concrete technique and synthetic sounds	
1974	Sequential Circuits	Sequential Circuits is founded by Dave Smith	US company
1975	Fairlight	Fairlight is founded by Kim Ryrie and Peter Vogel	Australian company
1979	First Digital LPs	First LPs produced from digital recordings made in Vienna	A mix of analogue playback and digital recording technology
1980	Electronic Dream Plant	Spider Sequencer for Wasp Synthesizer. One of the first low-cost digital sequencers	252 note memory, and used the Wasp DIN plug interface
1981	Yamaha	Yamaha R&D Studio opened in Glendale, California, USA	
1982	Philips/Sony	Sony launch CDs in Japan	First domestic digital audio playback device
1982	Robert Moog	First MIDI Specification announced by Robert Moog in his column in *Keyboard* magazine	
1983	Philips/Sony	Philips launch CDs in Europe	Limited catalogue of CDs rapidly expands
1983	Roland	Roland launches the TR-909, the first MIDI equipped drum machine	
1983	Sequential Circuit	Sequential Circuit's Prophet 600 is first synthesizer to implement MIDI	Prophet 600 is marred by awful membrane switch keypad
1983	Yamaha	'Clavinova' electronic piano launched	
1984	Yamaha	Marketing of custom LSIs begins	Yamaha begin to market their in-house expertise to the world market
1985	Yamaha	Yamaha R&D Studio opened in Tokyo, Japan	
1986	Steinberg	Steinberg's Pro 16 software for the Commodore C64	The start of the explosion of MIDI-based music software
1987	DAT	DAT (Digital Audio Tape) is launched. The first digital audio recording system intended for domestic use	Worries over piracy severely restrict its mass marketing
1987	Roland	MT-32 brings multi-timbral S&S synthesis in a module	The start of the 'keyboard' and 'module' duality
1987	Yamaha	Yamaha R&D Studio opened in London, England	
1989	Breakaway	The Breakaway Vocaliser 1000 is a pitch-to-MIDI device that translates singing into MIDI messages and sounds via its on-board sampled sounds	Somewhat marred by a disastrous live demonstration on the BBC's 'Tomorrow's World' programme

2 Analogue synthesis

2.1 Analogue and digital

The word 'analogue' means that a range of values are presented in a continuous rather than a discrete way. Continuous implies making measurements all the time, and also infinite resolution – although inherent physical limitations like the grain size on a magnetic tape, or the noise level in an electronic circuit will prevent any real-world system from being truly continuous. Discrete means that you use individual sample values taken at regular intervals rather than measuring all the time, with the assumption that the samples are a good representation of the original signal.

An analogue synthesizer is thus usually defined as one that uses voltages and currents to directly represent both audio signals and any control signals which are used to manipulate those audio signals. In fact, 'analogue' can also refer to any technology in which sound is created and manipulated in any way where the representation is continuous rather than discrete. Analogue computers were used before low-cost digital circuitry became widely available, and they used voltages and currents to represent numbers. They were used to solve complex problems in navigation, dynamics and mathematics.

Analogue electronics happens to be a convenient way of producing sound signals – but there are many other ways: mechanical, hydraulic, electrostatic, chemical, etc. For example, vinyl discs use analogue technology where the mechanical movement of the stylus is converted into sound. Tape recorders produce sound from analogue signals stored on magnetic tape.

In synthesizers, the use of the word 'analogue' usually implies voltage controlled oscillators (VCOs) and filters (VCFs). These have a characteristic set of audio characteristics: VCOs often have tuning stability problems, for example; and analogue filters can break into self-oscillation or may distort the signal passing through them. These features of the analogue electronics which are used in the design can contribute to the overall 'tone quality' of the instruments.

Analogue synthesizers are commonly regarded as being very useful for producing for bass, brass and the synthesizer 'cliché' sounds, but not a very good choice for simulating 'real' sounds. The typical clichéd sound is usually a 'synthy' sound consisting of slightly detuned oscillators beating against each other, with a resonant filter swept by a decaying envelope.

In contrast, digital synthesizers use numerical representations of the audio and control signals. They are thus capable of reproducing pre-recorded samples of real instruments with a very high fidelity. They also tend to be very precise and predictable, with none of the uncertainty of analogue instruments. Some of the many digital synthesis techniques are described in Chapter 5.

The difference between analogue and digital representations can be likened to an experiment to measure the traffic flow through a road junction. The actual passage of cars can be observed and the numbers of cars passing a specific point in a given time interval are noted down. The movement of the cars is analogue in nature, since it is continuous, whereas the numbers are digital, since they only provide numbers at specific times.

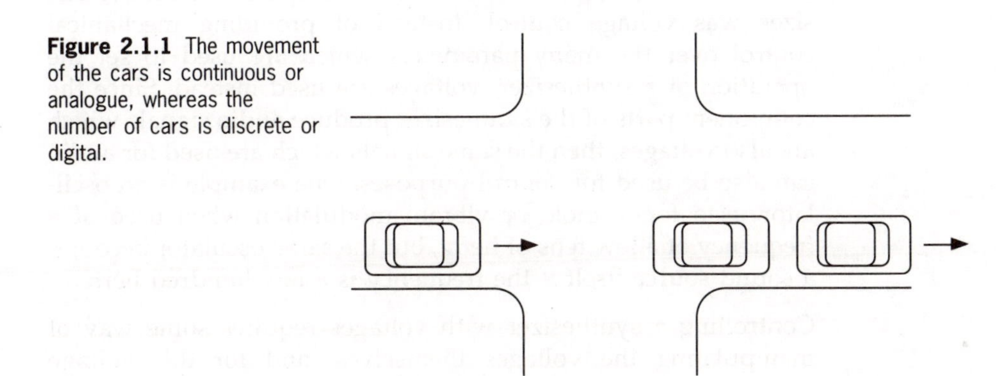

Figure 2.1.1 The movement of the cars is continuous or analogue, whereas the number of cars is discrete or digital.

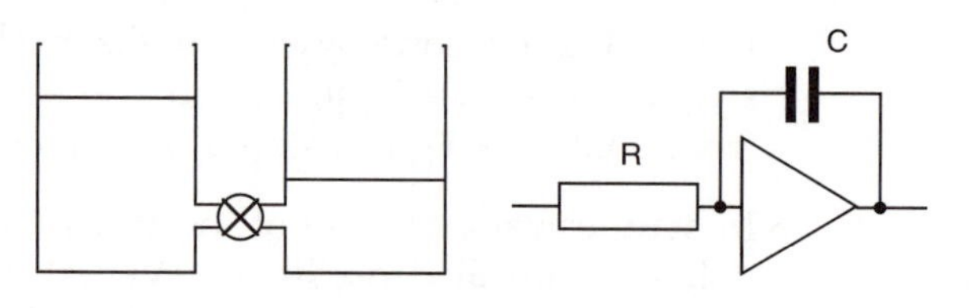

Figure 2.1.2 Two connected buckets can model an integrator circuit.

This link between a physical experiment and the numbers which can be used to describe it, is also significant because the first analogue synthesizers, and in fact, the first computers, were analogue, not digital. An analogue computer is a device which is used to solve mathematical problems by providing an electrical circuit which behaves in the same way as a real system, and then observing that happens when some of the parameters are changed. A simple example is what happens when two containers filled with water are connected together. This can be modelled by using an integrator circuit: a capacitor in a feedback loop. A step voltage applied to the integrator input simulates pouring water into one container – the voltage at the output of the integrator will rise steadily until the voltage is the same as the applied voltage, and then stops. If the integrator time constant is made larger, which is equivalent to reducing the flow of water between the containers (or making the second container larger), then the integrator will take longer to reach a steady state after a step voltage has been applied.

More sophisticated situations require more complex models, but the basic idea of using linear electronic circuits to simulate the behaviour of real-world mechanical systems can be very successful.

2.1.1 Voltage control

One of the major innovations in the development of the synthesizer was voltage control. Instead of providing mechanical control over the many parameters which are used to set the operation of a synthesizer, voltages are used instead. Since the component parts of the synthesizer produce audio signals which are also voltages, then the same signals which are used for audio can also be used for control purposes. One example is an oscillator used for tremolo or vibrato modulation when used at a frequency of a few tens of hertz, but the same oscillator becomes a sound source itself if the frequency is a few hundred hertz.

Controlling a synthesizer with voltages requires some way of manipulating the voltages themselves, and for this voltage controlled amplifiers (VCAs) are used. These use a control voltage to alter the gain of the amplifier, and can be used to control the gain of audio signals or control voltages. Using VCAs means that a synthesizer can provide a single common gain control element.

Although not all analogue synthesizers contain the same elements, many of the parts are common, and the method of control is the same throughout. Voltage control requires two main parts: sources and destinations.

Voltage control sources include the following:

- low frequency oscillators, for vibrato, tremolo and other cyclic effects
- envelope generators, which produce multi-segment control voltages, where the time and slope of each segment can be controlled independently
- pitch control: typically a pitch bend wheel or lever provides a control voltage where the amount of pitch bend is proportional to the voltage
- keyboard control: the output from a music keyboard provides a control voltage where the pitch is proportional to the voltage
- voltage controlled filters (VCFs) which can self-oscillate and so provide control signals
- voltage controlled oscillators (VCOs), which can be used as part of FM or ring modulation sounds.

Voltage controlled destinations include:

- low frequency oscillators – where the voltage is used to control the frequency or the waveshape
- envelope generators – where the voltages can be used to control the times or slopes of each of the segments
- voltage controlled filters (VCFs) – where the voltage is used to control the cut-off frequency of the filter, and perhaps the Q of the filter
- voltage controlled oscillators (VCOs) – where the voltage is used to control the frequency of the oscillator, or sometimes the shape or pulse width of the output waveform
- voltage controlled pan – where the voltage is used to control the stereo positioning of the sound
- voltage controlled amplifiers – where the voltage is used to control the gain of the amplifier.

Each of these modules will be explained in more depth in this chapter.

2.1.2 Tape and models

Not all analogue synthesizers have to be voltage controlled. The use of tape manipulation and real physical instruments to synthesize sounds might be regarded as the ultimate in 'analogue' synthesis, since it is actually possible to interact with the actual sounds directly and continuously. Despite this, the word 'analogue' usually implies the use of electronic synthesizers.

The 'source and modifier' model is often applied to analogue synthesizers, where the VCOs are the source of the raw audio, and the VCF, VCA and ADSR envelopes form the modifiers. But the same model can be applied to S&S synthesizers, or even to physical modelling. Even real world musical instruments tend to have a source (for a violin you vibrate the string using the bow) and modifier structure (for a violin it is the resonance of the body that gives the final 'tone' of the sound).

The controls of the sound source and the modifier can be split into two parts: performance controls which are altered during the playing of the instrument; and fixed parameter controls which tend to remain unchanged whilst the instrument is played.

Because it came first, many of the terminology, models, and metaphors of analogue synthesis are re-used in the more recent

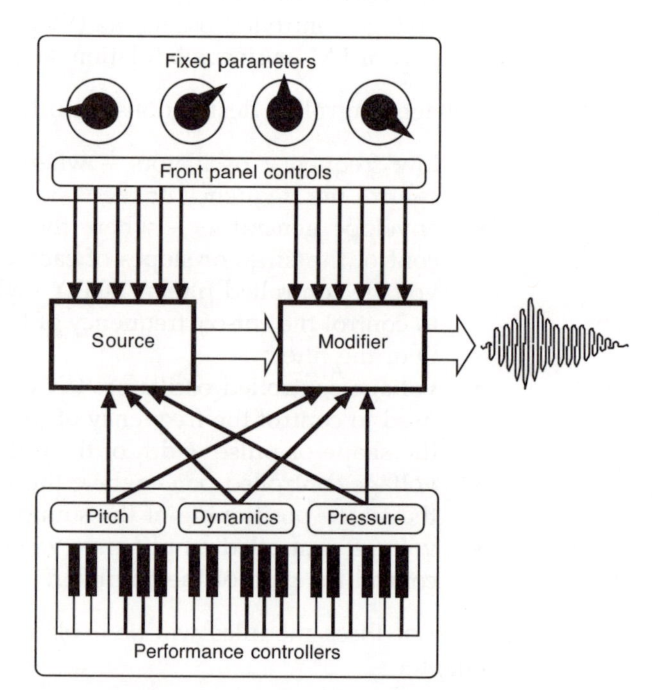

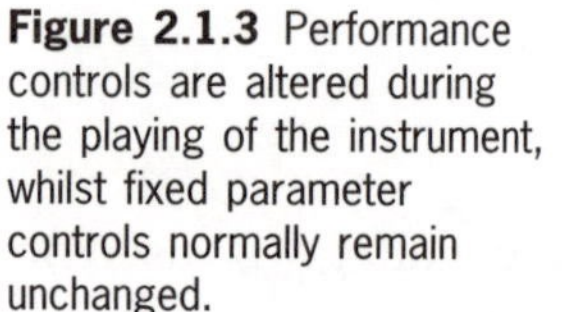

Figure 2.1.3 Performance controls are altered during the playing of the instrument, whilst fixed parameter controls normally remain unchanged.

digital methods. Although this serves to improve the familiarity for anyone who has used an analogue synthesizer, it does not help a more conventional musician who has never used anything other than a real instrument.

2.2 Subtractive synthesis

Subtractive synthesis is often mistakenly regarded as the ONLY method of analogue sound synthesis. Although there are other methods of synthesis, the majority of commercial analogue synthesizers use subtractive synthesis. Because it is often presented with a user interface consisting of a large number of knobs and switches, it is well suited to educational purposes. It can also be used to illustrate a number of important principles and models which are used in acoustics and sound theory.

Theory: source and modifier

Subtractive synthesis is based around the idea that real instruments can be broken down into three major parts: a source of sound, a modifier, which processes the output of the source, and some controllers which act as the interface between the performer and the instrument. This is most obviously apparent in many wind instruments, where the individual parts can be examined in isolation.

For example: a clarinet, where a vibrating reed is coupled to a tube, can be taken apart and the two parts investigated independently. On its own, the reed produces a harsh, strident tone, whilst the body of the instrument is merely a tube which can be shown to have a series of acoustic resonances related to its length, the diameter of the longitudinal hole, and other physical characteristics – in other words, it behaves like a series of resonant filters. Put together, the reed produces a sound which is then modified by the resonances of the body of the instrument to produce the final characteristic sound of the clarinet.

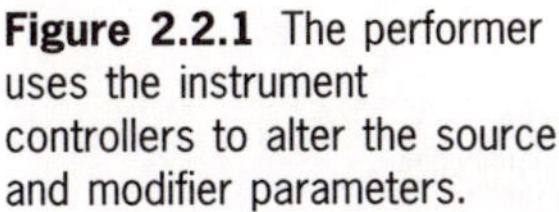

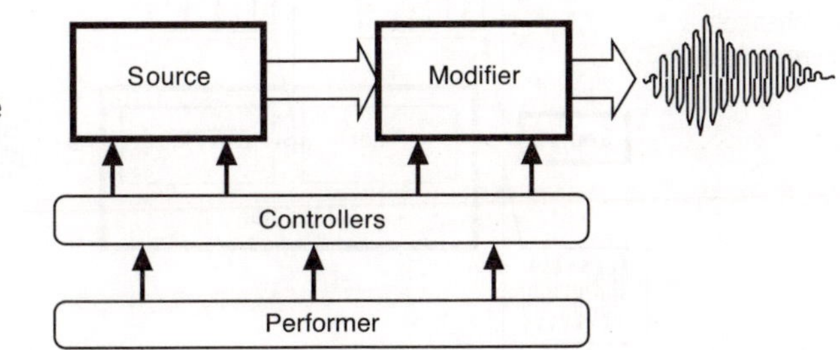

Figure 2.2.1 The performer uses the instrument controllers to alter the source and modifier parameters.

Although this model is a powerful metaphor for helping to understand how some musical instruments work, it is by no means a complete or unique answer. Attempting to apply the same concept to an instrument like a guitar is more difficult, since the source of the sound appears to be the plucked string, and the body of the guitar must therefore be the modifier of the sound produced by the string. Unfortunately, in a guitar, the source and modifier are much more closely coupled, and it is much harder to split them into separate parts. For example, the string cannot be played in isolation in quite the same way as the reed of a clarinet can, nor can the resonances of the guitar body be determined without the strings being present.

Despite this, the idea of modifying the output of a sound source is easy to grasp, and it can be used to produce a wide range of synthetic and imitative timbres. In fact, the underlying idea of source and modifier is a common theme in most types of sound synthesis.

Subtractive synthesis

Subtractive synthesis uses a sub-set of this generalized idea of source and modifier, where the source produces a sound which contains all the required harmonic content for the final sound, whilst the modifier is used to filter out any unwanted harmonics and shape the sound's volume envelope. The filter thus 'subtracts' the harmonics which are not required – hence the name of the synthesis method.

2.2.1 Sources

The sound sources used in analogue subtractive synthesizers tend to be based on mathematics. There are two basic types: waveforms and random. The waveforms are typically simple

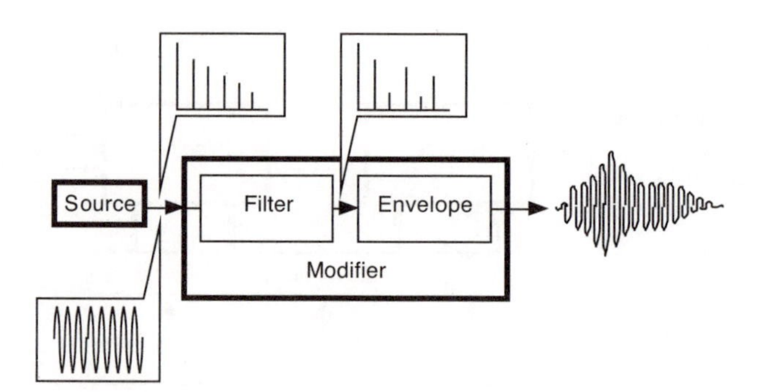

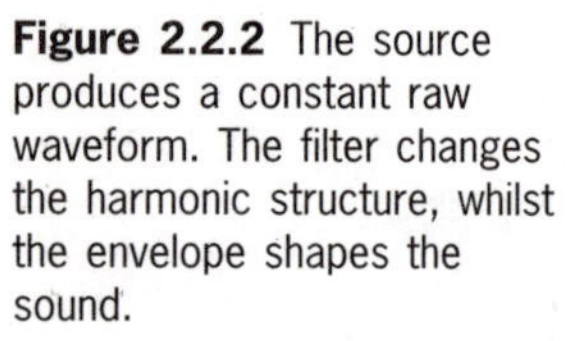
Figure 2.2.2 The source produces a constant raw waveform. The filter changes the harmonic structure, whilst the envelope shapes the sound.

waveshapes: sawtooth, square, pulse, sine and triangle are the most common. The shapes are ones which are easy to describe mathematically – and also to produce electronically. Random waveshapes produce noise, which contains a constantly changing mixture of all frequencies.

Oscillators are related to one of the component parts of analogue synthesizers: function generators. A function generator produces an output waveform, and this can be of arbitrary shape, and can be continuous or triggered. An oscillator which is intended to be used in a basic analogue subtractive synthesizer normally produces just a few continuous waveshapes, and the frequency needs to be controlled by a voltage.

Voltage controlled oscillators

Voltage controlled oscillators, or VCOs, provide voltage control of the frequency or pitch of their output. Some VCOs also provide voltage control inputs for modulation (usually FM) and for varying the shape of the output waveforms (usually the pulse width of the rectangular waveshape, although some VCOs allow the shape of other waveforms to be altered as well). Many VCOs have an additional input for another VCO audio signal, to which the VCO can be synchronized. Hard synchronization forces the VCO to reset its output to keep in sync with the incoming signal, which means that the VCO can only operate at the same or multiple frequencies of the input frequency. This produces a characteristic harsh sound. Other 'softer' synchronization schemes can be used to produce timbral changes in the output rather than locking of the VCO frequency.

A typical VCO has controls for the coarse (semitones) and fine (cents) tuning of its pitch, some sort of waveform selector (usually one of sine, triangle, square, sawtooth, and pulse), a pulse width control for the shape of the pulse waveform, and an output level control. Sometimes more than one simultaneous

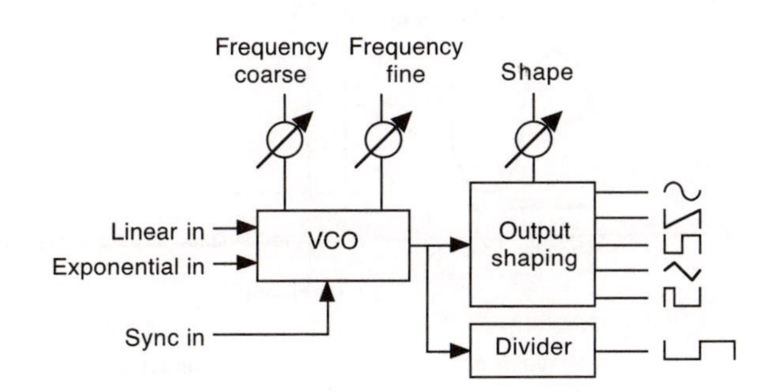

Figure 2.2.3 A block diagram of a typical VCO.

output waveform is available, and some VCOs also provide 'sub-octave' outputs which are one or two octaves lower in pitch. A control voltage for the pulse width allows the shape of the pulse waveform (and sometimes other waveforms as well), to be altered. This is called pulse width modulation, or shape modulation.

Harmonic content of waveforms

The ordering of waveforms on some early analogue synthesizers was not random. The waveforms are deliberately arranged so that the harmonic content increases as the rotary control is twisted.

The simplest waveshape is the sine wave (Figure 2.2.4). This is a smooth, rounded waveform based on the mathematical sine function. A sine wave contains just one 'harmonic': the first or fundamental. This makes it somewhat unsuitable for subtractive synthesis since it has no harmonics to be filtered.

A triangle waveshape has two linear slopes (Figure 2.2.5). It has small amounts of odd numbered harmonics, which gives it enough harmonic content for a filter to work on.

A square wave contains only odd harmonics (Figure 2.2.6). It has a distinctive 'hollow' sound and a very synthetic feel.

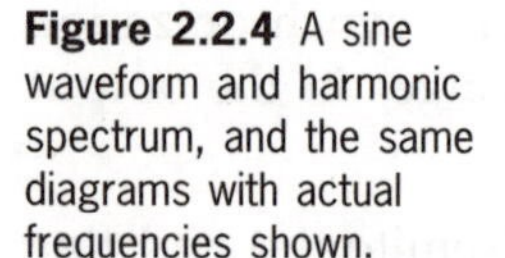

Figure 2.2.4 A sine waveform and harmonic spectrum, and the same diagrams with actual frequencies shown.

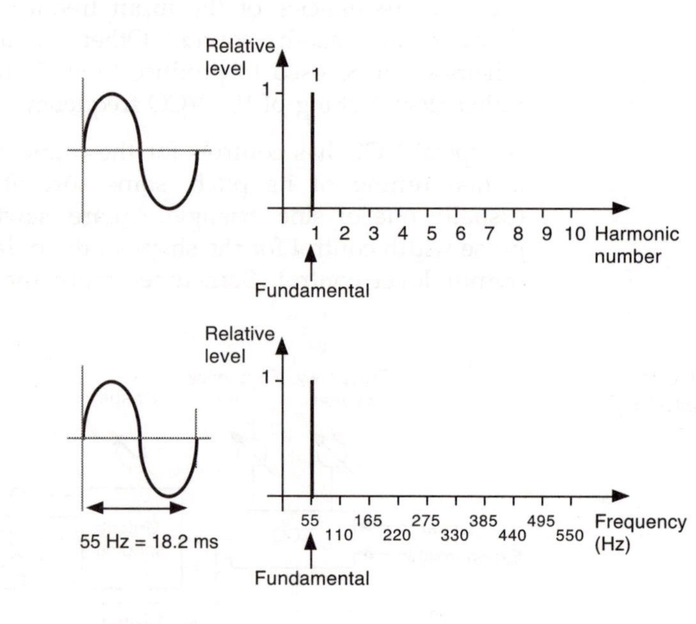

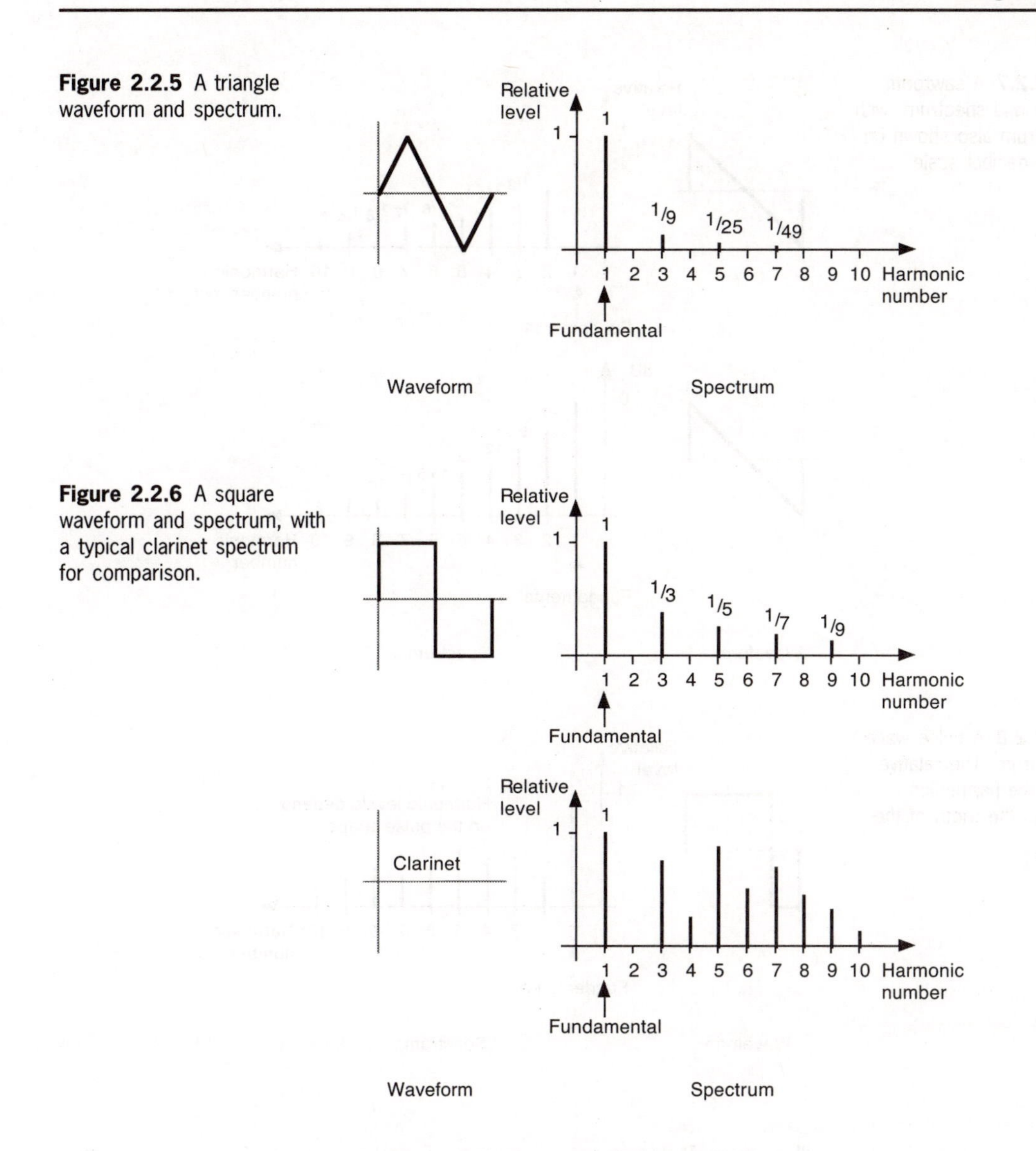

Figure 2.2.5 A triangle waveform and spectrum.

Figure 2.2.6 A square waveform and spectrum, with a typical clarinet spectrum for comparison.

A sawtooth wave contains both odd and even harmonics (Figure 2.2.7). It sounds bright, although many pulse waves can actually have more harmonic content. 'Super-sawtooth' waveshapes replace the linear slope with exponential slopes, as well as gapped sawtooths: these can contain greater levels of the upper harmonics than the basic sawtooth.

Depending on the ratio between the two parts (known as the mark-space ratio, shape, duty cycle or symmetry), pulse waveforms (Figure 2.2.8) can contain both odd and even harmonics, although not all of the harmonics are always present.

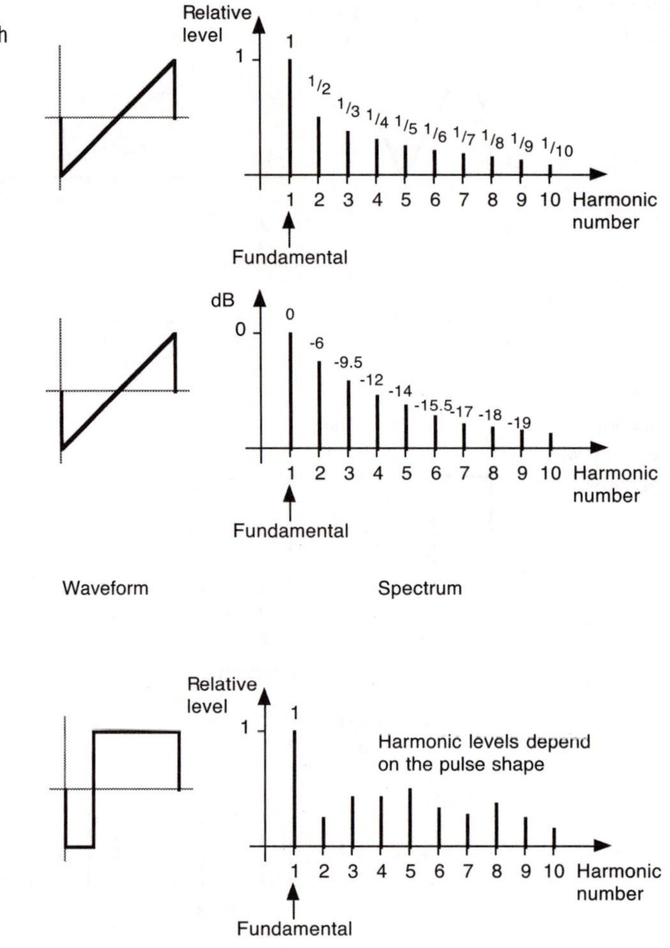

Figure 2.2.7 A sawtooth waveform and spectrum, with the spectrum also shown on a vertical decibel scale.

Figure 2.2.8 A pulse wave and spectrum. The relative levels of the harmonics depend on the width of the pulse.

The overall harmonic content of pulse waves increases as the pulse width narrows – although if a pulse gets too narrow it can completely disappear (the depth of pulse width modulation needs to be carefully adjusted to prevent this). A special case of a pulse waveshape is the square wave, where the even harmonics are not present. Pulse width modulated pulse waveforms are known as PWM waveforms, and their harmonic content changes as the width of the pulse varies. PWM waveforms are normally controlled with an LFO or an envelope, so that the pulse width changes with time. The audible effect when a PWM waveform is cyclically changed by an LFO is similar to two oscillators beating together.

All of the waveshapes and harmonic contents shown above are idealized. In the real world the edges are not as sharp, the shapes are not so linear, and the spectra are not as mathematically precise. For example, the 'sine' wave output on many VCOs is usually produced by shaping a triangle wave through a non-linear amplifier which rounds off the top of the triangle so that it looks like a true sine wave. The resulting waveform resembles a sine wave, although it will have additional harmonics – but for the purposes of subtractive synthesis, it is perfectly adequate. Section 2.3 on additive synthesis shows what real-world waveforms look like when they are constructed from simpler waveforms, rather than the perfect cases shown above.

2.2.2 Modifiers

There are two major modifiers for audio signals in analogue synthesizers: filters and amplifiers. Filtering is used to change the harmonic content or timbre of the sound, whilst amplification is used to change the volume or 'shape' of the sound. Both types of modifier are typically controlled by envelope generators, which produce complex control voltages which change with time.

2.2.3 Filters

A filter is an amplifier whose gain changes with frequency. It is usually the convention to have filters whose maximum gain is one, and so it is more correct to say that for a filter, the attenuation changes with frequency. A voltage controlled filter, or VCF, is one where one or more parameters can be altered using a control voltage. Filters are powerful modifiers of timbre, because they can change the relative proportions of harmonics in a sound.

Filters come in many different forms. One classification method is based on the shape of the attenuation curve. If a sine wave test signal is passed through a filter, then the output represents the attenuation of the filter at that frequency – this is called the frequency response of the filter. An alternative method injects a noise signal into the filter and then monitors the output, but the sine wave method is easier to carry out. The major types of frequency response curve are:

- Low-pass
- Band-pass
- High-pass
- Notch

Low-pass

A low-pass filter has more attenuation as the frequency increases. The point at which the attenuation is 3 dB is called the cut-off frequency, since this is the frequency at which the attenuation first becomes apparent. It is also the point at which half of the power in the audio signal has been lost, and so it is sometimes called the half-power point. Below the cut-off frequency, a low pass filter has no effect on the audio signal, and it is said to have a flat response (the attenuation does not change with frequency). Above the cut-off frequency then the attenuation increases at a rate which is called a slope. The slope of the attenuation varies with the design of the filter. Simple filters with one resistor and capacitor (RC) will have slopes of 6 dB per octave, which means that for each doubling of frequency, the attenuation increases by 6 dB. Each pair of RC elements is called a pole, and the slope increases as the number of poles increases. A two-pole filter will have an attenuation of 12 dB/octave, whilst a four-pole filter will have 24 dB/octave. Audibly, a four-pole filter has a more 'synthetic' tone, and makes much larger changes to the timbre of the sound as the cut-off frequency is changed. A two-pole filter is usually associated with a more 'natural' sound and more subtle changes to the timbre.

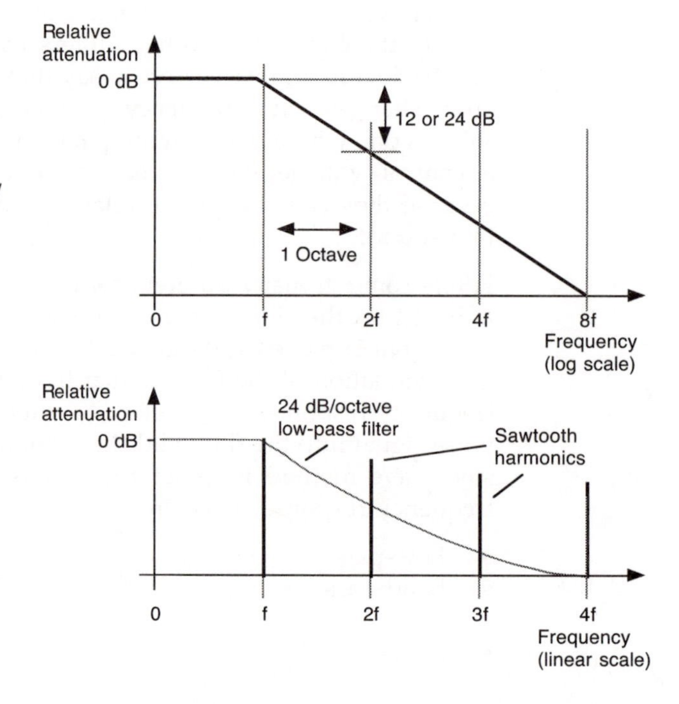

Figure 2.2.9 Filter responses are normally shown on a log frequency scale since a dB/octave cut-off slope then appears as a straight line. But harmonics are based on linear frequency scales, and on these graphs the filter appears as a curve.

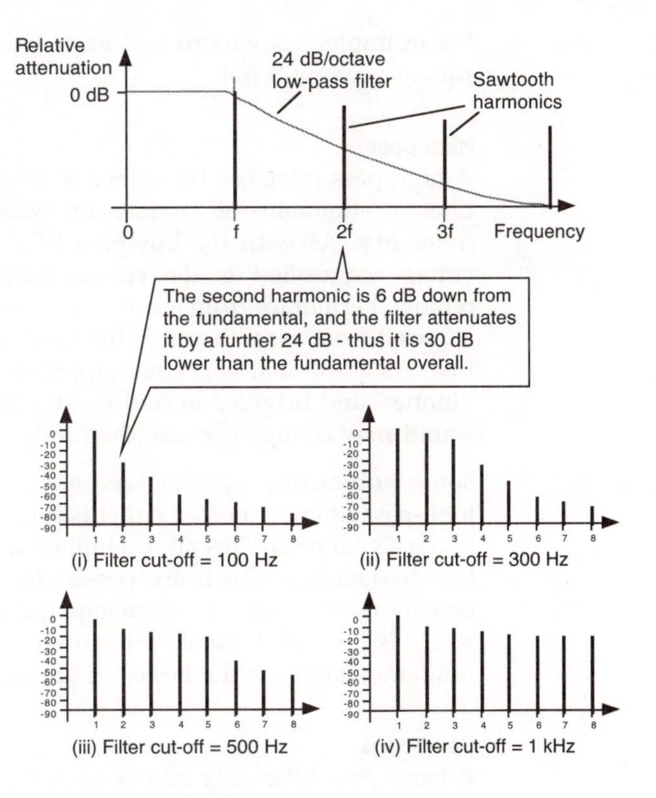

Figure 2.2.10 Low-pass filtering a sawtooth waveform with the cut-off frequency set to three different values: (i) At 100 Hz the filter cut-off frequency is the same as the fundamental frequency of the sawtooth waveform. The second harmonic is 30 dB below the fundamental and so the ear will hear an impure sine wave at 100 Hz. (ii) At 300 Hz the first three harmonics are in the pass-band of the filter, and the output will sound considerably brighter. (iii) At 500 Hz, the first five harmonics are in the filter pass-band, and so the output will sound like a slightly dull sawtooth waveform. (iv) At 1 kHz, the first ten harmonics are all in the pass-band of the filter and the output will sound like a sawtooth waveform.

Voltage-controlled low-pass filters usually have the cut-off frequency as the main controlled parameter. A sweep of cut-off frequency from high to low frequencies makes any audio signal progressively 'darker', with less high frequencies present. A filter sweeping from high frequency to low frequency of cut-off is often referred to as changing from 'open' to 'closed'. When the cut-off frequency is set to maximum, and the filter is 'open', then all frequencies can pass through the filter.

As the cut-off frequency of a low-pass filter is raised from zero, the first frequency which is heard is usually the fundamental. As the frequency rises, each of the successive harmonics (if any) of the sound will be heard. The audible effect of this is an initial sine wave (the fundamental), followed by a gradual increase in the 'brightness' of the sound as any additional frequencies are allowed through the filter. If the cut-off frequency of a low-pass filter is set to allow just the fundamental to pass through the filter, then the resulting sine wave will be identical for any input signal waveform. It is only when the cut-off frequency is increased and additional harmonics are heard, that the differences between the different waveforms will become apparent.

For example, a sawtooth will have a second harmonic, whilst a square wave will not.

High-pass

A high-pass filter has the opposite filtering action to a low-pass filter: it attenuates all frequencies which are below the cut-off frequency. As with the low-pass VCF, the parameter which is voltage controlled is the cut-off frequency. High-pass filters remove harmonics from a signal waveform, but as the frequency is raised from zero, then it is the fundamental which is removed first. As additional harmonics are removed, the timbre becomes 'thinner' and brighter in timbre, and the perceived pitch of the sound may change because the fundamental is missing.

Some subtractive synthesizers have a non-voltage-controlled high-pass filter connected either before or after the low-pass VCF in the signal path. This allows limited additional control over the low frequencies which are passed by the low-pass filter. It is usually used to remove or change the level of the fundamental, which is useful for imitating the timbre of instruments where the fundamental is not the largest frequency component.

Band-pass

A band-pass filter only allows a set range of frequencies to pass through it unchanged – all other frequencies are attenuated. The range of frequencies which are passed is called the bandwidth, or more usually, the pass-band, of the filter. Voltage-controlled band-pass filters usually have control over the cut-off frequency and bandwidth.

A band-pass filter can be thought of as a combination of a high-pass and a low-pass filter, connected in series: one after the other in the signal path. By using the same control voltage to the cut-off frequency inputs of two voltage controlled filters (one high-pass, the other low-pass), then the cut-off frequencies will 'track' each other, and the effective bandwidth of the band-pass filter will stay constant as the cut-off frequencies are changed. The width of the band-pass filter's pass-band can be controlled by adding an extra control voltage offset to one of the filters. If the cut-off frequency of the low pass filter is set below the cut-off frequency of the high-pass filter, then the pass-band does not exist, and no frequencies will pass through the filter.

Band-pass filters are often described in terms of the shape of their pass-band response. Narrow pass-bands are referred to as 'narrow' or 'sharp', and they produce marked changes in the frequency content of an audio signal. Wider pass-bands have

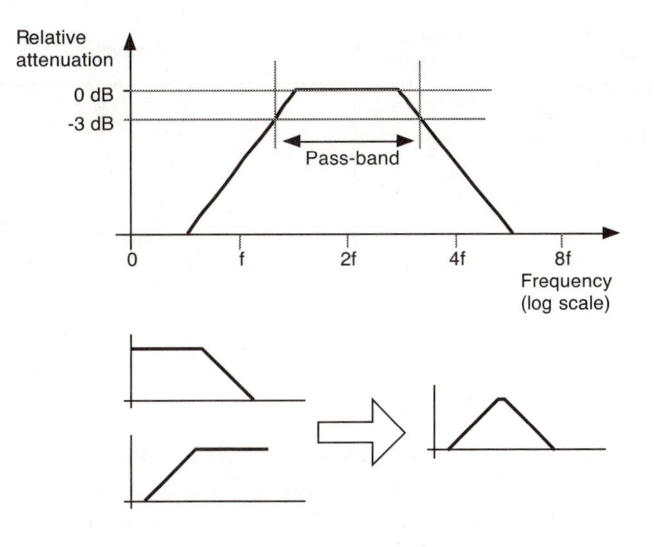

Figure 2.2.11 A band-pass filter only passes frequencies in a specific range. This is normally defined by the two points at which the filter attenuates by 3 dB. It can be thought of as a low-pass and a high-pass filter connected in series (one after the other). In the example shown, the lower cut-off frequency is about 1.3f (for the high-pass filter), whilst the upper cut-off frequency is about 2.7f (for the low-pass filter). The bandwidth of the filter is the difference between these two cut-off frequencies. Small differences are referred to as 'narrow', whilst large differences are known as 'wide'.

less effect on the timbre, since they merely emphasize a range of frequencies. The middle frequency of the pass-band is called the centre frequency.

Very narrow band-pass filters can be used to examine a waveform and determine its frequency content. By sweeping through the frequency range, each harmonic frequency will be heard as a sine wave when the centre frequency of the band-pass filter is the same as the frequency of the harmonic.

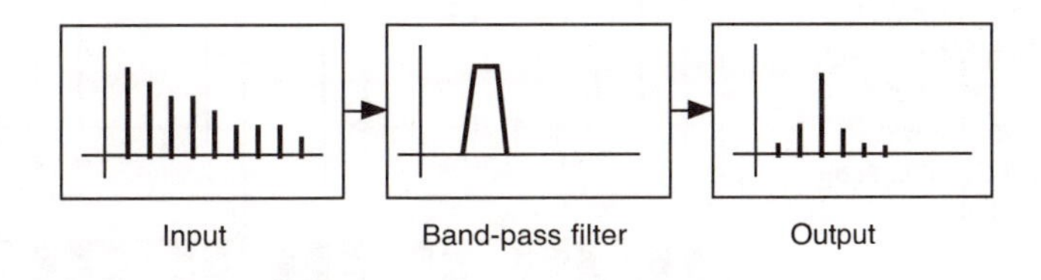

Figure 2.2.12 If a narrow band-pass filter is used to process a sound which has a rich harmonic content, then the harmonics which are in the pass-band of the filter will be emphasized, whilst the remainder will be attenuated. This produces a characteristic resonant sound. If the band-pass filter is moved up and down the frequency axis, then a characteristic 'wah-wah' sound will be heard – this is sometimes used on electric guitar sounds.

Notch

A notch filter is the opposite of a band-pass filter. Instead of passing a band of frequencies, it attenuates just those frequencies and allows all others to pass through unaffected. Notch filters are used to remove or attenuate specific ranges of frequencies, and narrow 'notches' can be used to remove single harmonic frequencies from a sound. Voltage-controlled notch filters usually provide control over both the cut-off and bandwidth of the filter.

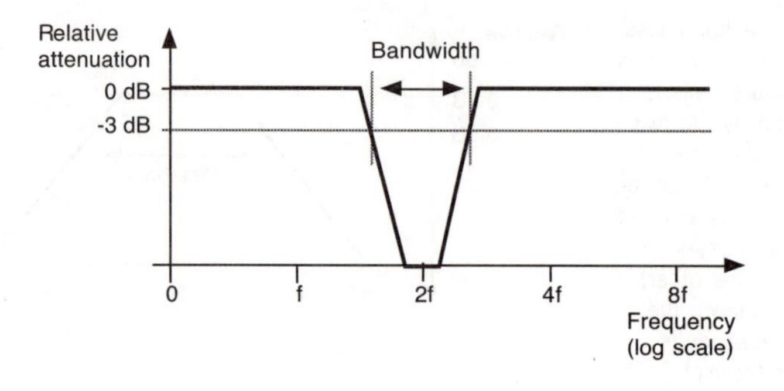

Figure 2.2.13 A notch filter is the opposite of a band-pass filter. It attenuates a band of frequencies. It can also be formed from a series combination of a low-pass and a high-pass filter.

Scaling

If the keyboard pitch voltage is connected to the cut-off frequency control voltage input of a VCF, then the cut-off frequency can be made to track the pitch being played on the keyboard. This means that any note played on the keyboard is subjected to the same filtering, since the cut-off frequency will follow the pitch being played. This is called pitch tracking or keyboard scaling.

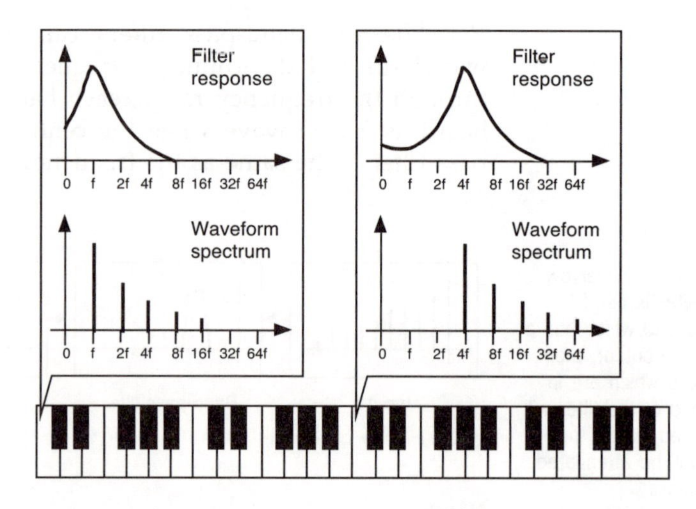

Figure 2.2.14 Filter scaling, tracking or following is the term used to describe changing the filter cut-off so that it follows changes in the pitch of a sound. This allows the spectrum of the sound produced to stay the same. In the example shown, the filter peak tracks the changes in the pitch of the sound when notes two octaves apart are played – the peak coincides with the fundamental frequency in each case. With no filter scaling then the note with a fundamental of 4f two octaves up would be strongly attenuated if the filter cut-off frequency did not change from the peak at a frequency of f.

Resonance

Low-pass and high-pass filters can have different response curves depending on a parameter called resonance or Q (short for 'quality', but rarely referred to as such). Resonance is a peaking or accentuation of the frequency response of the filter at a specific frequency. For band-pass filters, the Q figure is given by the formula:

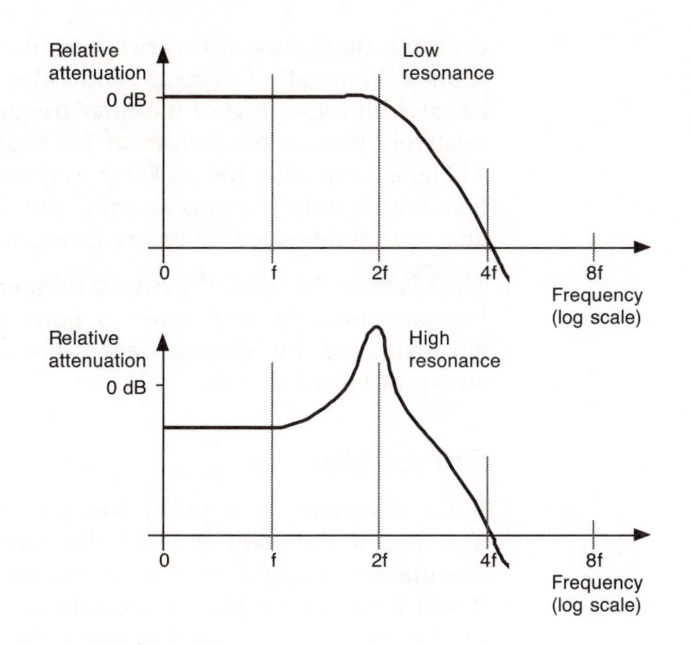

Figure 2.2.15 Resonance changes the shape of a low-pass filter response most markedly at the cut-off frequency, but it also affects the shape below the cut-off frequency. The result is a smooth transition from a low-pass to something like a narrow band-pass filter.

$$Q = \text{Centre frequency}/\text{Bandwidth (or pass-band)}$$

This formula is often also used for the resonance in the low-pass and high-pass filters used in synthesizers. For these low-pass and high-pass filters the resonance is usually at the cut-off frequency, and it forms a 'peak' in the frequency response.

In many VCFs, feedback is used to produce resonance. By taking some of the output signal and adding it back into the input of the filter, the response of the filter can be emphasized at the cut-off frequency. See Section 2.7 for more information on the implementation of filters.

Most subtractive synthesizers implement only low-pass and band-pass filtering, where the band-pass is often produced by increasing the Q of the low-pass filter so that it is a 'peaky' low-pass rather than a true band-pass filter. This phenomenon of a peak of gain in an otherwise low-pass (or high-pass) response is called 'corner peaking'. Some models of synthesizer also have an additional simple high-pass filter, whilst notch filters or band reject are very uncommon.

There are two types of filter: constant Q, and constant bandwidth. Constant-Q filters do not change their Q as the frequency of the filter is changed. This means that they are good for applications where the filter is used to produce a sense of pitch from an unpitched source like noise. Since the Q is constant, the bandwidth varies with the filter frequency and so sounds 'musical'. Constant-bandwidth filters have the same bandwidth regardless of the filter frequency. This means that a relatively narrow bandwidth of 100 Hz for a filter frequency of 4 kHz, is very wide for a 400 Hz frequency: the Q of a constant-bandwidth filter changes with the filter frequency. Most analogue synthesizer filters are constant Q.

The effect of changing the cut-off frequency of a highly resonant low-pass filter in 'real time' is quite distinctive, and can be approximated by singing 'eee-yah-oh-ooh' as a continuous sweep of vowel sounds.

Filter oscillation

If the resonance of a peaky low-pass or a band-pass VCF is increased to the point at which the filter plus its feedback has a cumulative gain of more than one at the cut-off frequency, then it will break into self-oscillation. In fact, this is one method of producing an oscillator – you put a circuit with a narrow band-pass frequency response into the feedback loop of an amplifier or op-amp. The oscillation produces a sine wave – often much purer than the 'sine' waves produced by the VCOs!

2.2..4 Envelopes

An envelope is the overall 'shape' of the volume of a sound, plotted against time (Figure 2.2.17). In an analogue synthesizer, the volume of the sound output at any time, is controlled by a voltage-controlled amplifier (see VCA), and the voltage which is used is called an envelope. Envelopes are produced by envelope 'generators', and have many variants. Envelope generators are categorized by the number of controls which they provide over the shape of the envelope. The simplest provide control only

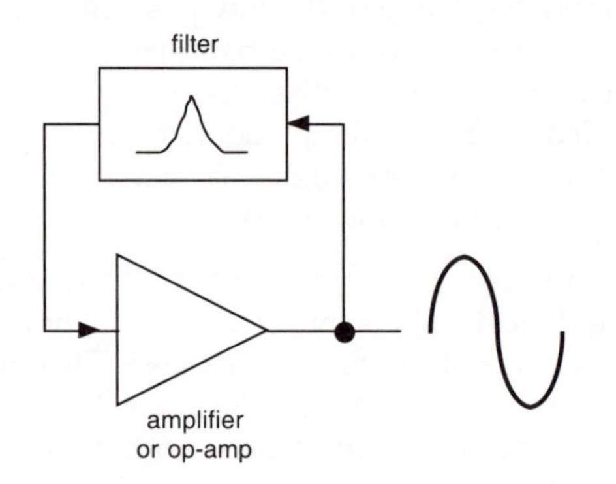

Figure 2.2.16 If a filter with a strong resonant peak in its response is connected around an amplifier, then the circuit will tend to oscillate at the frequency with the highest gain – at the peak of the filter response. This can be easily demonstrated (perhaps too easily) with a microphone and a public address system.

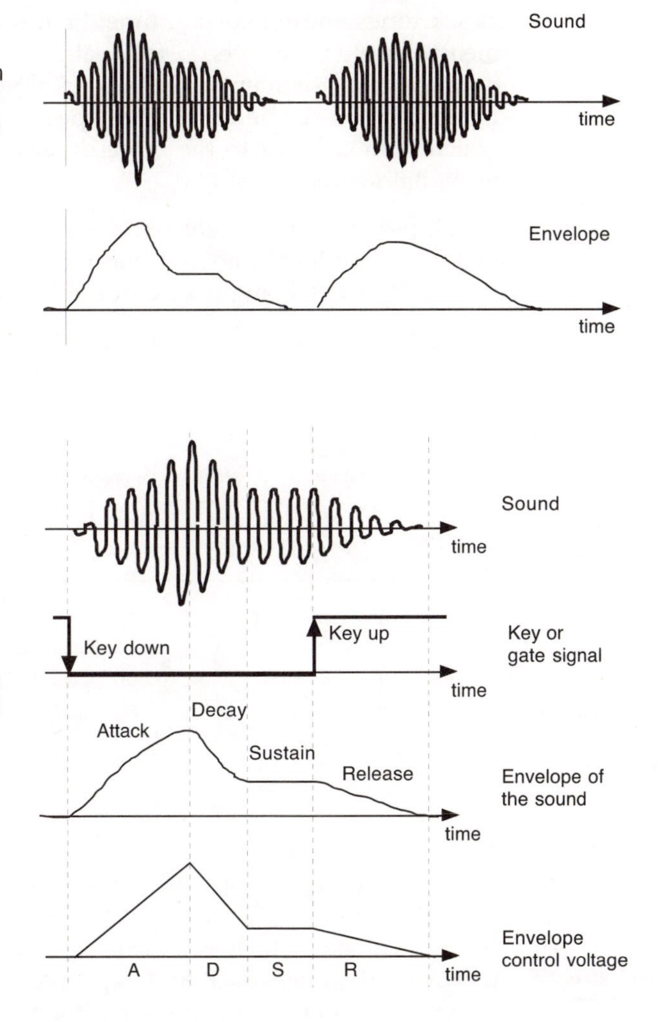

Figure 2.2.17 The 'envelope' of a sound is the overall shape – the change in volume with time. The shape of an envelope often forms a distinctive part of a sound.

Figure 2.2.18 Envelopes are divided into segments depending on their position. The start of the sound is called the 'attack segment'. After the loudest part of the sound, the fall to a steady 'sustain' segment is called the 'decay' segment. When the sound ends, the fall from the sustain segment is called the 'release' segment.

over the start and end of a sound, whilst the most complex may have a very large number of parameters.

Envelopes are split into segments or parts (Figure 2.2.18). The time from silence to the initial loudest point is called the attack time, while the time for the envelope to decrease or decay to a steady value is called the decay time. For instruments that can produce a continuous sound, like an organ, the decay time is defined as the time for the sound to decay to the steady-state 'sustain' level, whilst the time that it takes for the sound to decay to silence when it ends is called the release time.

Bowed stringed instruments can have long attack, decay and release times, whilst plucked stringed instruments have shorter attack times and no sustain time. Pianos and percussion instruments can have very fast attack times and complex decay/sustain segments. There is an almost standardized set of names for the segments of envelopes in analogue synthesizers, which is in contrast to the more diverse naming schemes used in digital synthesizers.

Envelopes are usually referred to in terms of the control voltage that they produce, and it is normally assumed that they are started by a key being pressed on a keyboard.

The following are some of the common types of envelope generator.

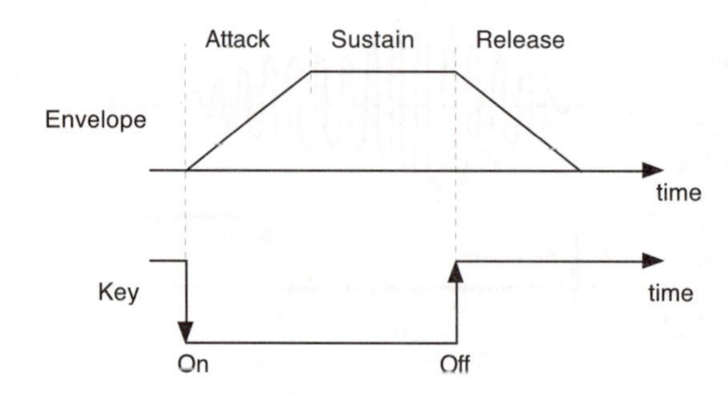

Figure 2.2.19 In an AR (attack release) envelope the pressing down of a key (or a similar gating device on a synthesizer that does not use keys) starts the attack segment. When the peak level has been reached, then the envelope stays at this level until the key is released (or the gating signal is removed) and the the envelope falls in the release segment. If the key is released whilst the envelope is in the attack segment, then the envelope normally moves to the release segment, and need not reach the peak level (see Figure 2.2.27). Some synthesizers provide a control which forces the whole of the attack segment to be completed.

Attack release (AR)

Attack release envelopes only provide control over the start and end of a sound (Figure 2.2.19). The two-segment envelope control voltage which is produced rises up to the maximum level, and then falls back to the quiescent level, which is usually zero volts. AR envelopes are often found on 1970s vintage string machines: simple polyphonic keyboards which used organ 'master oscillator and divider' technology with simple filtering and chorus effects processing to give an emulation of an orchestral string sound (see Section 3.3 on DCOs for more information).

Attack decay (AD)

If the envelope decays as soon as the attack segment reaches its maximum level, and then the decay time sets how long it takes for the envelope to drop to zero, then only percussive envelope shapes can be produced (unless the decay time is set to be very long, as in some ADR envelopes). These two-segment AD envelopes are often found connected to the frequency control input of VCOs, where the envelope then produces a rapid change in pitch at the start of the note, known as a 'chirp'. This

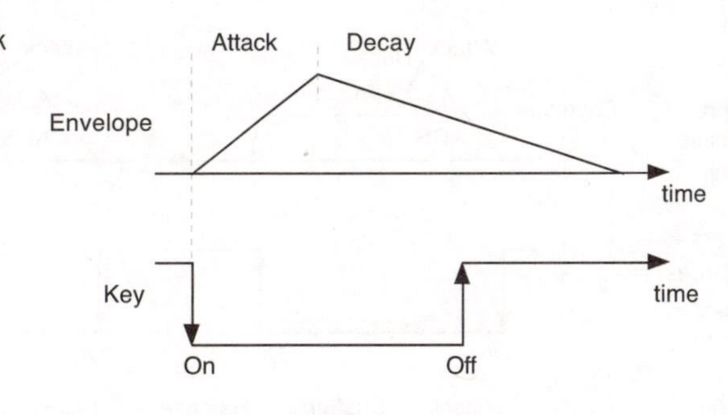

Figure 2.2.20 An AD (attack decay) envelope is similar to an AR envelope, except that there is no sustain segment. When the peak level is reached, the envelope decays, even if the key is held down.

can be effective for vocal and brass sounds. Inverting the envelope can produce downwards changes in pitch instead of upwards.

Attack decay release (ADR)

The ADR envelope uses long decay times to simulate a high sustain level, in which case the resulting envelope is very like an AR envelope, or else a percussive AD envelope by using shorter decay times (Figure 2.2.21).

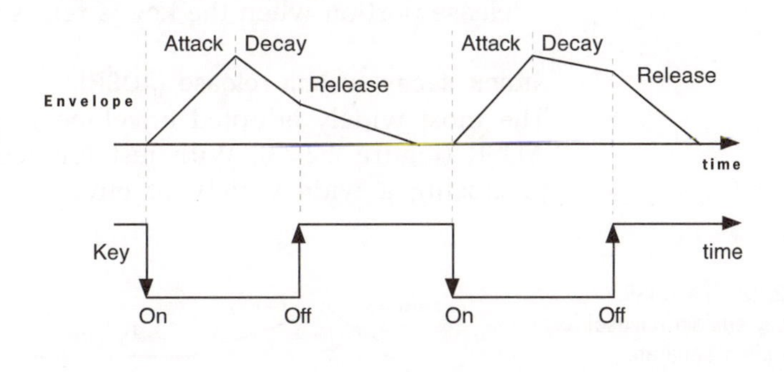

Figure 2.2.21 The ADR envelope provides control over separate decay and release segments. This allows more complex envelope shapes to be produced than is possible with AR or AD envelope generators. If the key or gate is released during the attack segment, then the envelope moves to the release segment and ignores the decay segment.

Attack decay sustain (ADS)

If a sustain level is added to an AD envelope, then the ADS envelope generator is the result (Figure 2.2.22). The attack segment reaches a maximum value, and the decay time then sets how long it takes for the envelope to reach the sustain level. Some ADS envelope generators have switches that make the release time the same as the decay time, or else have a very short release time. The type of envelope that is produced depends on the sustain level. If the sustain level is set to the maximum level (the same as the attack reaches) then two-segment AR type envelopes are produced. If the sustain level is set to zero, then only two-segment AD envelopes are produced. With the sustain level set mid-way, then four segment 'ADSR' type envelopes can

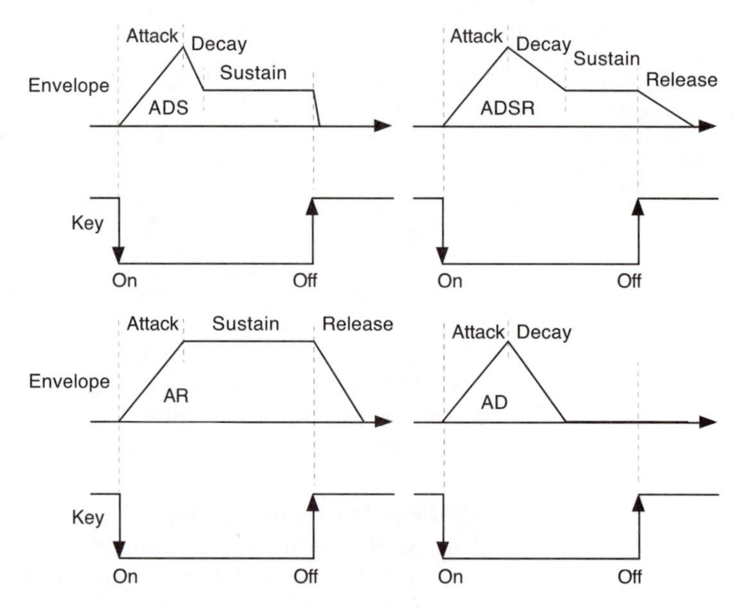

Figure 2.2.22 An ADS envelope adds a sustain segment at the end of the decay segment. The 'release' time is normally set to the same as the decay time, although some synthesizers provide a switch which forces a fast release time regardless of the setting of the decay time. An ADS envelope generator can be used to produce a wide variety of envelopes, including ones which have many of the characteristics of ADSR (see Figure 2.2.23), AR and AD envelopes.

be produced. These have an initial attack and decay portion, then the sustain portion whilst the key is held down, and then a release portion when the key is released.

Attack decay sustain release (ADSR)

The most widely adopted envelope generator is probably the ADSR (Figure 2.2.23). With just four controls, it is capable of producing a wide variety of envelope shapes, with only the

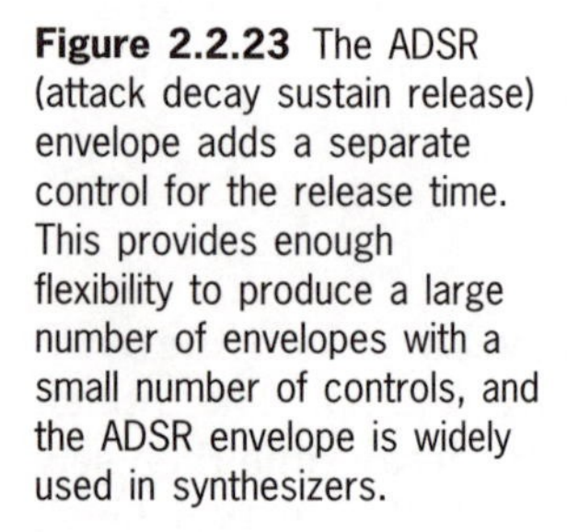

Figure 2.2.23 The ADSR (attack decay sustain release) envelope adds a separate control for the release time. This provides enough flexibility to produce a large number of envelopes with a small number of controls, and the ADSR envelope is widely used in synthesizers.

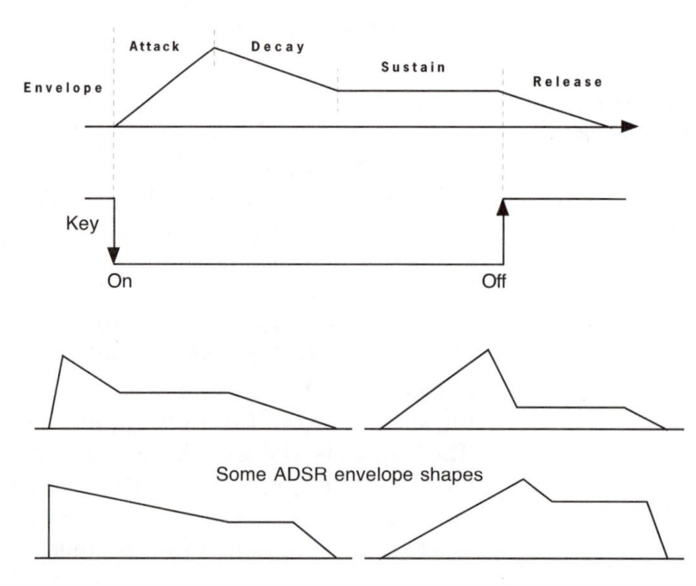

ADBDR dual-decay variant offering superior flexibility at the cost of one extra control. The ADSR envelope generator's main weakness is that the sustain segment is static: it is a fixed level. For this reason, ADSR-type envelopes are not particularly well suited to producing percussive piano-type envelopes, where the 'sustain' portion of the sound gradually decays to zero.

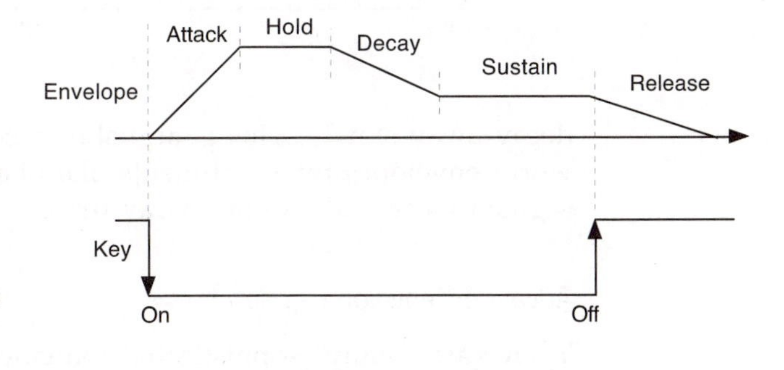

Figure 2.2.24 An AHDSR (attack hold decay sustain release) envelope adds a 'hold' segment at the end of the attack segment – rather like the sustain segment, but the length is set by a time rather than when the key or gate is released. As with other envelope shapes, if the key is released before the sustain segment, then the envelope moves to the release segment.

Attack hold decay sustain release (AHDSR)

Some envelopes force the envelope to stay at the maximum or peak level for a fixed time when the attack segment has finished, and before the decay segment can start (Figure 2.2.24). This is useful when a percussive envelope is set with very rapid attack and decay times, and the minimum length of the envelope needs to be controlled. For some sounds, an AD envelope with fast times (less than 10 ms) can be too short to be audible.

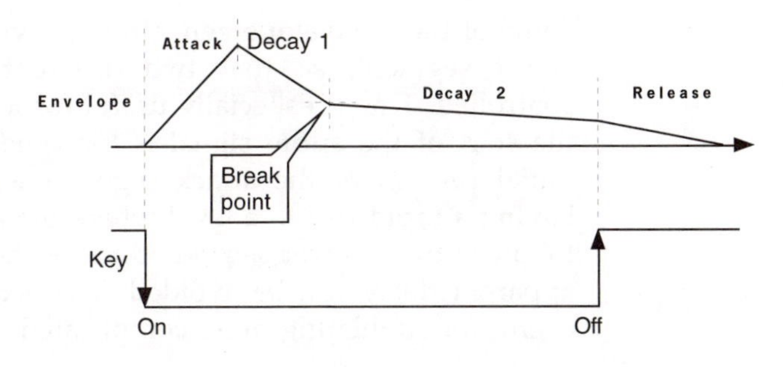

Figure 2.2.25 The ADBDR (attack decay 1 break decay 2 release) envelope has two decay segments, and the transition from one decay is set by a variable level control – rather like a sustain level control. By setting the decay time to a long value, then they can be used as pseudo-sustain segments, and so an ADBDR envelope can produce similar envelopes to an ADSR type.

Attack decay 1 break decay 2 release (ADBDR)

By splitting the decay segment into two portions, with a 'break point' level controlling when one decay portion finishes and the other starts, then a wide range of envelope shapes can be produced (Figure 2.2.25). By setting the second decay to very long times, then it can be used in much the same way as a sustain segment, although it has the advantage that it can still

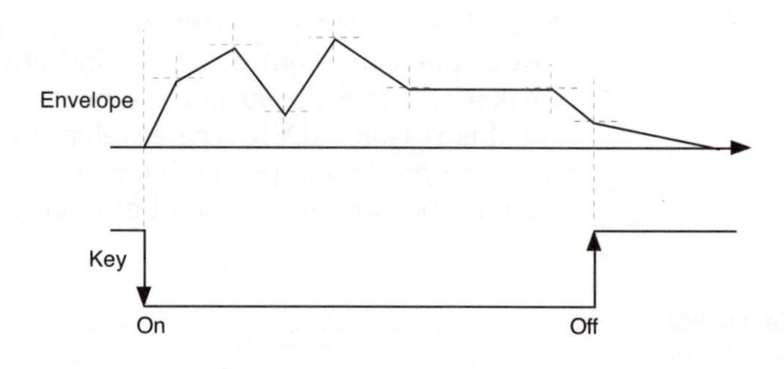

Figure 2.2.26 Multi-segment envelopes can have several attack, decay and release segments, as well as hold and sustain segments. Break points can also be used to split a segment into smaller segments.

decay away slowly. This is arguably a better emulation of real world envelopes for instruments like pianos, where the sustain segment is actually a long decay time.

Advanced envelope generators

There are many sophisticated enhancements of the basic analogue ADSR envelope generator (Figure 2.2.26). Most of these are ADSRs with the addition of initial time delay, break-points in the attack or decay segments, and times for the peak and sustain levels. Although the extra controls provide more possibilities for envelope shapes, they also greatly increase the complexity of the user interface. Delayed envelopes (denoted by an initial 'D' in the abbreviation: DADSR for delayed ADSR) are used when the start of the envelope needs to be delayed in time without the need for using a long attack time, or where the attack needs to be rapid after the delay time.

Some of these envelope generators provide a break point in the attack segment, so that two different attack times can be controlled. This is especially useful for long attack times, where the start of the audio signal is too quiet to be heard, and the initial portion of the attack segment is heard as a delay. By having a rapid rise to a level where the audio signal is audible, followed by a slower second attack portion, then this unwanted apparent delay can be avoided. This extra break point is also useful for simulating more complicated attack curves.

Break points are not always explicitly named as such. The interaction between the gate signal and the envelope often has implied break points at the transitions between attack, decay, sustain and release. These are frequently not documented in the manufacturer's product information. The usual method of operation is as follows. If the key is only held down for a short time, and the envelope is still in the attack segment when the key is released, then the envelope will go into the release segment. In

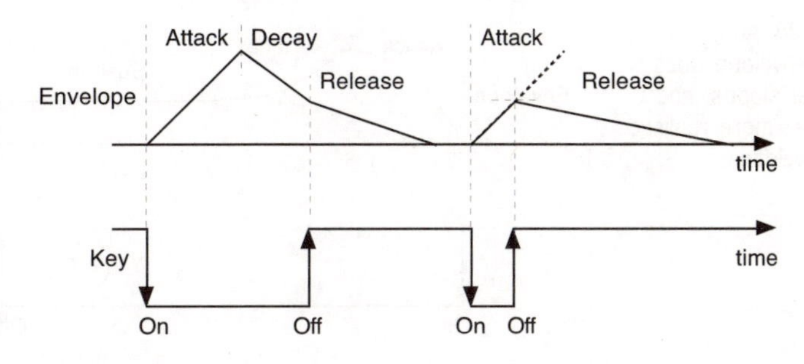

Figure 2.2.27 The transition from the attack segment to the release segment when the key or gate is released can be thought of as adding in a break point to the attack segment.

this case the envelope may not reach the maximum level, although some envelope generators always rise to the maximum level. If there is a hold time associated with the maximum level, then this is usually not affected by the key being released. If the envelope has reached the decay segment, then when the key is released, the envelope will go into the release segment.

If the initial, final, peak and sustain levels are all controllable, then the envelope flexibility can become approximately equivalent to the multi-segment envelopes often found in digital synthesizers, although the terminology is normally very different. See Chapter 5 for more details on digital envelopes.

Some analogue synthesizers only have one envelope generator, which is then used to control both the VCF and VCA. If two envelopes are available, then patching one to the filter and the other to the amplifier provides independent control over the volume and timbre. A third envelope could be used to control the pitch of the VCOs, or perhaps the stereo position of the sound using two VCAs arranged as a pan control.

Linear or exponential?

Many real-world quantities change in a non-linear way. This can be because of the process involved, or because of the way that the change is perceived. For example, the theoretical population growth curve of many animal species shows an exponential or power-law growth because the initial two animals produce two new individuals, who then eventually join the breeding population, and then these four individuals produce four new offspring. The doubling of the population in each successive generation produces a rapidly increasing population curve. Similarly, because human ears perceive sound in a non-linear way, each doubling of the apparent volume level requires about ten times the energy in the sound. Again, the relationship connecting the two variables is a non-linear one.

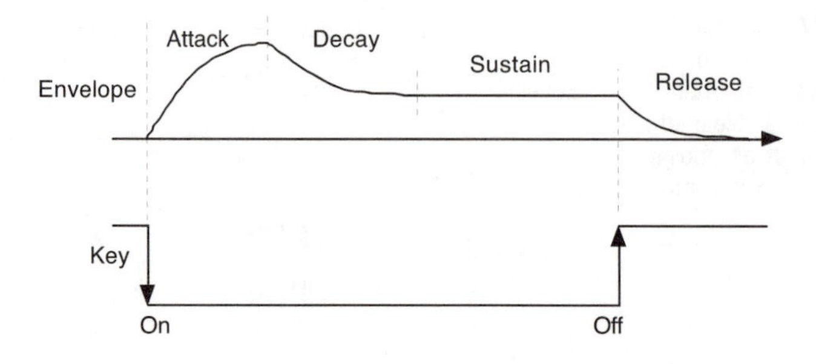

Figure 2.2.28 An exponential envelope does not use linear slopes, and often provides more realistic sounding envelopes.

Many natural sound envelopes have non-linear curves. Changes are usually rapid at first, and gradually slow down (Figure 2.2.28). This is particularly apparent with the attack segment of envelopes, where a linear rise in volume sounds too slow at first, whereas an exponential rise in volume sounds 'correct' – in fact, it sounds 'linear' to the human ear! Some envelope generators enable a switched selection between linear and exponential curves. Envelope generators with break points in the attack, decay and release segments can produce similar effects to exponential curves, albeit with a crude approximation.

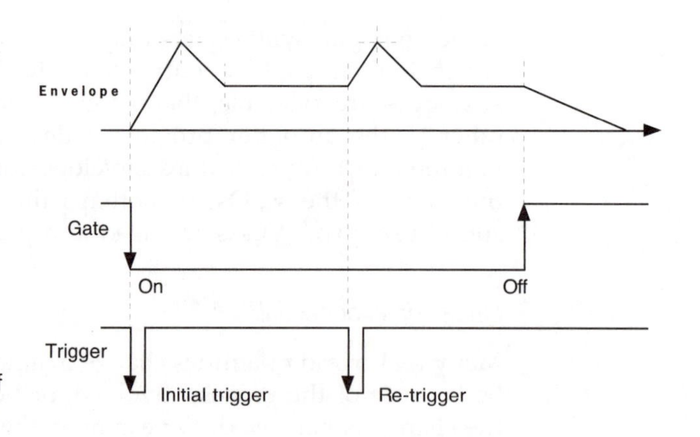

Figure 2.2.29 The re-triggering of an envelope generator can sometimes be used to add in a break point and start a new attack, normally from the level which had been reached by the envelope. The overall length of the envelope is controlled by the key being pressed down, or a similar gate control in synthesizer which are not controlled by a keyboard. The re-triggering of the envelope is controlled by a trigger signal which is generated by the start of each new note. This is normally found on monophonic synthesizers, where the gate is produced globally from any keys which are being held down, whilst the triggers are produced individually by each key.

Triggering

The initiation of an envelope generator is often assumed to be caused by a key being pressed on a music keyboard. Although this is the way that many synthesizers are set up, it is not the only way that envelopes can be started – an LFO or VCO could provide a trigger which will start the envelope generator. In this case, the envelope is not tied to the keyboard, and can be used when a complex repeated control voltage is required.

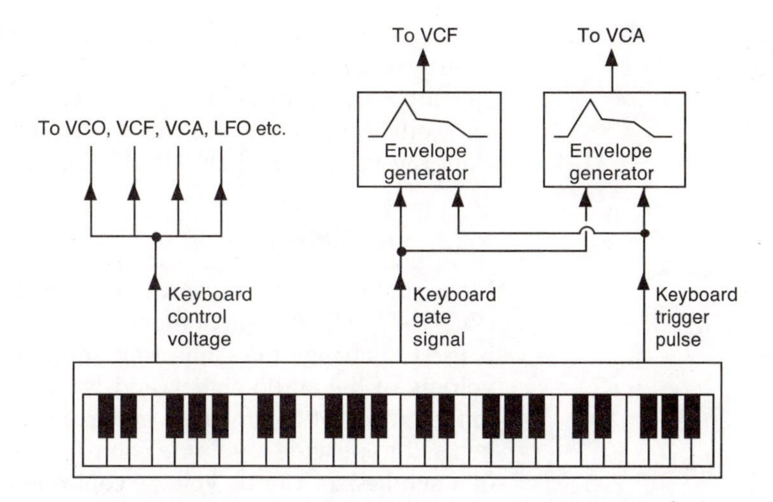

Figure 2.2.30 The gate and trigger routeing from a keyboard to the envelope generation is normally fixed, whilst the keyboard control voltage can be routed to a number of destinations.

When the keyboard is used to start an envelope, two separate signals are produced. The 'gate' signal indicates when the key is up or down, whilst the start of the key depression is shown by a 'trigger' pulse. The response of an envelope generator to these two signals depends on how the envelope generator is configured.

'Single trigger' envelope generators start when they receive a gate and a trigger, and progress through the envelope, entering the release segment when the gate signal ends to indicate that the key is no longer being held down.

'Multi trigger' envelope generators start when they receive a gate signal and a trigger pulse, but additional trigger pulses will restart part of the attack segment and the decay segment. These extra trigger pulses are normally only produced by monophonic synthesizers (one note at once) when a key is held down, and another key is pressed.

'LFO trigger' or 'txternal trigger' envelope generators normally ignore the trigger pulse, and treat the input signal as a gate. The width of the LFO waveform, or the length of the external signal set the length of the gate signal.

Whereas sources of audio signals or control voltages can be routed to almost any destination in a synthesizer, the routeing of trigger and gate signals is often much more restricted – usually they are hard wired from the keyboard.

Voltage-controlled parameters

Some envelope generators provide voltage control of the segment times and levels. This enables the shape of the envelope to be changed with one or more control voltages. One use of this facility is for 'scaling', where the length of all the times in the envelope are changed to imitate variations in envelope shape with pitch, in which case the control voltage would be derived from the keyboard pitch control voltage.

2.2.5 Amplifiers

Most analogue synthesizers have a voltage controlled amplifier (VCA) as the final stage of the modifier section. The control voltage is used to change the gain of an amplifier. The VCA controls the volume of the audio signal, and is sometimes connected directly to the output of an envelope generator. An offset voltage can also be used to provide a volume control – so even the output volume of a synthesizer can be voltage controlled (Figure 2.2.31).

There are two types of input to VCAs:

- Linear inputs are used for tremolo and AM. They are also used with exponential curve envelopes.
- Exponential inputs are used for volume changes and linear curve envelopes.

The combination of linear and exponential envelopes with linear and exponential VCAs provides much scope for confusion. Using an exponential curve envelope with an exponential VCA produces a result which has sudden or abrupt changes rather than steady transitions.

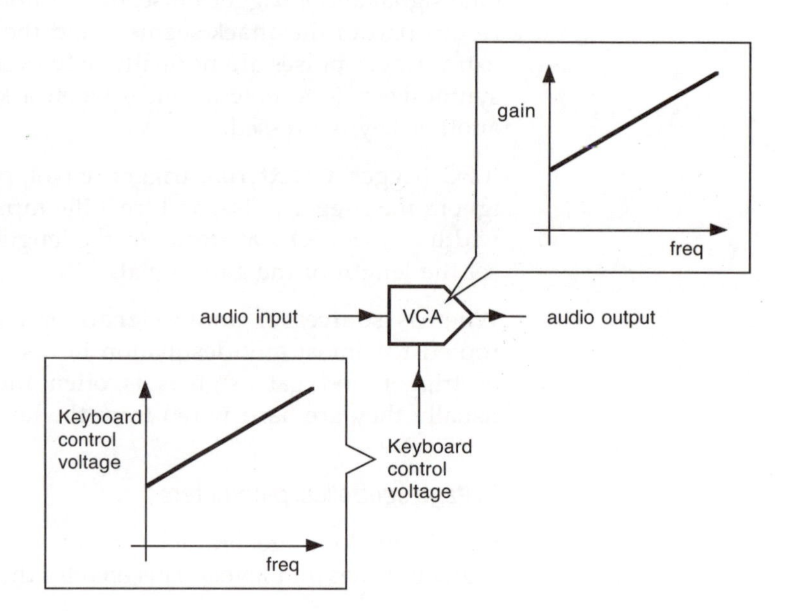

Figure 2.2.31 A VCA can be used to produce control of volume which follows the keyboard by routeing the keyboard control voltage to the VCA gain control. This is similar to the tracking of a filter, and produces a coarse high-pass filtering effect, where higher notes are attenuated less than lower notes.

Tremolo is a cyclic variation in the volume of a sound. It is produced by using an LFO control voltage to alter the gain of a VCA. Tremolo normally uses a sine or triangle waveform at frequencies between 5 and 20 Hz. Higher frequencies from an LFO or a VCO produce amplitude modulation (AM), where the output of the VCA is a combination of the audio signal and the LFO or VCO frequency. See section 2.4 for more details on AM.

Apart from their normal use as volume controlling devices, VCAs can also be used to provide 'filtering' effects. By connecting the keyboard pitch voltage to the control voltage input of a VCA, the gain of the VCA is then dependent on the pitch control voltage from the keyboard. Since the keyboard pitch voltage normally rises as the keyboard note position rises, then the VCA will act as a high-pass filter, since low notes will be be at a lower volume than higher notes. By inverting the keyboard pitch voltage, then a low-pass 'filter' effect can be produced. This coupling of the VCA to the keyboard pitch voltage is called 'scaling', since the output of the VCA is scaled according to the pitch.

2.2.6 Other modifiers

LFOs

Low-frequency oscillators are used to produce low frequency control voltages. They are in two forms: VCOs or special purpose oscillators. VCO-based LFOs can have their frequency controlled with an external control voltage, whilst special purpose oscillators cannot.

Unlike audio frequency VCOs, LFOs need to produce waveforms where the shape is normally more important than the harmonic content. So, in addition to the sine, square, pulse and sawtooth waveforms, additional shapes like an inverted sawtooth are also provided. These might be used when the LFO is connected to a source like a VCO, and is controlling the pitch of the VCO. The basic sawtooth, or ramp-up waveform, would then produce a pitch that rose slowly and dropped quickly. The inverted shape, although still called a sawtooth, would now be a ramp-down waveform, and would give a pitch that rose quickly and dropped slowly.

Two specialized LFO waveform outputs are often found on LFOs: sample and hold, and arbitrary.

Sample and hold is the name given to a random or repetitive sequence of control voltages which are produced by using the

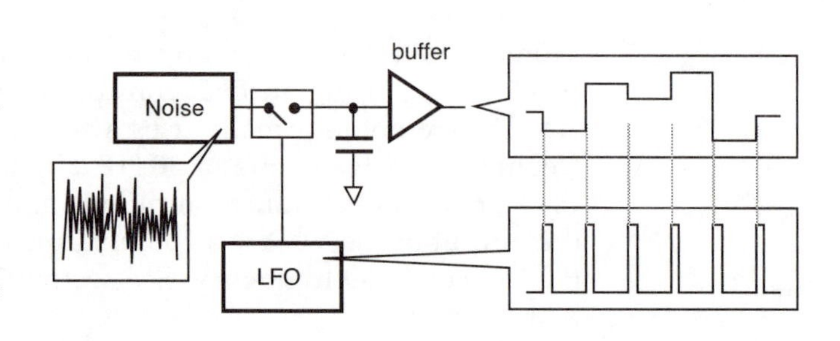

Figure 2.2.32 Sample and hold circuits take regular 'samples' of a noise (or other waveform) and then maintain that level until the next sample is taken. The rate of the samples is normally controlled by an LFO. The output consists of a series of steady voltages with rapid transitions, but whose level is not predictable. If the noise source is replaced with a repetitive waveform, then the output levels depend on the timing relationships between the sample LFO and the waveform being sampled.

LFO to repeatedly take the value of another voltage source, and then keeping that value until the next time that it measures the value again (Figure 2.2.32). This process is called 'sampling' the value, and that value is then 'held' until the next sample is taken. The technique is thus called sample and hold.

If the voltage source that is sampled is noise, then the sample values will be random in level. This produces a series of values which do not repeat, and are not regular or predictable. The regular timing from the periodic sampling is the only known quantity. If another LFO or VCO is sampled, then one of two results is possible. If the second LFO or VCO is not synchronized to the sampling LFO, then the output of the sample and hold will be a series of values which are partly random and partly repetitive – the exact pattern depends on the relative frequencies and the LFO/VCO waveform. If the sampling LFO and the second LFO/VCO are synchronized so that they are locked together with the LFO/VCO being a multiple or fraction of the sampling LFO, then the output pattern will repeat.

Sample and hold is often used to control the cut-off frequency of a resonant low-pass filter. This is an effective way of providing 'interest' and 'movement' in a sound when it is in the sustain segment of an ADSR envelope. Unfortunately, the rhythmic random changing timbres that this produces have become an over-used cliché.

Arbitrary waveforms are ones which are constructed from a series of simpler waveform segments. There are many variations possible:

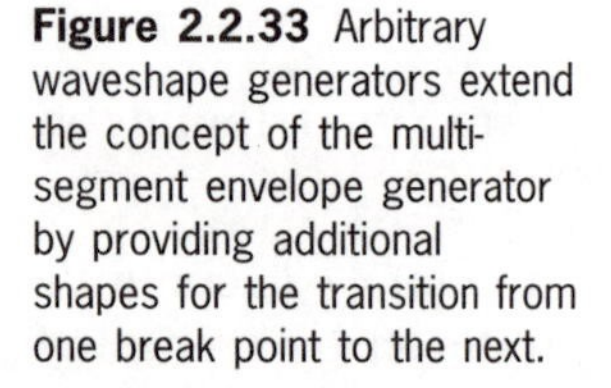

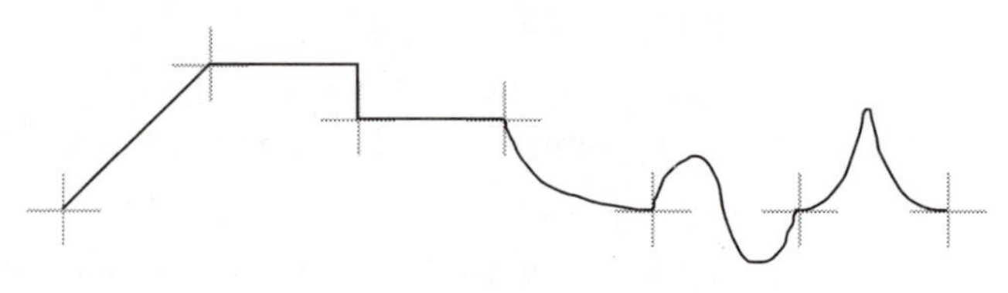

Figure 2.2.33 Arbitrary waveshape generators extend the concept of the multi-segment envelope generator by providing additional shapes for the transition from one break point to the next.

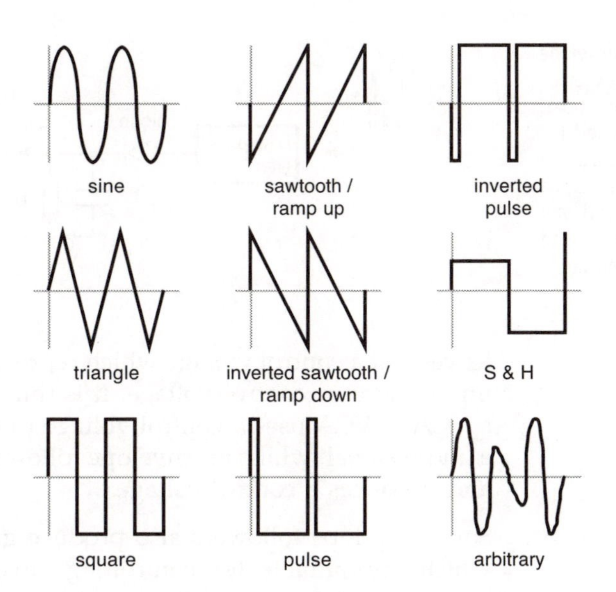

Figure 2.2.34 LFO outputs are normally provided in a variety of shapes to give additional control possibilities – although in practice, the sine wave is almost always used for vibrato or tremolo, and the square wave is almost exclusively used for trills. The other shapes are often presented in normal and inverted forms, and are often used for special effects sounds.

- two or more levels (rather like a simple sequencer)
- two or more straight-line slopes (much like an envelope)
- two or more curves (exponential, linear, sine, power law, etc.).

Arbitrary waveform generators are also called function generators. They can be used to replace envelope generators, control panning and effects settings, and even act as simple sequencers to produce a series of pitched notes.

LFO output waveforms are frequently available simultaneously, so that a sine wave can be used at the same time as a square waveform. The common outputs are:

- sine
- triangle
- square
- sawtooth/ramp up
- inverted sawtooth/ramp down
- pulse
- inverted pulse (100%-pulse width)
- S&H
- arbitrary.

Envelope follower

An envelope follower takes an audio signal or control voltage, converts it to just positive values, and then low-pass filters it with a filter which has a very low cut-off frequency – a few hertz. This removes any high frequencies from the input, and

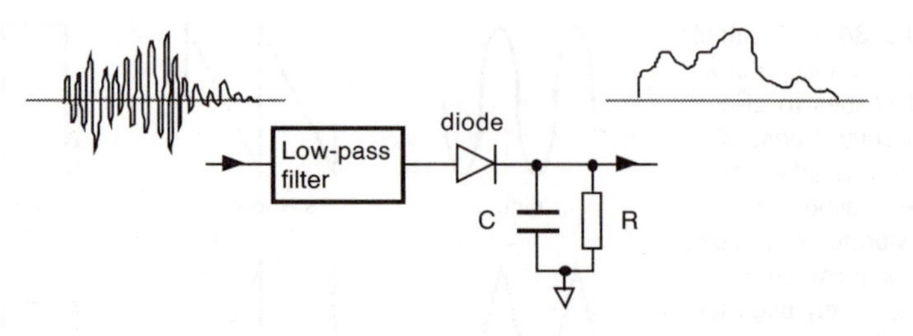

Figure 2.2.35 An envelope follower is used to 'extract' the envelope from an audio signal. This can be used to process external signals in a synthesizer. The audio signal is low-pass filtered, and then a diode-pump circuit is used to provide the final output voltage.

leaves just a control voltage which represents the envelope of the input audio or control voltage. It is thus almost the opposite of a VCA: a VCA uses a control voltage to change the envelope of a audio signal, whilst an envelope follower takes an audio signal and produces a control voltage.

Some envelope followers also produce gate and trigger outputs which are suitable for controlling envelope generators – the envelope follower is then a complete module for interfacing an external audio signal with an analogue synthesizer. If the envelope follower is used to process source control voltages, then it can be used to 'smooth' rapidly changing waveforms which have sharp transitions, or even produce portamento effects if the keyboard pitch control voltage is processed.

Externally triggered sample and hold

If a sample and hold circuit has an external sample clock input, then it can be used to sample voltage sources at non-periodic intervals. One suitable sample clock source is the keyboard gate or trigger signals. Using the keyboard to control the sample and hold produces an output which changes only when a new key is pressed on the keyboard. By using an envelope follower to produce gate or trigger signals from an external audio input, then the sample and hold can be driven from an external audio signal. In this way, any audio signal can be used as a source of control voltages.

Waveshaper

Although rarely implemented on analogue synthesizers, the waveshaper is a non-linear amplifier which allows control over the relationship between the input and output signals. Any non-linearity in this relationship changes the shape of the waveform passing through the waveshaper, and this changes the harmonic content of the signal. Chapter 5 contains more information on the use of waveshaping in digital synthesizers.

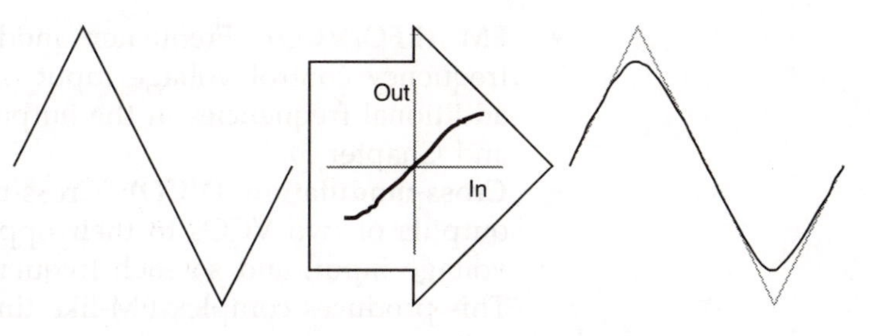

Figure 2.2.36 A waveshaper uses a non-linear transfer function to change the shape of a waveform. This is often used to convert a triangle waveform into an approximation of a sine wave, and is adequate for shaping LFO and VCO outputs.

Another interpretation of an analogue waveshaper is that it adds distortion to the signal, and so it is best used for monophonic signals. A more familiar waveshaper is the 'fuzz box' used by guitarists, where the passing of polyphonic audio signals through a clipping circuit produces large amounts of distortion.

Modulation

Modulation is another type of modifier. Any parameter which can be voltage controlled is a potential means of modulation. Although VCAs are available from the front panels of many analogue synthesizers, they are also used inside to allow control voltages to act as modulators – anywhere where a control voltage is used to change the amplitude or level of a signal or control voltage.

Some of the many possible ways that sources can be modified using modulation are:

- **LFO (LFO/envelope/keyboard)**: LFO modulation changes the rate or frequency of the LFO. This can be used to produce vibrato or tremolo whose rate is not fixed.
- **VCO mod (LFO/envelope/keyboard)**: LFO modulation of a VCO produces vibrato. Envelope modulation produces pitch sweeps. Keyboard modulation changes the scaling of the VCO: it can change the keyboard so that an octave on the keyboard represents any pitch interval to the VCO.
- **Filter mod (LFO/envelope/keyboard)**: LFO modulation of a filter produces cyclic timbre changes. Envelope modulation produces dynamic timbral changes during the course of a single note. Keyboard modulation controls how the filter 'tracks' the note on the keyboard.
- **PWM (LFO/envelope/keyboard)**: PWM modulation changes the timbre of the source waveform.
- **AM (LFO/VCO)**: Amplitude Modulation with low frequencies produces tremolo. At higher frequencies it adds extra frequencies to the audio signal. (See Section 2.4).

- **FM (LFO/VCO)**: Frequency modulation uses the linear frequency control voltage input of the VCOs. It produces additional frequencies in the output signal. (See Section 2.4 and Chapter 5).
- **Cross-modulation (VCO)**: Cross-modulation connects the outputs of two VCOs to their opposites' frequency control voltage input, and so each frequency modulates the other. This produces complex FM-like timbres, but it can be difficult to control and keep in tune.
- **Pan (LFO/VCO/envelope/keyboard)**: LFO modulation of the stereo pan position produces 'auto-pan', where the audio signal moves cyclically from one side of the stereo image to the other. VCO modulation can spread individual harmonics across the stereo image. Envelope modulation moves the image with the note envelope. Keyboard modulation places notes in the stereo image dependent upon their position on the keyboard.
- Many other sources and modifiers can be modulated. The effects section of many analogue synthesizers allows parameters like the reverberation time, flange speed, and others to be controlled.

Controllers

In conventional instruments, the control of the sound production is often a mechanical linkage between the performer and the instrument. A saxophone player uses a number of levers to control the opening and closing of the holes which determine the effect length of the saxophone. Control over the timbre can be accomplished by how the lips grasp the mouthpiece and the reed, as well as the use of the tongue. Further expression comes from the lungs with control over air pressure.

The interfacing between the performer and the synthesizer sound generation circuitry is accomplished by one or more controller devices. The main note pitch controller is usually a modified organ type keyboard, although sometimes weighted action piano-type keyboards are used. Changes in pitch are normally produced with a rotary control called a pitch bend wheel, and a similar control is used to add in modulation effects like vibrato or tremolo. Control over volume and timbre can be accomplished by using a foot pedal – as used in organs for volume.

Keyboard

The familiar music keyboard with its patterned combination of black and white keys is widely used as the main discrete pitch control for note selection, as well as initiating envelopes. Although normally connected together, the pitch selection and envelope triggering functions can be separated.

Pitch bend

Continuous control over the pitch is achieved by using a 'pitch bend' controller. These are normally rotating wheels or levers, and usually change the pitch of the entire instrument over a specified range (often a semitone or a fifth). They produce a control voltage whose value is proportional to the angle of the control. Pitch bend controls normally have a spring arrangement which always returns the control to the centre 'zero' position (no pitch change) when it is released. This central position is often also mechanically detented, so that it can be felt by the operator, since it will require force to move it away from the centre position.

Modulation

Modulation is controlled using rotary wheels or lever, where the control voltage is proportional to the angle of the control. Modulation controllers are not normally sprung so that they return to the centre position. Some instruments allow pressure on the keyboard to be used as a modulation controller.

There have been some attempts to combine the functions of pitch bend and modulation into a single 'joystick' controller, but the most popular arrangement remains the two wheels: pitch bend and modulation.

Foot controllers

Foot controllers are pedals which provide a control voltage which is proportional to the angle of the pedal. Although associated with volume control, they can be used as modulation controls, or even as pitch bend controls.

Foot switches

Foot switches are foot-operated switches which normally only have two values (some multi-valued variants are produced, but these are rare). They are used to control parameters such as sustain, portamento, etc.

See Chapter 7 for more details on controllers.

2.2.7 Using analogue synthesis

Learning how to make the best use of the available facilities provided by an analogue synthesizer requires time and effort.

Although there are a number of 'standard' configurations of VCO, VCF, VCA and envelopes, the key to making the most of an analogue synthesizer is understanding how the separate parts work: both in isolation and in combination. Roland (1978, 1979) and De Furia (1986) are excellent references for further reading on this subject.

As a brief introduction to some of the techniques of using an analogue synthesizer, the remainder of this section shows how a subtractive analogue synthesizer can be an excellent learning tool for exploring some of the principles of audio and acoustics. Here are some of the demonstrations which can be carried out using a subtractive synthesizer.

Harmonic content of waveforms

The harmonic content of different waveshapes can be audibly demonstrated by using a low-pass VCF with high resonance (set just below self-oscillation), or a narrow band-pass filter. Each VCO waveform is connected to the filter input, and the filter cut-off frequency is slowly increased from zero to maximum (Figure 2.2.37). As the resonant peak passes the fundamental, the filter output will be a sine wave at that frequency. As the cut-off frequency is increased further, the fundamental sine wave will disappear, and the next harmonic will be heard as the cut-off frequency matches the frequency of the harmonic. The audible result is a series of sine waves, whose frequency matches the frequencies of the harmonics.

If noise is passed through the filter, then the output will be sine waves whose frequencies will be within the pass-band of the resonant peak, and whose levels will change randomly. The audible result is rather like whistling.

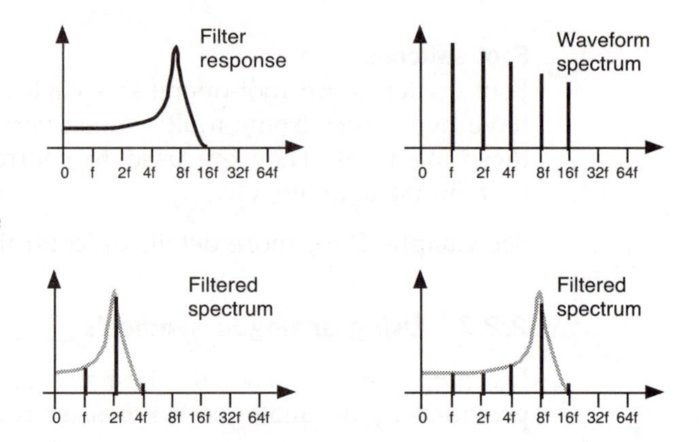

Figure 2.2.37 By varying the cut-off frequency of a resonant low-pass filter, the harmonic content of a waveform can be heard. As each of the harmonics which are present in the spectrum pass through the peak of the filter, they will be clearly heard. The frequency of the harmonic can be determined by noting the frequency of the filter when the harmonic is heard.

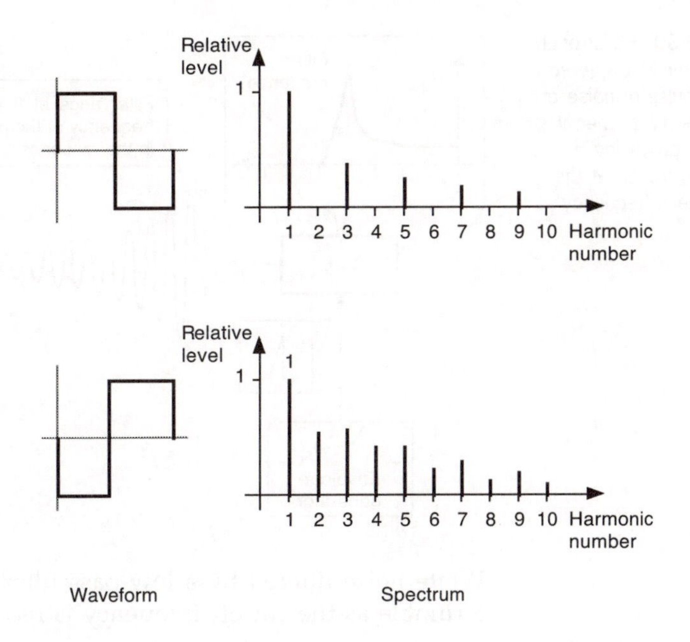

Figure 2.2.38 The harmonic content of a square wave and a rectangular wave are different – especially the even harmonics. The second harmonic is not present in a square wave, and yet can be clearly heard in a rectangular waveform. This can be used to produce square waves from a VCO which provides control over the width of the pulse. By adjusting the pulse width control and listening for the disappearance of the second harmonic, a square wave can be produced.

Harmonic content of pulses

The harmonic content of different pulse widths of pulse waveforms can be demonstrated by listening to the pulse waveform and changing the pulse width manually. At a pulse width of 50%, the sound will be noticeably hollow in timbre: this is a square wave. The square wave position can be heard because the second harmonic which is one octave above the fundamental will disappear. Using the resonant filter technique described in the previous example, individual harmonics can be examined – tuning the filter to the harmonic which disappears for a square wave can be used to emphasize this effect.

As the pulse width is reduced, then the timbre will become brighter and brighter, and with very small pulse widths, the sound may disappear entirely. (This is a problem with the VCO circuitry, and not an acoustic effect!) Conversely, increasing the pulse width from 50% produces the same changes in the timbre, and, again at very large pulse widths, may result in the loss of the sound.

Filtering

Many resonance and ringing filter effects can be demonstrated by connecting a percussive envelope to a VCF control voltage input, and turning up the resonance. Just below self-oscillation, the filter can be made to oscillate for a short time by using the envelope to trigger the oscillation.

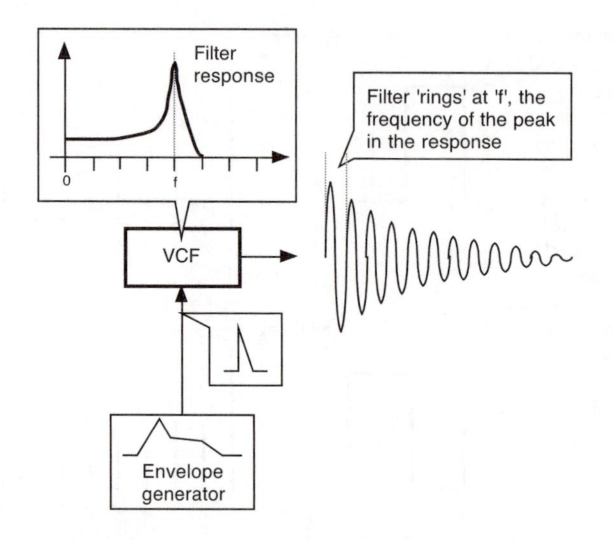

Figure 2.2.39 If a strongly resonant filter is 'triggered' by a brief pulse of noise or an envelope pulse, then it can 'ring' – producing a decaying oscillation at the cut-off or peak frequency.

White noise filtered by a low-pass filter changes from a hiss to a rumble as the cut-off frequency is reduced.

Beats

Beats occur when two VCOs or audio signals are detuned relative to each other. The interference between the two signals produces a cyclic variation in the overall level as they combine or cancel each other out repeatedly (Figure 2.2.40). The time between the cancellations is related to the difference in frequency. between the two audio signals or VCOs. Using two VCOs with a beat frequency of 1 Hz or less produces a lively, rich and interesting sound.

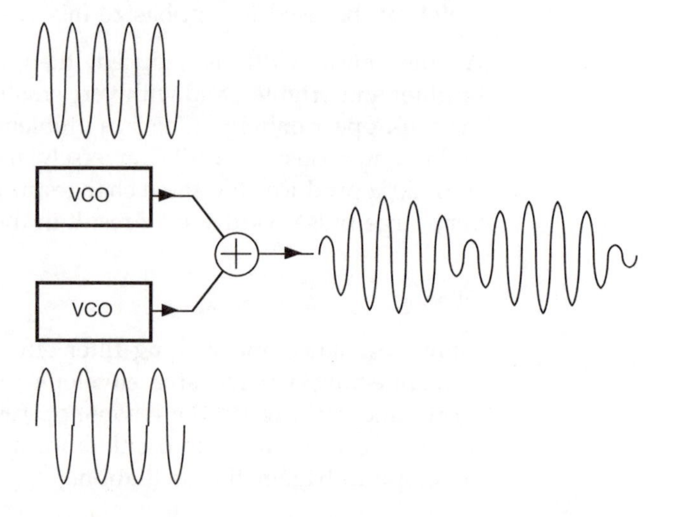

Figure 2.2.40 Beats can be demonstrated by mixing together the outputs of two VCOs which have slightly different frequencies. The two waveforms will cyclically add together or subtract, and so produce an output that varies in level. The audible effect is an interesting 'chorus' type of sound for frequency differences of less than 2 Hz, and vibrato for 2 to 20 Hz.

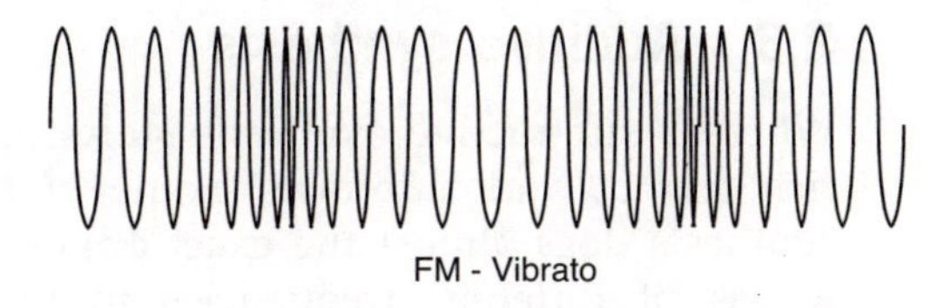

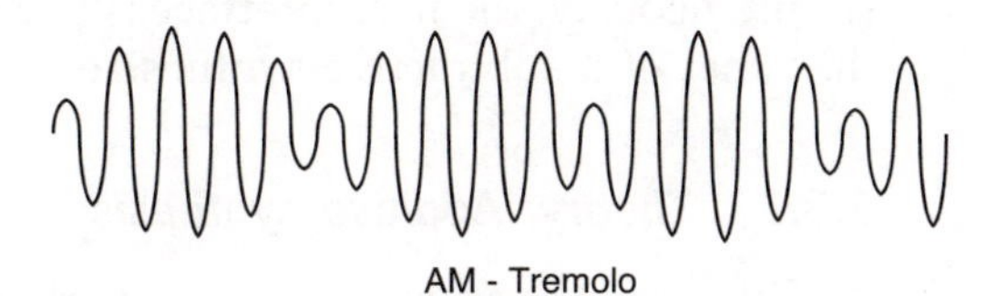

Figure 2.2.41 Vibrato is a cyclic variation in the frequency of a sound, whilst tremolo is a cyclic variation in the level of a sound.

Pulse width modulation, or PWM, uses an LFO to cyclically change the width of a pulse waveform from a single VCO. The result has many of the audible characteristics of two VCOs beating together.

Vibrato versus tremolo

Vibrato is frequency modulation (FM). The frequency of tne audio signal is changed. Using an LFO to modulate the frequency of a VCO produces vibrato.

Tremolo is amplitude modulation (AM). The level of the audio signal is changed. Using an LFO to modulate the level of an audio signal using a VCA produces tremolo.

Table 2.2.1 Envelope segments

Symbol	*Segment*	*Description*	*Type*
D	Delay	The time from the start of the envelope to the start of the attack segment.	Time
I	Initial	The first level of the envelope. The quiescent level.	Level
A	Attack	The time taken for the envelope to rise from the initial level to the maximum (peak) level.	Time
H	Hold	The time that the envelope stays at the maximum (peak) level.	Time
P	Peak	The level to which the envelope rises at the end of the attack time.	Level
D	Decay	The time for the envelope to fall from the maximum (peak) level, to the sustain or final level.	Time
B	Break-point	The level at which one decay segment changes to another.	Level
D	Decay	The time for the second decay segment to fall from the break point level, to the sustain or final level.	Time
S	Sustain	The level at which the envelope stays whilst the key is held down (gate signal on).	Level
S	Sustain	The time for which the sustain segment lasts (often the minimum time).	Time
R	Release	The time for the envelope to fall, from the sustain level, to the final level.	Time
F	Final	The final level of the envelope (usually the same as the initial).	Level

2.3 Additive synthesis

Whereas subtractive synthesis starts out with a harmonically-rich sound, and 'subtracts' some of the harmonics, additive synthesis does almost the exact opposite. It adds together sine waves of different frequencies to produce the final sound. Because large numbers of parameters need to be controlled simultaneously, the user interface is usually much more complex than that of a subtractive synthesizer.

2.3.1 Theory: Additive synthesis

Additive synthesis is based on work produced by Fourier, a French mathematician from the seventeenth century. In 1807, Fourier showed that the shape of any repetitive waveform could be reproduced by adding together simpler waveforms, or alternatively, that any periodic waveform could be described by specifying the frequency and amplitude of a series of sine waves. The restriction that the waveshape must repeat is imposed to keep the mathematics manageable. Without the restriction it is still possible to convert any waveform into a series of sine waves, but since the waveform is not constant, then the sine waves which make it up are not constant either.

One useful analogy is to think of trying to describe writing to someone, who has never seen it, over the telephone. You might start out by describing how the words are broken up into letters, and these letters are made up out of lines, dots and curves. This works perfectly well as long as the words you might try to describe stay fixed, but if they change, then you would have to keep updating your description. You could still convey the information about the shape of the letters which make up the words, but you would have to provide lots more detailed description as the letters changed.

The simplest example of synthesizing a waveform using Fourier synthesis is a sine wave. A sine wave is made up of just one sine wave, at the same frequency! In terms of harmonics, a sine wave contains just one frequency component, at the repetition rate of the sine wave: its frequency.

More complicated waveshapes can be made by adding additional sine waves. The simplest method involves using simple integer multiples of the fundamental frequency. So, if the fundamental is denoted by f, then the additional frequencies will be $2f$, $3f$, $4f$, etc. These are the frequencies which occur in some of the basic waveshapes: sawtooth, square, etc., and are known

Table 2.3.1 Harmonics and frequencies

Frequency	*Harmonic*	*Overtone*
f	fundamental	fundamental
2f	2	1
3f	3	2
4f	4	3
5f	5	4
6f	6	5
7f	7	6
8f	8	7
9f	9	8
10f	10	9

as harmonics. Because the numbering of the harmonics is based around their position above the fundamental or first harmonic, with a frequency of f, then the second harmonic has a frequency of $2f$. The second harmonic is also sometimes called the first overtone.

2.3.2 Harmonic synthesis

So far additive synthesis seems to be based around producing a specific waveform from a series of sine waves. In practice, the 'shape' of a waveform is not a good guide to its harmonic content, since minor changes to the shape can produce large changes in the harmonic content. Conversely, simple changes of phase for the harmonics can produce major changes in the shape of the waveform. In fact, although the human ear is mainly concerned with the harmonic content, the relative phase of the harmonics can be very important at low frequencies. For frequencies above 440 Hz, you can change the phase of a harmonic and thus alter the resulting shape of the waveform, but the basic timbre will sound the same. Control over phase is thus useful under some circumstances, and is found in some additive synthesizers.

The harmonic content of waveshapes is a useful starting point for examining this relationship between shape and perception.

The simplest waveshape is the sine wave. Sine waves sound clean and pure, and perhaps even a little bit boring. Adding in small amounts of odd numbered harmonics produces a triangular waveshape, which has enough harmonic content to stop it sounding quite as pure as the sine wave.

A square wave contains only odd harmonics. It has a characteristic 'hollow' sound, and the absence of the second harmonic is

Figure 2.3.1 (i) A triangle waveform constructed from six sine wave harmonics is very different from a sine wave, even though the fundamental is by far the strongest component. (ii) A combination of equal amounts of the first 12 harmonics produces a waveform which looks (and sounds) like a pulse waveshape.

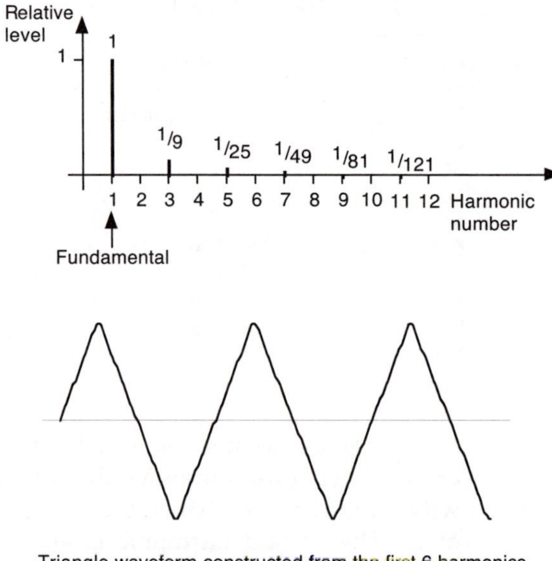

(i)

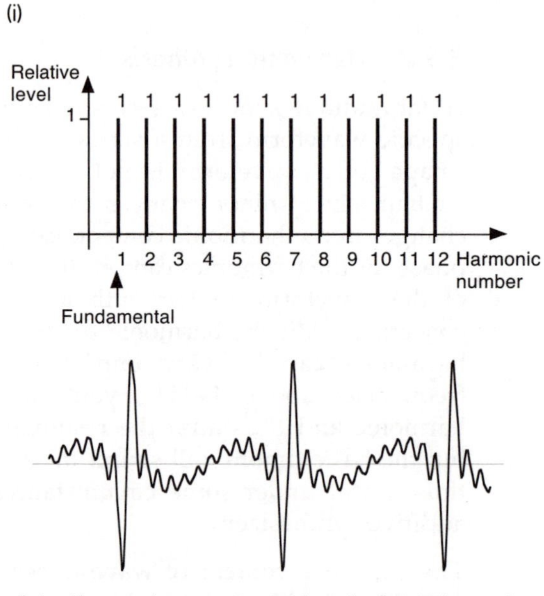

(ii)

particularly noticeable if a square wave is compared with a saw-tooth wave. A square wave which has been produced with a phase change in the second harmonic no longer looks like a 'square' wave, and yet the harmonic content is the same (Figure 2.3.2).

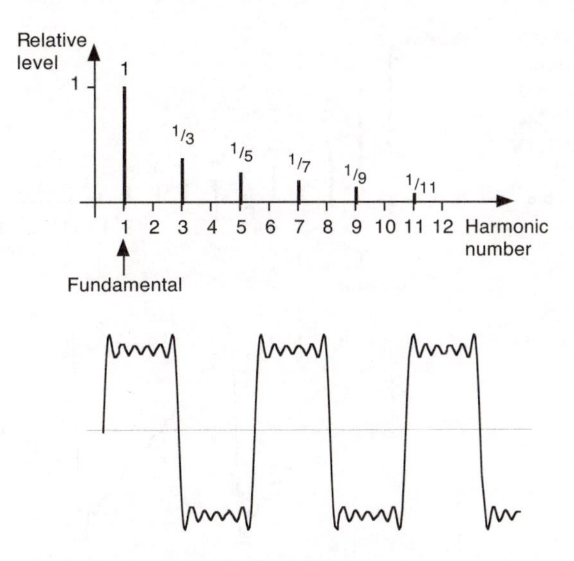

(i)

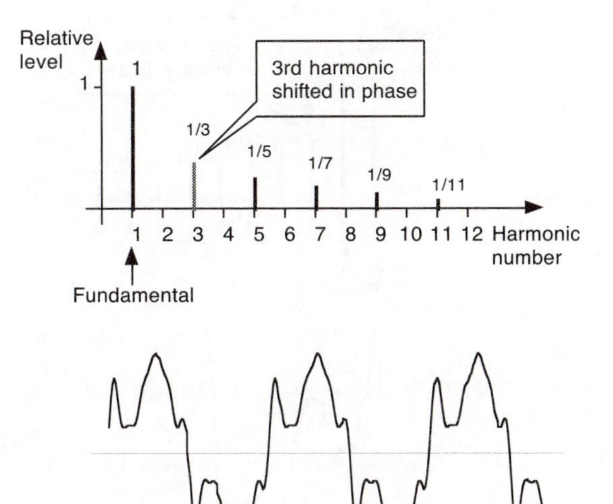

(ii)

Figure 2.3.2 (i) A square waveform constructed from six sine wave harmonics has a close approximation to the ideal waveshape. (ii) Changing the phase of the third harmonic radically alters the shape of the waveform.

A sawtooth wave contains both odd and even harmonics. It sounds bright, although many pulse and 'supersawtooth' waveshapes can contain greater levels of harmonics. Again, a sawtooth wave with a phase change in the second harmonic

Figure 2.3.3 (i) A sawtooth waveform constructed from 12 sine wave harmonics has a close approximation to the ideal waveshape. (ii) Changing the phase of the second harmonic radicaly alters the shape of the waveform.

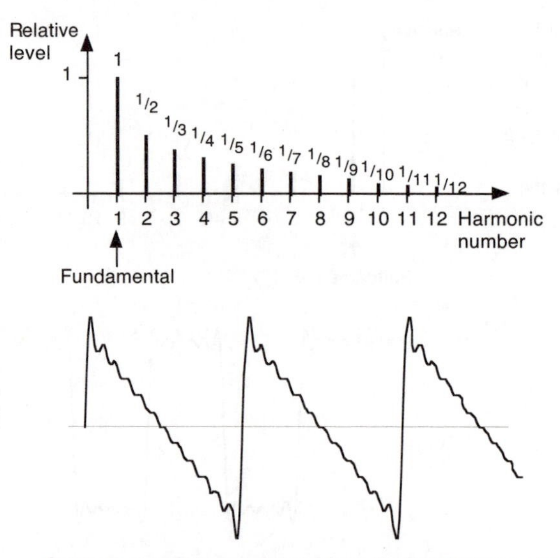

Sawtooth waveform constructed from the first 12 harmonics

(i)

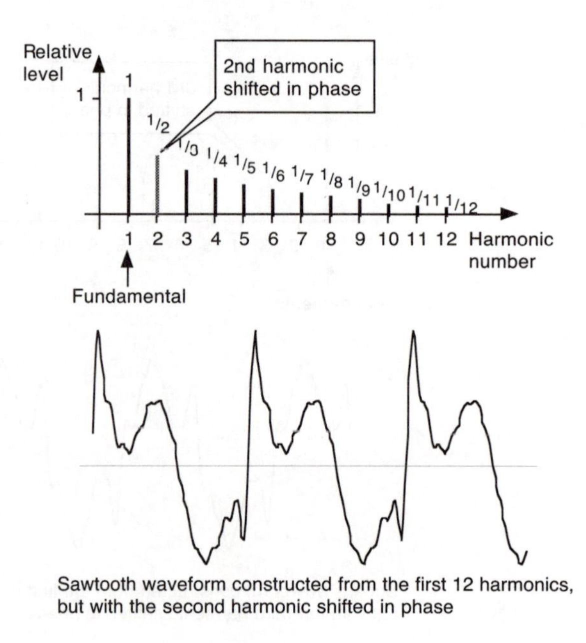

Sawtooth waveform constructed from the first 12 harmonics, but with the second harmonic shifted in phase

(ii)

does not look like a sawtooth, although it still sounds like one to the ear (Figure 2.3.3).

Pulse waves contain more and more harmonics as the pulse width narrows (or widens). A 10% pulse has the same spectrum as a 90% pulse, and it also sounds the same to the ear. One special case is the square wave, where the even harmonics are missing completely. Pulse widths of anything other than 50% include the second harmonic – and this can usually be clearly heard as the pulse width is varied away from the 50% value.

Finally, there is the 'even harmonic' wave. If a sawtooth contains both odd and even harmonics, and a square wave contains just the odd harmonics – then what does a wave containing just the even harmonics look like? Actually, it is just another square wave, but one octave higher in pitch!

In practice, adding together sine waves produces waveforms which have some of the characteristics of the mathematically perfect ideal waveforms – but not all. Producing square edges on a square wave would require large numbers of harmonics – an infinite number for a 'perfect' square wave. Using just a few harmonics can produce waveforms which have enough of the harmonic content to produce the correct type of timbre, even though the shape of the waveform may not be exactly as expected.

2.3.3 Harmonic analysis

In order to produce useful timbres, an additive synthesizer user really needs to know about the harmonic content of real instruments, rather than mathematically derived waveforms. The main method of determining this information is Fourier analysis, which reverses the concept of making any waveform out of sine waves and uses the idea that any waveform can be split into a series of sine waves.

The basic concept behind Fourier analysis is quite simple, although the practical implementation is usually very complicated. If an audio signal is passed through a very narrow band-pass filter that sweeps through the audio range, then the output of the filter will indicate the level of each band of frequencies which are present in the signal. The width of this band-pass filter determines how accurate the analysis of the frequency content will be: if it is 100 Hz wide, then the output can only be used to a resolution of 100 Hz, whereas if the band-pass filter has a 1 Hz bandwidth, then it will be able to indicate individual frequencies to a resolution of 1 Hz.

For simple musical sounds which contain mostly harmonics of the fundamental frequency, the resolution required for Fourier analysis is not very high. The more complex the sound, the higher the required resolution. For sounds which have a simple

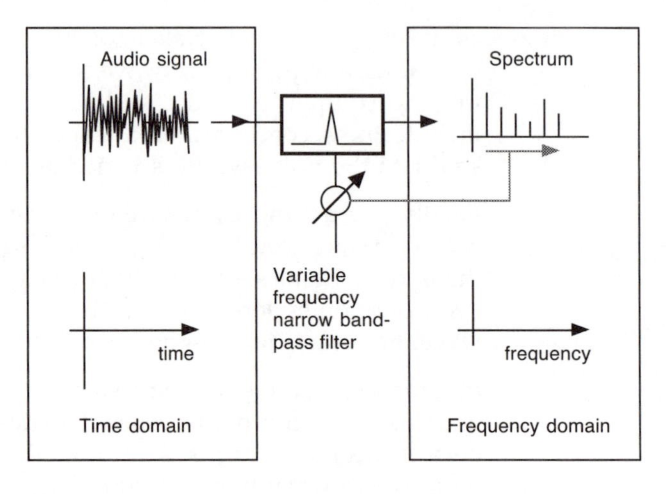

Figure 2.3.4 Sweeping the centre frequency of a narrow band-pass filter can convert an audio signal into a spectrum: from the time domain to the frequency domain.

structure consisting of a fundamental and harmonics, then a rough 'rule of thumb' is to make the bandwidth of the filter less than the fundamental frequency, since the harmonics will be spaced at frequency intervals of the fundamental frequency.

Having 1 Hz resolution in order to discover that there are five harmonics spaced at 1 kHz intervals is extravagant. Smaller bandwidths require more complicated filters, and this can increase the cost, size and processing time, depending on how the filters are implemented. Fourier analysis can be achieved using analogue filters, but it is frequently carried out by using digital technology. (See Section 5.7 on analysis-synthesis techniques).

Numbers of harmonics

How many separate sine waves are needed in an additive synthesizer? Supposing that the lowest fundamental frequency which will be required to be produced is an low A at 55 Hz, then the harmonics will be at 110 Hz, 165 Hz, 220 Hz, 275 Hz, 330 Hz, 385 Hz, 440 Hz. . . The 32nd harmonic will be at 1760 Hz, and the 64th harmonic is at 3 520 Hz.

An A at 440 Hz has a 43rd harmonic of 19 800 Hz. Most additive synthesizers seem to use between 32 and 64 harmonics.

Harmonic and inharmonic content

Real-world sounds are not usually deterministic: they do not contain just simple harmonics of the fundamental frequency. Instead, they also have additional frequencies which are not

Table 2.3.1 Additive frequencies and harmonics

Frequency				Harmonic
55	110	220	440	Fundamental
110	220	440	880	2
165	330	660	1320	3
220	440	880	1760	4
275	550	1100	2200	5
330	660	1320	2640	6
385	770	1540	3080	7
440	880	1760	3520	8
495	990	1980	3960	9
550	1100	2200	4400	10
605	1210	2420	4840	11
660	1320	2640	5280	12
715	1430	2860	5720	13
770	1540	3080	6160	14
825	1650	3300	6600	15
880	1760	3520	7040	16
935	1870	3740	7480	17
990	1980	3960	7920	18
1045	2090	4180	8360	19
1100	2200	4400	8800	20
1155	2310	4620	9240	21
1210	2420	4840	9680	22
1265	2530	5060	10120	23
1320	2640	5280	10560	24
1375	2750	5500	11000	25
1430	2860	5720	11440	26
1485	2970	5940	11880	27
1540	3080	6160	12320	28
1595	3190	6380	12760	29
1650	3300	6600	13200	30
1705	3410	6820	13640	31
1760	3520	7040	14080	32
1815	3630	7260	14520	33
1870	3740	7480	14960	34
1925	3850	7700	15400	35
1980	3960	7920	15840	36
2035	4070	8140	16280	37
2090	4180	8360	16720	38
2145	4290	8580	17160	39
2200	4400	8800	17600	40
2255	4510	9020	18040	41
2310	4620	9240	18480	42
2365	4730	9460	18920	43
2420	4840	9680	19360	44
2475	4950	9900	19800	45
2530	5060	10120	20240	46
2585	5170	10340	20680	47
2640	5280	10560	21120	48
2695	5390	10780	21560	49
2750	5500	11000	22000	50
2805	5610	11220	22440	51
2860	5720	11440	22880	52
2915	5830	11660	23320	53
2970	5940	11880	23760	54
3025	6050	12100	24200	55
3080	6160	12320	24640	56
3135	6270	12540	25080	57
3190	6380	12760	25520	58
3245	6490	12980	25960	59
3300	6600	13200	26400	60
3355	6710	13420	26840	61
3410	6820	13640	27280	62
3465	6930	13860	27720	63
3520	7040	14080	28160	64

simple integer multiples of the fundamental frequency. There are several types of these unpredictable 'inharmonic' frequencies:

- noise
- beat frequencies
- sidebands
- inharmonics

Noise has, by definition, no harmonic structure, although it may be present only in specific parts of the spectrum: coloured noise. So any noise which is present in a sound will appear as random additional frequencies within those bands, and whose level and phase are also random.

Beat frequencies arise when the harmonics in a sound are not perfectly in tune with each other. 'Perfect' waveshapes are always assumed to have harmonics at exact multiples of the fundamental, whereas this is not always the case in real-world sounds. If a harmonic is slightly detuned from its mathematically 'correct' position, then additional harmonics may be produced at the beat frequency, so if a harmonic is 1 Hz too high in pitch relative to the fundamental, then a frequency of 1 Hz will be present in the spectrum.

Sidebands occur when the frequency stability of a harmonic is imperfect, or when the sound itself is frequency modulated. Both cases result in pairs of frequencies which mirror around the 'ideal' frequency. So a 1 kHz sine wave which is frequency modulated with a few hertz will have a spectrum which contains frequencies on either side of 1 kHz, and the exact content will depend on the depth of modulation and its frequency.

Inharmonics are additional frequencies which are structured in some way, and so are not noise, but which do not have the simple integer multiple relationship with the fundamental frequency. Timbres which contain inharmonics typically sound like a 'bell' or 'gong'.

Many additive synthesizers only attempt to produce the harmonic frequencies, with perhaps a simple noise generator as well. This deterministic approach limits the range of sounds which are possible, since it ignores the many stochastic, random, elements which make up real-world sounds.

2.3.4 Envelopes

The control of the level of each harmonic over time uses envelope generators and VCAs. Ideally, one envelope generator and one VCA should be provided for each harmonic. This would

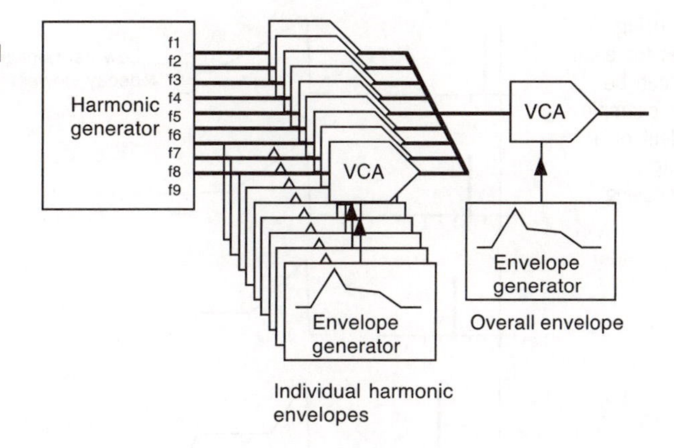

Figure 2.3.5 Individual envelopes are used to control the harmonics, but an overall envelope allows easy control over the whole sound which is produced.

mean that the overall envelope of the final sound was the result of adding together the individual envelopes for each of the harmonics, and so there would be no overall control over the envelope of the complete sound. Adding an overall envelope generator and VCA to the sum of the individual harmonics allows quick modifications to be made to the final output.

In order to minimize the number of controls and the complexity, the envelope generators need to be as simple as possible without compromising the flexibility. Delayed ADR envelopes are amongst the easiest of envelope generators to implement in discrete analogue circuitry, since the gate signal can be used to control a simple capacitor charge and discharge circuit to produce the ADR envelope voltage. Delayed ADR envelopes also require only four controls (delay time, attack time, decay time, and release time), whereas a delayed ADSR would require five controls and more complex circuitry. If integrated circuit envelope generators are used, then the ADSR envelope would probably be used, since most custom synthesizer chips provide ADSR functionality.

Control grouping and ganging

With large numbers of harmonics, having separate envelopes for each harmonic can become very unwieldy and awkward to control. The ability to assign a smaller number of envelopes to harmonics can reduce the complexity of an additive synthesizer considerably. This is only effective if the envelopes of groups of harmonics are similar enough to allow a 'common' envelope to be determined. Similarly, ganging together controls for the level of groups of harmonics can make it easy to make rapid changes to timbres – altering individual harmonics can be very time-consuming. Simple groupings like 'all of the odd', or 'all of the even' harmonics, can be useful starting points for this technique.

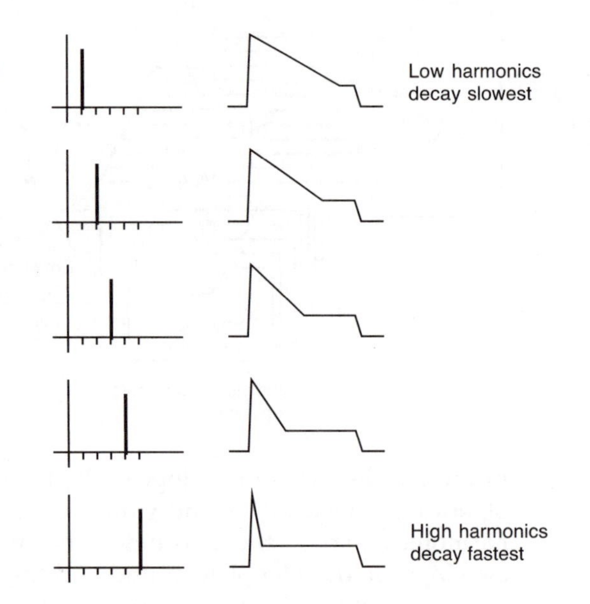

Figure 2.3.6 By using different envelopes for each harmonic, a filter can be 'synthesized'. This example shows the equivalent of a low-pass filter being controlled by a decaying envelope.

Filter simulation/emulation

Filters modify the harmonic content of a sound. In the case of an additive synthesizer, there are two ways that this can be carried out: with a filter; or with a filter emulation. As with the overall envelope control mentioned above, there are advantages to having a single control for the combined harmonics, and a VCF could be added just before the VCA. Such a filter would only provide crude filtering of the sound, in exactly the same way as in subtractive synthesis.

Filter emulation uses the individual envelope generators for the harmonics to 'synthesize' a filter by altering the envelopes. For example, if the envelopes of higher harmonics are set to have progressively shorter decay times, then when a note is played, the high harmonics will decay first. This has an audible effect which is very similar to a low-pass filter being controlled by a decaying envelope. The difference is that the 'filter' is the result of the action of all the envelopes, rather than one envelope. Consequently, individual envelopes can be changed, which then allows control over harmonics that would not be possible using a single VCF.

As with the envelope control ganging and grouping, similar facilities can be used to make filter emulation easier to use, although the implementation of this is much easier in a fully digital additive instrument.

2.3.5 Practical problems

Analogue additive synthesis suffers from a number of design difficulties. Generating a large number of stable, high purity sine waves simultaneously can be very complex, especially if they are not harmonically related. Providing sufficient controls for the large number of available parameters is also a problem. Depending on the complexity of the design, an additive synthesizer might have the following parameters repeated for each harmonic:

- frequency (fixed harmonic or variable inharmonic)
- phase
- level
- envelope (DADR, DADSR or multi-segment: four or more controls)

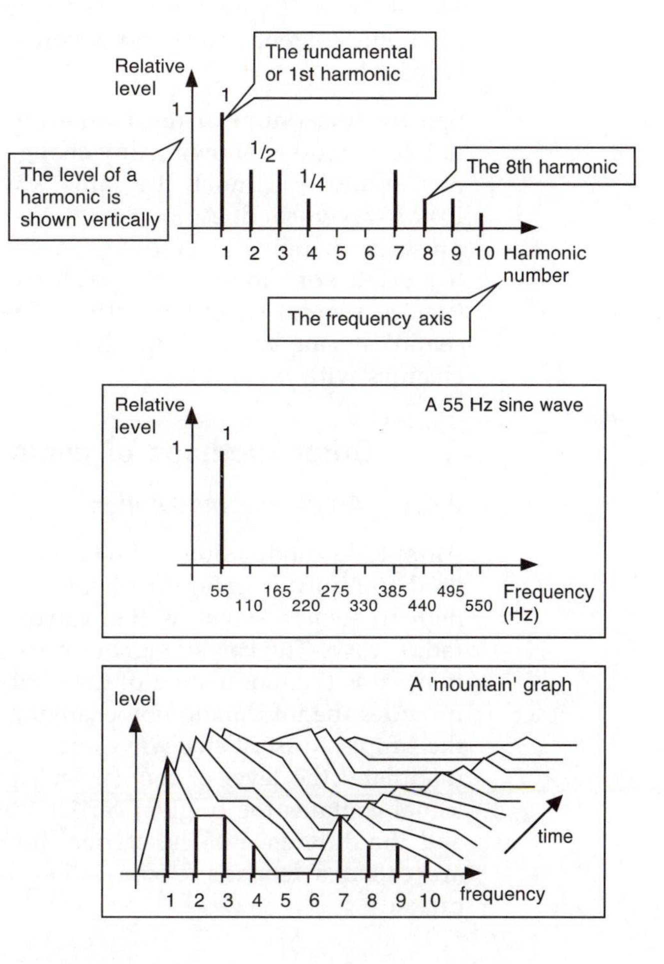

Figure 2.3.7 A spectrum is a plot of frequency against level. It thus shows the harmonic content of an audio signal. In most of the examples in this book, the horizontal axis is normally shown with harmonic numbers instead of frequencies – the 55 Hz sine wave spectrum shows the correspondance with frequency. When a spectrum changes with time, then a 'mountain' graph may be used to show the changes in the shape.

For a 32-harmonic additive synthesizer, these eight parameters give a total of just over 250 separate controls, ignoring any additional controls for ganging and filter emulation. Although it is possible to assemble an additive synthesizer using analogue design techniques, practical realizations of additive synthesizers have tended to be digital in nature, where the generation and control problems are much more easily solved.

Spectrum plots

The subtractive and additive sections in this chapter have both shown plots of the harmonic content of waveforms, showing a frequency axis plotted against level. This 'harmonic content' graph is called a spectrum, and it shows the relative levels of the frequencies in an audio signal. Whereas a waveform is a way of showing the shape of a waveform as its value changes with time, a spectrum is a way of showing the harmonic content of a sound. The shape of a waveform is not a very good indication of the harmonic content of a sound, whereas a spectrum is – by definition.

Spectra (the plural of the Latin-derived word 'spectrum') are not very good at showing any changes in the harmonic content of a sound – in much the same way that a single cycle of a PWM waveform does not convey the way that the width of the pulse is changing over time. To show changes in spectra, a 'waterfall' or 'mountain' graph is used, which effectively 'stacks' several spectra together. The resulting 3D-like representation can be used to show how the frequency content changes with time.

2.4 Other methods of analogue synthesis

2.4.1 Amplitude modulation

Amplitude modulation, or AM, is a variation on one method used to transmit radio broadcasts. AM radio works by using a high frequency signal as the 'carrier' of the audio signal as a radio wave. The carrier signal on its own conveys no information – it is the modulation of the carrier by the audio signal that provides the information by changing the level of the carrier. In the simplest case, a sine wave audio signal is used to change (or modulate) the level of the carrier signal. The resulting output signal contains the original carrier frequency, but also the sum and the difference of the carrier and audio frequencies – these are called sidebands, because they are on either side of the carrier.

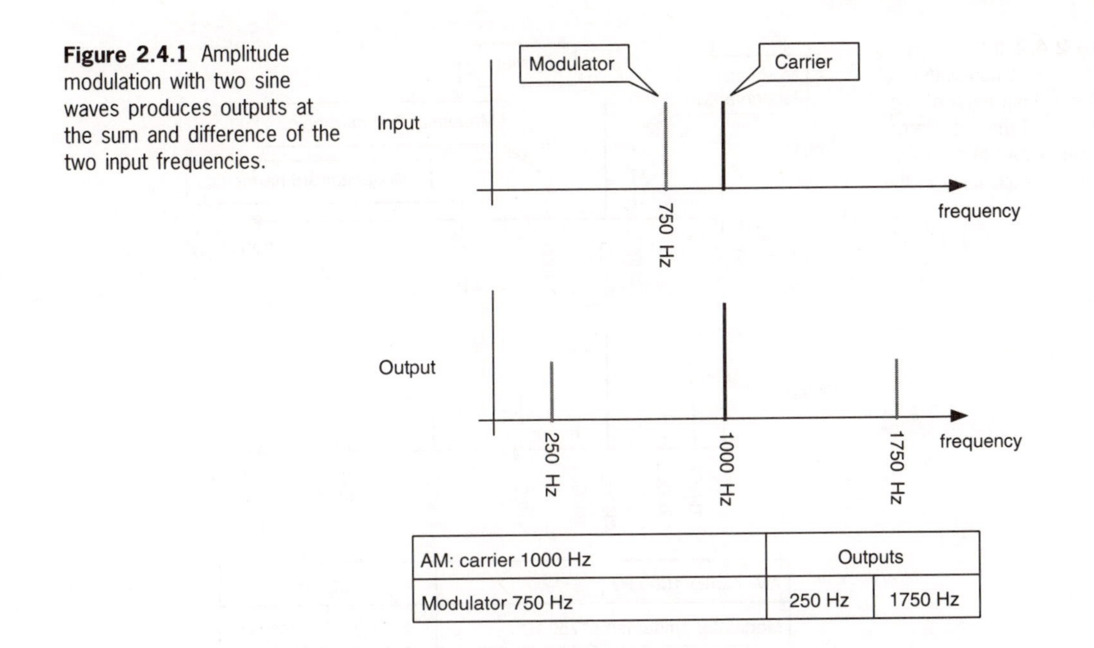

AM: carrier 1000 Hz	Outputs	
Modulator 750 Hz	250 Hz	1750 Hz

Figure 2.4.1 Amplitude modulation with two sine waves produces outputs at the sum and difference of the two input frequencies.

For audio AM, the two frequencies are both in the audio range, but the same principles apply – the output consists of the carrier frequency and the sum and difference frequencies. So with a carrier of 1000 Hz and a modulator of 750 Hz, the output sideband frequencies will be 1000, 1750 and 250 Hz. Notice that the modulating frequency is not present in the output. For 100% modulation, the sidebands have half the amplitude of the carrier.

For AM with waveforms other than sine waves, each component frequency is treated separately. So for a sine carrier and a non-sinusoidal wave modulator, there are actually the equivalent of several modulator frequencies: one for each harmonic in the modulator. For a sawtooth modulator wave this means that there will be integer multiples of the modulator frequency, at decreasing levels. Each of these harmonics will produce sidebands around the carrier. The carrier frequency of 1000 Hz will also be present in the output. Again, with 100% modulation, the sidebands will have half the amplitude of the carrier.

With a non-sinusoidal carrier of 1000 Hz, and a sine wave modulator of 750 Hz, it is the equivalent of several carrier frequencies, and each carrier produces its own set of sidebands from the modulation frequency. For a sawtooth carrier, this means that there will be the equivalent of a carrier at each integer multiple of the carrier frequency, and each will produce sidebands from the modulator frequency. With 100% modula-

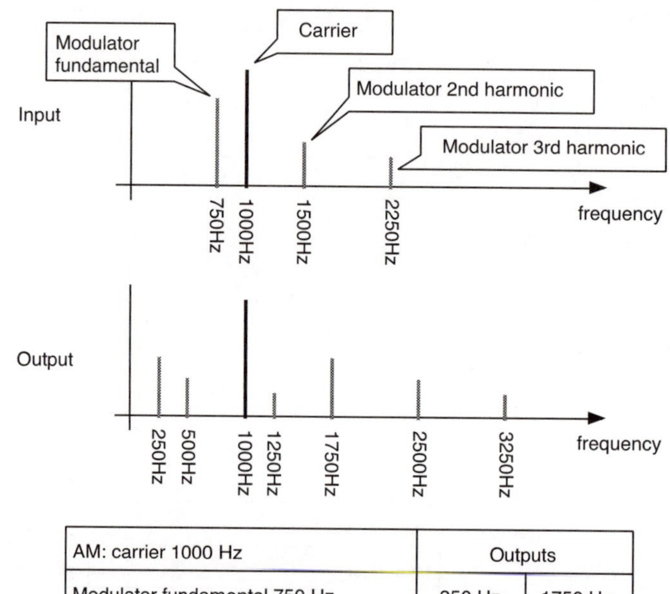

AM: carrier 1000 Hz	Outputs	
Modulator fundamental 750 Hz	250 Hz	1750 Hz
Modulator 2nd harmonic 1500 Hz	500 Hz	2500 Hz
Modulator 3rd harmonic 2250 Hz	1250 Hz	3250 Hz

Figure 2.4.2 If the modulator is a non-sinusoidal waveform, then each of the harmonics of the modulator produces a pair of sum and difference frequencies in the output.

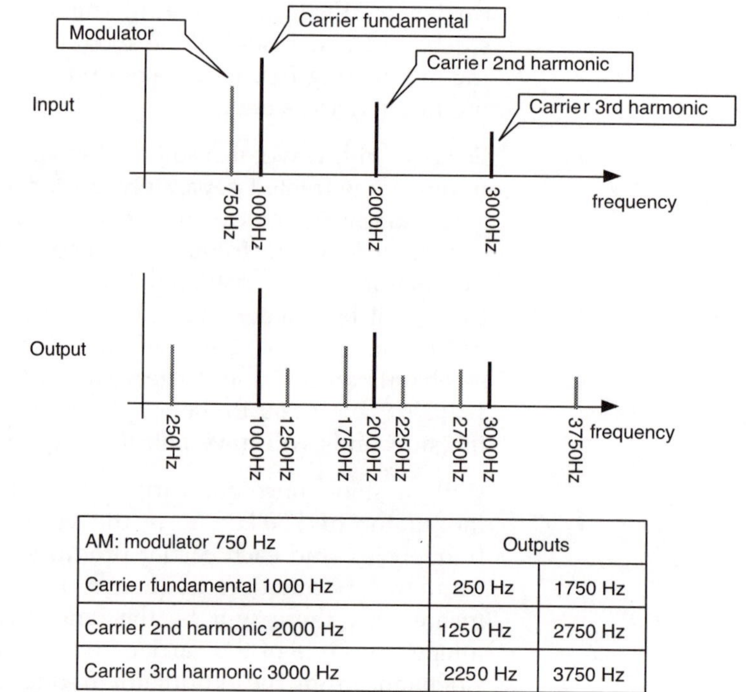

AM: modulator 750 Hz	Outputs	
Carrier fundamental 1000 Hz	250 Hz	1750 Hz
Carrier 2nd harmonic 2000 Hz	1250 Hz	2750 Hz
Carrier 3rd harmonic 3000 Hz	2250 Hz	3750 Hz

Figure 2.4.3 If the carrier is a non-sinusoidal waveform, then each carrier harmonic appears in the output, and also produces a pair of sum and difference frequencies.

tion the sidebands will have half the amplitude of the carrier. All of the harmonics in the carrier wave will also be present in the output.

For the case of two non-sinusoidal waves, AM produces a set of sidebands for each carrier harmonic, using each modulator harmonic. AM is thus a simple way of producing complex sounds with a number of harmonics which are not related to the fundamental (inharmonics).

(If the modulating frequency is lower than about 25 Hz, then AM is called tremolo, and it is perceived as a rapid cyclic change in the amplitude.)

2.4.2 Frequency modulation

Frequency modulation, or FM, also employs another method which is normally used for the transmission of radio broadcasts. FM radio again uses a high frequency signal as the 'carrier' of the audio signal. The modulation of the carrier signal by the audio signal 'carries' the information by changing the frequency of the carrier. The simplest case is where a sine wave audio signal is used to change (or modulate) the frequency of the carrier signal. The amount of frequency change is called the deviation, Δf_c, and instead of producing just one pair of sideband frequencies, FM can produce many sidebands, where the extra sidebands are similar to the harmonics in the sawtooth AM case described in Section 2.4.1 – and this is just for sine wave carrier and modulator frequencies.

The number of sidebands which are produced can be determined by using the modulation index, which is a measure of the amount of modulation which is being applied to the carrier. The modulation index is given by dividing the deviation by the modulator frequency, f_m.

$$\text{Modulation index} = \Delta f_c / f_m$$

Notice that the modulation index is dependent not only on how much the carrier frequency is changed, but also on the modulator frequency. The resulting output signal contains the original carrier frequency, but also the sum and difference sidebands for each of the multiples of the modulator frequency.

For audio FM with two sine waves, the output consists of the carrier frequency and sidebands made up from the sum and difference frequencies of the carrier and multiples of the modulator frequency. The number of sidebands depends on the modulation index, and a rough approximation is that there are

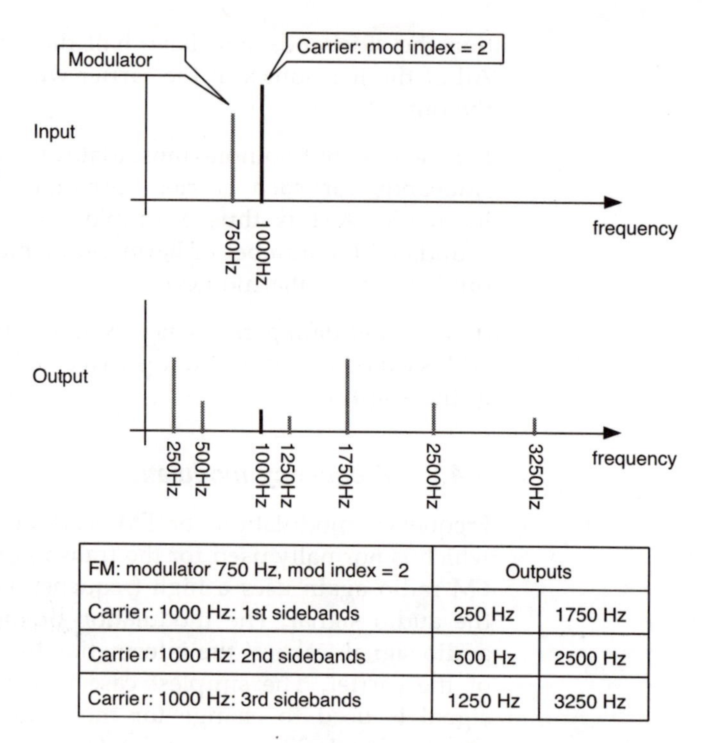

FM: modulator 750 Hz, mod index = 2	Outputs	
Carrier: 1000 Hz: 1st sidebands	250 Hz	1750 Hz
Carrier: 1000 Hz: 2nd sidebands	500 Hz	2500 Hz
Carrier: 1000 Hz: 3rd sidebands	1250 Hz	3250 Hz

Figure 2.4.4 Frequency modulation depends on the depth of modulation as well as the input frequencies. The number of sidebands which are produced depends on the modulation index.

two more than the modulation index. The modulating frequency is not present in the output. The amplitudes of the sideband frequencies are determined by a set of curves called Bessel functions (Chowning and Bristow, 1986).

For FM with waveforms other than sine waves, each component frequency is treated separately. So for a sawtooth carrier and a sine wave modulator the output is similar to the sawtooth AM case, but there are many more sidebands produced. FM is thus a very powerful technique for producing complex spectra, but in an analogue synthesizer it suffers from problems related to the frequency stability of the carrier and modulator VCOs, and the response of the carrier VCO to frequency modulation at audio frequencies.

(If the modulating frequency is lower than about 25 Hz, then FM is known as vibrato – a cyclic change in pitch.)

2.4.3 Ring modulation

Ring modulation takes two audio signals and combines them together in a way that produces additional harmonics. It uses a circuit known as a 'balanced modulator' to produce a single

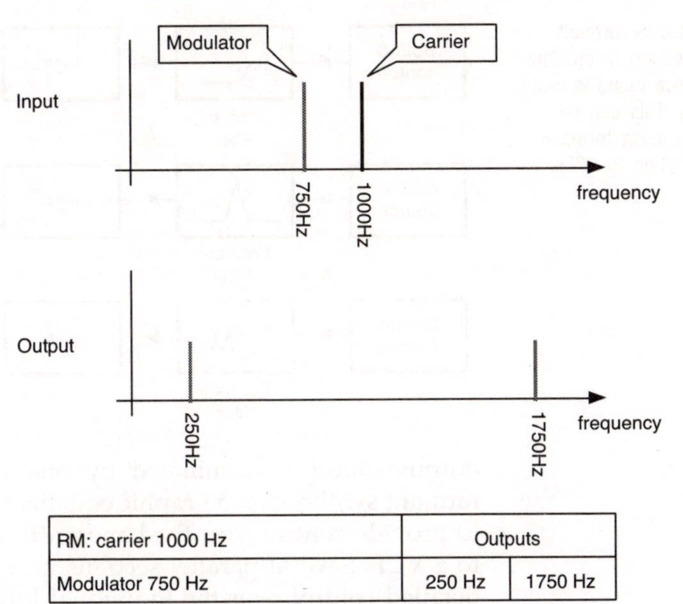

RM: carrier 1000 Hz	Outputs	
Modulator 750 Hz	250 Hz	1750 Hz

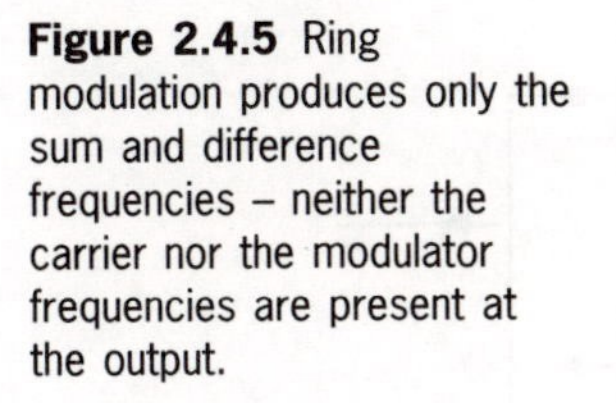
Figure 2.4.5 Ring modulation produces only the sum and difference frequencies – neither the carrier nor the modulator frequencies are present at the output.

output from two inputs: the output consists of the sum of the two input frequencies, and the difference between the two input frequencies. The original inputs are not present in the output signal.

This is similar to AM, except that it is only the additional frequencies that are generated which are present at the output: only the sidebands are heard, not the carrier or the modulator. This means that ring modulation can be useful where the original pitch information needs to be lost, which makes it useful for pitch transposition, especially where one of the sets of extra frequencies can be filtered out.

Modulation summary

Table 2.4.1

AM	Carrier in output	Simple sidebands for sine waves
FM	Carrier in output	Multiple sidebands for sine waves
RM	No carrier in output	Simple sidebands for sine waves

2.4.4 Formant synthesis

Formant synthesis is intended to emulate the strong resonant structure of many real instruments, where the spectrum of the

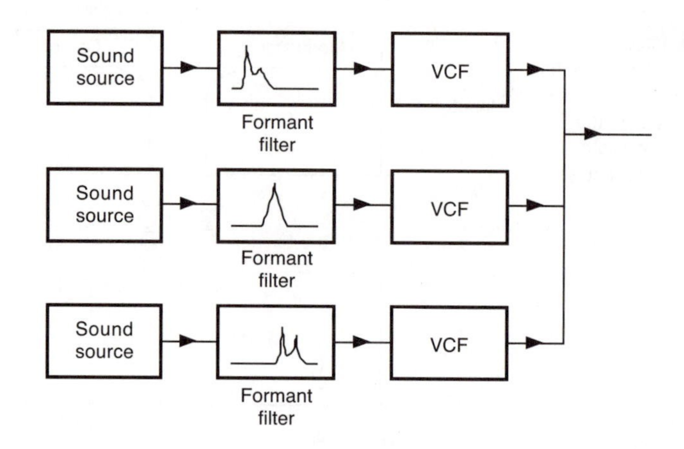

Figure 2.4.6 A formant synth is intended to emulate the resonance found in real instruments. This can be achieved by using formant filters in addition to VCFs.

output sound is dominated by one or more formants. In a formant synthesizer, a graphic equalizer or complex filter is used to provide control over the bandwidth of the sound in addition to a VCF. Several parallel sections may be used to enable more detailed control over the individual formant areas of the sound (see also Section 5.1).

2.4.5 Damped oscillators & ringing filters (drum sounds)

Circuits which have a strong resonance at a specific frequency can be made to oscillate if a sudden input causes them to self-oscillate. This 'ringing' is usually a sine wave and it dies away at a rate which is dependent on how close to self-oscillation the circuit is. The nearer it is to oscillating, the longer the ringing will last. Some VCFs can be made to self-oscillate if their Q or resonance is high enough, and at Q values just below this, they will ring. Conversely, an oscillator can be 'damped' so that it does not self-oscillate, but it will then ring. Filters and oscillators are just different applications of resonant circuits.

Decaying sine waves are very useful for producing percussive sounds, and many of the drum sounds produced by rhythm machines in the 1970s and early 1980s were produced by using ringing circuits.

2.4.6 Organ technologies

Most traditional organs are based around additive synthesis techniques, where a large number of sine waves are produced from a master oscillator, and then individual notes select mixes of sine waves via drawbar or other controls for the harmonic content. Unlike additive synthesizers, until the middle of the 1980s, organs tended not to have envelope control over the

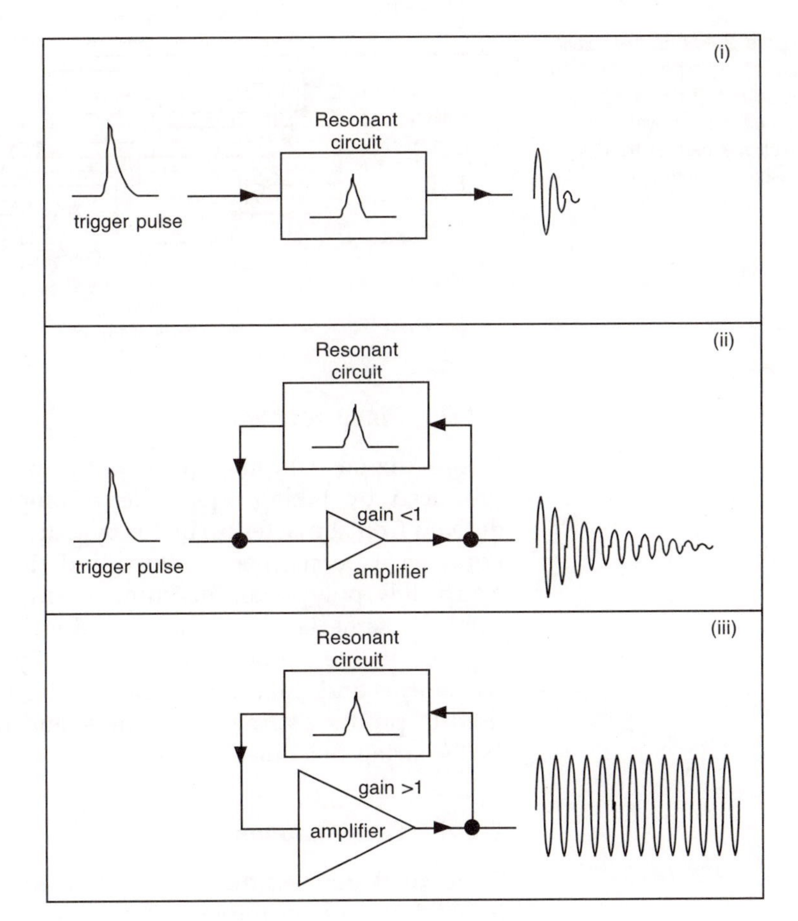

Figure 2.4.7 A resonant circuit can produce some ringing when a trigger pulse is applied. (ii) When a resonant circuit is placed in the feedback loop of an amplifier with a gain of less than one, then the ringing of the resonant circuit is enhanced. (iii) If the gain of the amplifer is greater than one, then the circuit will oscillate at the frequency of the least attenuation in the resonant circuit.

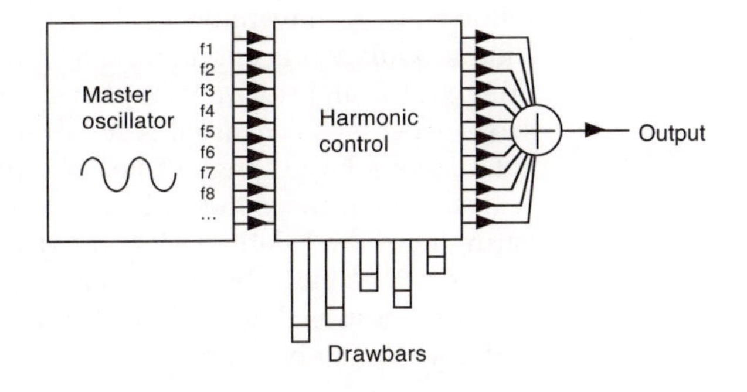

Figure 2.4.8 Organs produce sounds by the addition of sine waves. The methods of producing the sine waves can be mechanical, electro-mechanical and electronic.

individual harmonics which make up the sounds. The advent of digital technology and sampling has made organs much more closely related to S&S synthesizers. Chapter 3 gives further details of digital master oscillators, whilst Chapter 4 describes S&S synthesis in more detail.

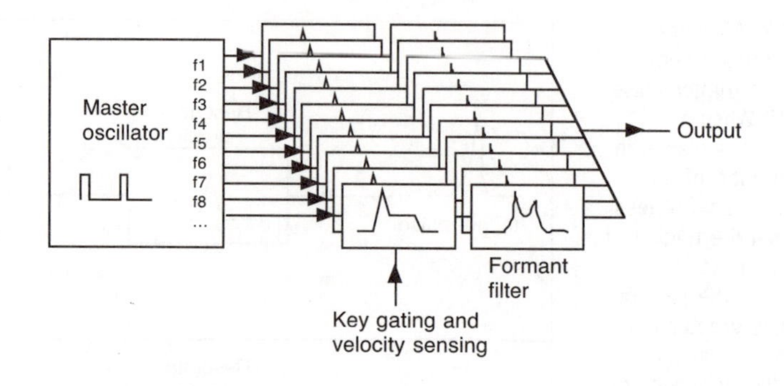

Figure 2.4.9 Simple 'piano' and 'string' type sounds can be produced by gating and filtering pulse waveforms which are derived from a master oscillator.

2.4.7 Piano technologies

Before digital sampling technology, piano type sounds were produced by taking square or rectangular waveforms, often derived from a master oscillator by a divider technique, and then applying a percussive envelope and filtering. This produces a completely polyphonic instrument, although the sound suffers from the same lack of dynamic individual harmonic control as organs of the same time period. By using narrow pulse waveforms and different envelopes, the same techniques can be used to produce string-like sounds, and this was used in many 1970s 'string machines'.

2.4.8 Combinations

Some analogue synthesizers use a combination of synthesis techniques, for example where several oscillators are used (additive-style) to provide the sound source, although this is then followed by a conventional subtractive synthesis modifier section. Ring modulation is another method which sometimes appears in otherwise straightforward implementations of subtractive synthesizers – perhaps because it is relatively simple to implement, and yet allows a large range of bell-like timbres which contrast well with the often more melodic subtractive synthesis timbres. Some 'string machines' in the 1970s added a VCF and ADSR EG section to provide 'synth brass' capabilities. Such combinations can provide additional control and creative potential, although their additions rarely become adopted generally.

2.4.9 Tape techniques

Perhaps the most straightforward method of analogue synthesis is the use of the tape recorder. By recording sounds onto magnetic tape, they can be stored permanently for later modification and manipulation. The raw sounds used can be either natural or synthetic. Chapter 4 details the use of tape as a recording medium, whilst Chapter 1 outlines some of the creative possibilities of using tape as a synthesis tool.

2.4.10 Optical techniques

Whilst tape offers a large number of possibilities for manipulating sound once it has been recorded on the tape, it does not allow the user to generate or control a sound directly. The audio signals are recorded onto the tape as changes to the magnetic fields stored on the iron oxide coating of the plastic tape, and so cannot be seen or changed, other than by recording a new sound over the top.

In contrast, by using the optical sound tracks which are often used in film projectors, it is possible to directly input the raw sound itself. Modern film projectors can use magnetic or digital techniques as well, but the basic method uses a light source and an optical sensor on either side of the film. When the sound track is clear, all the light passes through the film to the sensor, and conversely, when the film is dark, then no light passes to the sensor. By varying the amount of light that can pass through the film to the light sensor, then the output of the sensor can be controlled. If the film sound track varies at a fast enough rate, then audio signals can be produced at the output of the sensor. Film soundtracks usually control the amount of light by altering the width of the clear part of the film – the wider the gap, the more light passes through to the sensor. The part of the film used to record this 'sound' track is by the side of the picture, and looks much like an oscilloscope view of an audio signal, except that it is mirrored around the long axis (Figure 2.4.10).

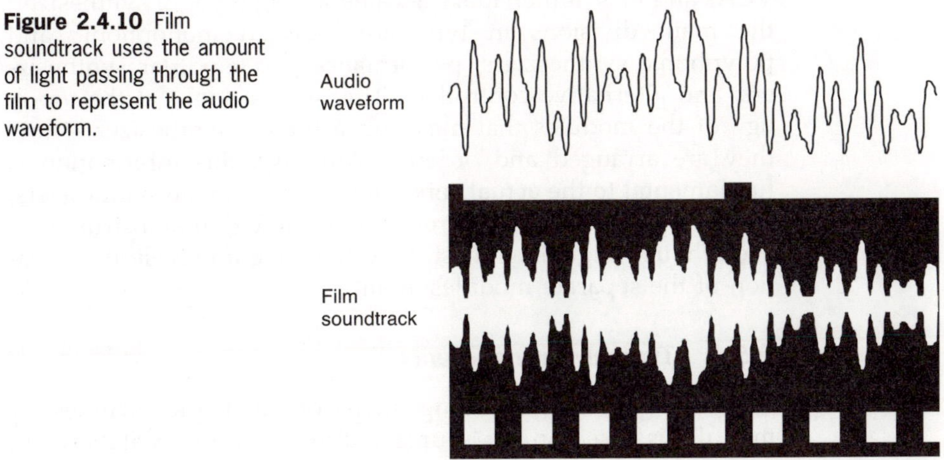

Figure 2.4.10 Film soundtrack uses the amount of light passing through the film to represent the audio waveform.

By taking film which has no sound recorded onto it, and then drawing onto the film sound track with an opaque ink, it is possible to create sounds which will only be heard when the film is played. Sounds can thus be drawn or painted directly onto film. Although this sounds like an effective marriage of art and science, it turns out that the process of drawing sounds by hand is a slow and tedious one, and the precision required to obtain consistent timbres is very high.

Combining the drawing skills of optical sound creation with the tape manipulation processes of music concrete can offer a much more versatile technique. In this case, only short segments of film soundtrack need to be drawn, since the resulting short sounds can be recorded onto tape, copied many times to provide longer sounds, and then manipulated using tape techniques.

2.4.11 Sound effects

Perhaps the ultimate 'analogue' method of synthesizing sounds is the work of the 'sound effects' team in a film or television studio. Using a floor covered with squares containing various surfaces, and a large selection of props, 'Foley' artistes produce many of the everyday sounds that accompany film and television programmes. For more unusual 'spot' effects, specialized props or pre-recorded sound effects may be used. Choreographing the sound effects for a detailed scene can be a very complex and time-consuming task – very similar to controlling an orchestra!

2.5 Topology

How do the component parts of a synthesizer fit together? This section starts by looking at typical arrangements of VCOs, VCFs, VCAs and EGs. It then looks at categorizing types of synthesizer: the main divisions in type are between monophonic and polyphonic synthesizers, performance and modular synthesizers, and alternative controllers. This section looks at the topology of the modules that make up a typical synthesizer – how they are arranged and ordered. Although this information is fundamental to the actual construction of analogue synthesizers, the theory behind it is also relevant to some digital instruments, even though digital synthesizers often have no physical realization of the separate modules at all.

2.5.1 Typical arrangements

The most common arrangement of analogue synthesizer modules is based on the 'source and modifier' or 'excitation and

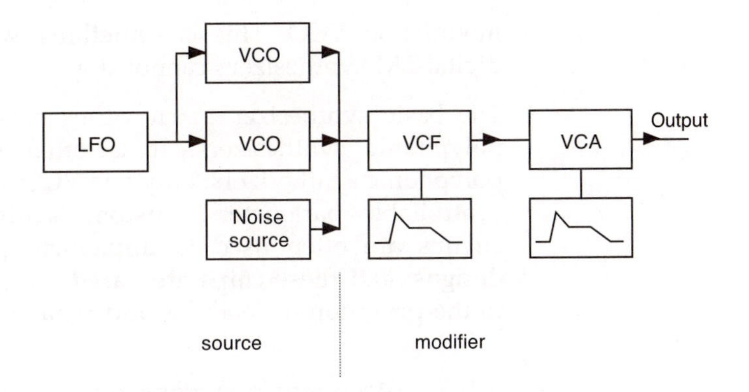

Figure 2.5.1 The basic synthesizer patch uses one or more VCOs and a noise generator as the sound source, with an LFO to provide vibrato modulation. The modifier section comprises a VCF and VCA, both controlled by one or more Envelope Generators.

filter' model. This uses one or more VCOs plus a noise generator as the sources of the raw timbre. It then uses a VCF and VCA controlled by one or more EGs to shape and refine the final timbre. An LFO is used to provide cyclic modulation: usually of the VCO pitch.

This basic arrangement of modules is used so often by manufacturers that it has become permanently hard-wired into many designs, even some modular systems! It is also used in many digital designs where there is no need for a rigid arrangement of modules because they are implemented in software.

One alternative method of subtractive synthesis replaces the single VCF with several. This enables more specific control of portions of the sound spectrum, and is often associated with the use of band-pass rather than low-pass filters. Because having separate filters for the oscillators enables them to be used as components of the final sound, rather than as a single source processed through a single modifier, this paralleling of facilities can be much more flexible in its creative possibilities. It is often used in the so-called 'Formant synthesis' technique, where the aim is to emulate the peaks in frequency response which characterize many real-world instruments, and particularly the human voice. Additive synthesis is an extension of this formant synthesis technique, where additional VCOs, VCFs and VCAs are added as required. Ganging of EGs by using voltage control of the EG parameters can make the control easier.

By using one VCO to modulate another, FM synthesis can be used, although the limitations of the VCO tuning stability and scaling accuracy restrict its use. By using VCFs to process the outputs of each VCO, then the FM can be dynamically changed from using sine waves to using more complex waveshapes by increasing the cut-off frequency of the VCF on the output of the modulation VCO. This is something which most commercial digital FM synthesizers cannot do!

The basic synthesizer patch varies between monophonic and polyphonic synthesizers. It is often simplified for use in polyphonic synthesizers: only one VCO and VCA, and often less controllable parameters. Custom 'synth-on-a-chip' integrated circuits are often used to implement polyphonic synthesizer designs, and these chips are based on a minimalistic approach to the provision of modules and parameters.

2.5.2 Monophonic synthesizers

Monophonic synthesizers tend to be performance-oriented instruments designed for playing melodies, solos or lead lines. Despite the name, many monophonic analogue instruments can actually play more than one note at once: many have a duophonic note memory which allows two different note pitches to be assigned to two VCOs. With only one or two notes capable of being played simultaneously, an assignment strategy is required so that any additional notes played can be dealt with in a predictable way. Two common schemes are last-note or low-note priority. Last-note priority is a time-based scheme: it always assigns the most recently played note to the synthesizer's voice circuitry, whilst low-note priority is a pitch-based scheme which always assigns the lowest pitched note to the voice circuitry. Low note priority can be a powerful performance feature – for example, the performer can play legato 'drone' notes with the thumb of their right hand and use the rest of the fingers to play runs on top, with staccato playing dropping back to the 'drone' note. This technique is most effective with envelopes which are not retriggerable – that is, they do not restart the attack segment each time a new key is pressed on the keyboard. (See Chapter 7 for more details on keyboard design and note assignment.)

Portamento is a gliding effect which happens between notes. On a monophonic synthesizer it is normally used as a performance

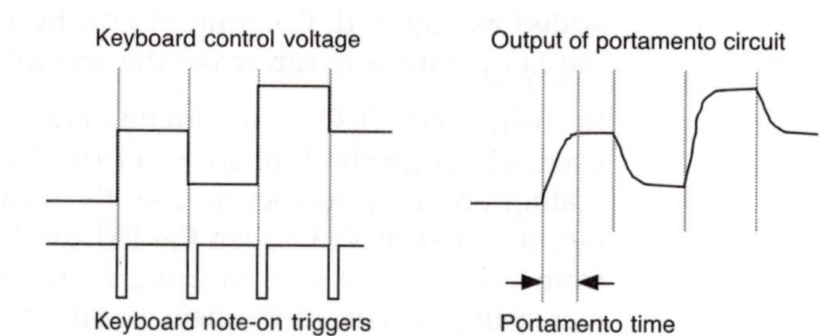

Figure 2.5.2 Portamento provides a smooth transition between successive pitches from the VCOs. The time taken for the keyboard control voltage to change from the previous value to the new value is called the portamento time.

effect to give a contrast between the sudden pitch transition between notes, and the slow change of a portamento. Portamento circuits in analogue synthesizers work by restricting the rate at which a control voltage can change. Normally, the pitch control voltage from a keyboard will change rapidly when a new note is selected. A portamento circuit changes the slope of the transition between the two voltages. It thus takes time for the note to move from the existing pitch to the new pitch.

Monophonic synthesizers normally arrange the front panel controls so that they form a logical arrangement, often mimicking the topology of the modules inside. The front panel is normally arranged so that sources and controllers are on the left, with modifiers and the final output on the right. Early analogue monophonic synthesizers, and most modular systems, do not have any form of memory for the positions and settings of the front panel controls, and so a clear and functional arrangement of controls can aid the user in remembering settings. The process of using such a synthesizer requires a lot of practice to become thoroughly familiar with the workings of the instrument. Recalling a sound is often achieved iteratively, with adjustments of the controls gradually homing in on the required sound. Individuals who have mastered a synthesizer in this way have many similarities to a classically trained instrumentalist, where the way to produce a sound from the instrument requires dexterity, skill and a degree of coaxing.

By the end of the 1970s, memory stores for the rapid recall of front panel settings had begun to appear, and by the end of the 1980s almost all monophonic synthesizers were equipped with memories. Front panels began to reflect this change by concentrating more on simplifying both the recall of memories, and of making simple minor edits to them. Many synthesizers are now simply replay machines for preset sounds, and for many users, their programming has changed from being part of the performance art to being an unwanted chore.

The performance controls on monophonic analogue synthesizers are mono-oriented: pitch-bend (often set to an interval of a fifth or an octave); octave switch (up or down one or two octaves: often to compensate for a small keyboard span); modulation (normally vibrato) and occasionally after-touch (normally controlling vibrato). For those instruments which do have front-panel controls, then they can be used as an additional method of control: real-time changes to sounds can be made 'live'. This usage of front panel and performance controls arises from the monophonic nature of the keyboard. For a right-handed player,

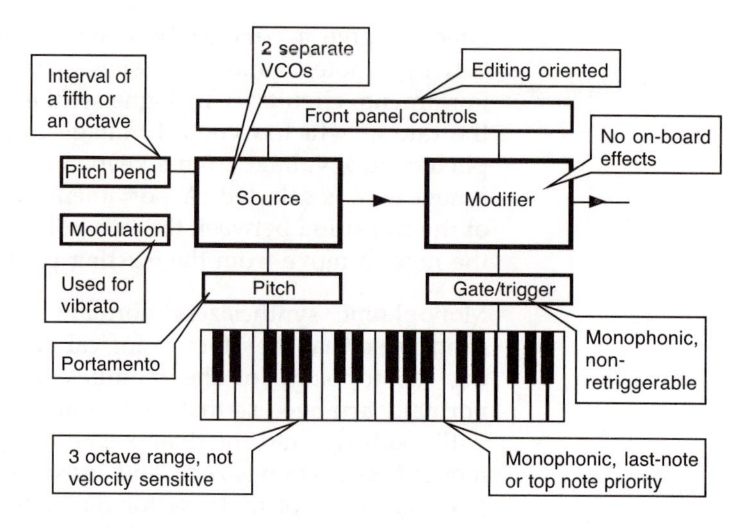

Figure 2.5.3 A summary of the main features of a typical analogue monophonic synthesizer.

the right hand is used to play the keyboard, whilst the left hand is used to provide additional expression by manipulating the pitch bend and modulation wheels. 'Classical' two-handed static position techniques for playing monophonic melodies are rarely seen, and instead, a flowing hand movement with lots of crossovers is used, thus freeing the left hand for the performance controls. Left handed versions of monophonic synthesizers are very rare indeed: the placement of the performance controls is invariably on the left side of the keyboard.

2.5.3 Polyphonic synthesizers

Polyphonic analogue synthesizers are often implemented as several monophonic synthesis 'engines' or 'voices' connected to a common polyphonic keyboard. Each of these 'voices' receives monophonic note pitch voltage, gate and trigger information, and performance controller information. It is usual for each voice to produce the same sound or timbre: multi-timbrality is normally only found on digital instruments. The assignment of the voices to the keys which are played on the keyboard is carried out by key assignment circuitry or software in the polyphonic keyboard. This deals with the reassignment of notes which are playing (note stealing) and the method of assigning notes to the voices (last note priority, etc.). (More details of keyboards can be found in Chapter 7.)

Controlling portamento on a polyphonic synthesizer is much more complex than on a monophonic synthesizer. The transitions between several notes can be made using several portamento algorithms. These are often named according to the effective polyphony which they produce, although in practice, only short portamento times are used to give a slight movement of pitch at the beginning of notes: this is frequently used for vocal, brass and string sounds. Longer portamento times do not suit polyphonic keyboard technique, except for special effect usage with block chords, and often a glissando is more musically useful – where all the notes in between the last note played and the next are played in sequence.

Memory stores seem to be more widespread in early polyphonic analogue synthesizers than in their monophonic equivalents. Initially, some manufacturers produced low-cost polyphonic synthesizers without memories, but these were not very popular in comparison to their more expensive memory-equipped versions (the Roland Juno-6 and Juno-60 memory version illustrate this well, since the follow-up model, the Juno 106, was only available with memories). The designers of polyphonic synthesizers seem to have placed more emphasis on the accessibility of the memory recall controls than on the front panel controls for programming the synthesizer voices. The Yamaha CS-80 demonstrates this principle: the programmable memories are hidden underneath a flap, and have tiny controls, whilst the large, colourful memory recall buttons are handily placed right at the front of the control panel.

The performance controls on polyphonic analogue synthesizers tend to be optimized for polyphonic playing techniques. Pitch bend is normally only a semitone, and can often be applied to only the top note, or the last note which has been played on the keyboard. The modulation wheel is often replaced or paralleled by a foot pedal, and often controls timbre via the VCF cut-off rather than vibrato. After-touch is almost invariably used to control vibrato or tremolo, and some instruments provide polyphonic after-touch pressure sensing instead of the easier-to-implement global version. Some instruments have an LFO which is common to all the voices, and so vibrato or tremolo modulation is applied at exactly the same frequency and phase to all the voices. In contrast, instruments which use separate LFOs for each voice circuit will have slightly different frequencies and phases, and this can greatly improve string and vocal sounds.

Real-time changes to the timbres are normally made using additional controllers: foot pedals, foot switches and breath controllers. Manipulating front panel controls whilst playing with both hands on a polyphonic keyboard seems to be unpopular, and if front panel controls are used, then the playing technique used often reverts to monophonic usage, as described

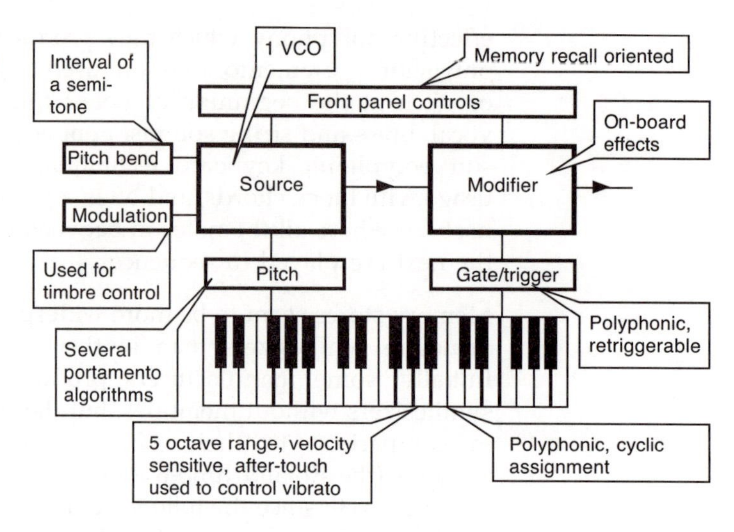

Figure 2.5.4 A summary of the main features of a typical analogue polyphonic synthesizer.

above – although polyphonic keyboards almost always have retriggered envelopes, which restricts some performance techniques. The performance controls are placed on the left hand side of the keyboard, just as with monophonic synthesizers.

2.5.4 Performance versus modular synthesizers

Performance synthesizers (monophonic or polyphonic) need a simplified and ordered control panel in order to make them usable in live performance. For this reason they usually have a fixed topology of modules: VCO, VCF, VCA.

Modular synthesizers are not arranged in a logical order because there is no way to anticipate what they will be used for, except for the simplest cases. The most usual arrangement has the oscillators and other sound sources grouped together, usually on the top or on the left, with the modifiers (filters, amplifiers) in the centre or middle, and the envelope generators on the right or bottom.

Performance instruments have memories which can be used to store and recall sounds or timbres quickly. They are often used as replay machines for a series of presets.

Modular synthesizers normally have no memory facilities, or very simple generic ones which do not have the immediacy of those found in polyphonic instruments.

It has been said that modular synthesizers are the ultimate synthesizers, and that it is only time that limits people's use of

Figure 2.5.5 (i) A performance-oriented synthesizer is designed to rapidly recall stored sounds and allow detailed performance effects to be applied with a range of specialized controllers. (ii) A modular synthesizer provides a wide range of modules which provide great flexibility, but at the expense of complexity and ease-of-use.

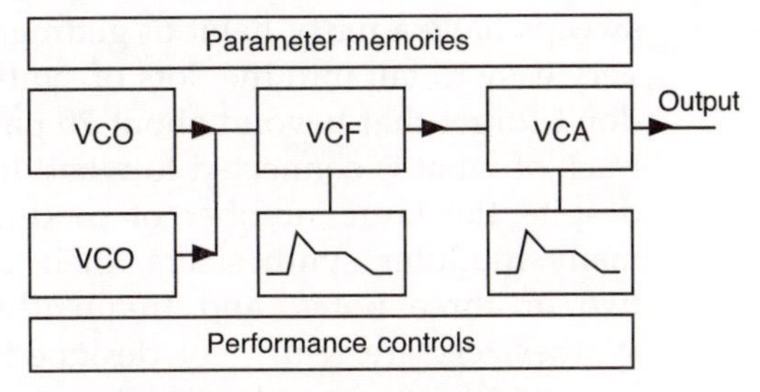

(i) A performance-oriented synthesizer

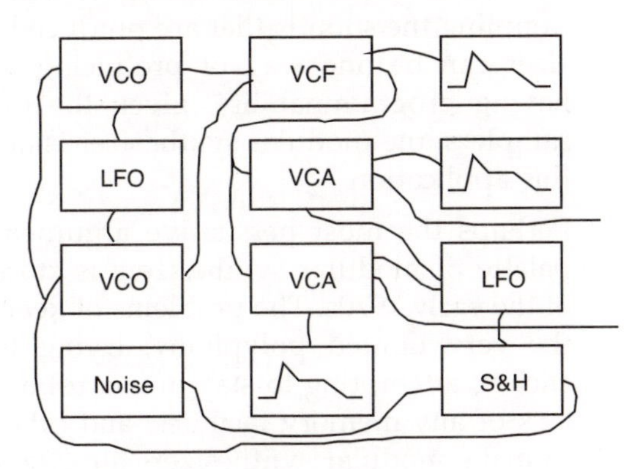

(ii) A modular synthesizer

them. Actually, modular synthesizers are severely limited, by a combination of the design and the user. The design is limited by the problems of trying to cope with patch-leads and lots of controls underneath. The user is fully occupied trying to hold everything that is happening in their head: a simple VCO-VCF-VCA setup with a couple of envelope generators can be spread over more than a dozen modules and 20 or more patch-leads. The limitations are all too evident: no programmability, a terrible user interface and lots of scribbled sheets noting down settings and patches. They are also write-only devices; once the user has produced a patch, coming back three months later and trying to figure out what is happening is almost impossible. It is often much faster to start all over again.

Modular synthesizers are very good for appearance. Large panels covered with knobs, switches and patch-cords can look very impressive on stage. In reality, modular synthesizers are very good at producing lots of variations on a very specific set of sounds, and not very much outside of that set. Some FM type sounds can be produced, but not very usefully, since the VCO modulation at audio frequencies is often less than ideal. Filter sweeps have a nasty habit of getting boring too – and it is very, very easy to fall into the 'lots of synth brass sounds' cliché. And don't forget that beyond about 20 patch-cords, most people lose track of what is connected to what! It is also often forgotten that despite the large number of modules which are available in many modular synthesizers, their polyphony is very limited: two or three notes, and frequently only one note! Modular synthesizers are really not designed for polyphonic use – and trying to keep several separate sets of modules with anything like the same parameter settings is almost impossible. Although sampling the sounds that are produced using a modular synthesizer can be one way of producing a polyphonic sound and having programmability, given the synthesis power of many samplers, the modular synthesizer is almost redundant even for this application.

Perhaps the most persuasive argument for the limited timbre palette of modular synthesizers is stored forever in recordings of the early 1970s. The problems of keeping track of patch-cords; the very limited polyphony; trying to avoid sweeping filter clichés; attempting to stay in control of the sound; the complete lack of any memory facilities; and other limitations all conspire to make modular synthesizers an expensive chore.

Of course, from a very different viewpoint, modular synthesizers are very collectable, and may well be very sought after in the future as 'technological antiques'.

2.5.6 Keyboards versus other controllers

Most synthesizers come with a keyboard. Most expander modules are equipped with MIDI input, which is a strongly keyboard-oriented interface. Many of the controls on a typical synthesizer are monophonic keyboard oriented: pitch bend, modulation, keyboard tracking, after-touch, key scaling. . .

Alternative controllers often have different parameters available which are not keyboard related. Stringed instruments like violin and cello have control over the pressure of the bow on the string in a way which is analogous to velocity and after-touch combined. Guitars enable the performer to use vibrato on specific notes: something which is very difficult on most keyboard-based synthesizers. Woodwind instruments have a number of performance techniques which do not have a keyboard equivalent – like pitch bending, changing the timbre, or producing harmonics, all by using extra breath pressure and lip techniques. (Additional information on controllers can be found in Chapter 7.)

2.6 Early versus modern implementations

Electronics is always changing. Components, circuits, design techniques, standards and production processes may become obsolete over time. This means that the design and construction of electronic equipment will continuously change as these new criteria are met. The continuing trend seems to be for smaller packaging, lower power, higher performance and lower cost – but at the price of increasing complexity, embedded software, difficulty of repair and rapid obsolescence. Over the last 25 years, the basic technology has changed from valves and transistors towards microprocessors and custom integrated circuits.

2.6.1 Tuning and stability

The analogue synthesizers of the late 1960s and early 1970s are infamous for their tuning problems. But then so are many acoustic instruments!

In fact, it was only the very earliest synthesizers which had major tuning problems. The first Moog VCOs were relatively simple circuits built at the limits of the available knowledge and technology – no-one had ever built analogue synthesizers before. The designs were thus refined prototypes which had not been subjected to the rigorous trials of extended serious musical use. It is worth noting that the process of converting laboratory prototypes into rugged, 'road-worthy' equipment is still very difficult, and at the time, valve amplifiers and electro-mechanical devices like tape echo machines were very much the technology. Modular synthesizers were the first 'electronic' devices to become musical instruments that actually left the laboratory.

The oscillators in early synthesizers were affected by temperature changes because they used diodes or transistors to generate the required exponential control law – and these change their characteristics with temperature (you can use diodes or transistors as temperature sensors!). Once the problem was identified it was quickly realized that there was a need for temperature compensation. A special temperature compensation resistor called a 'Q81' was frequently used – they have a negative temperature coefficient which exactly matches the positive temperature coefficient of the transistor.

Eventually circuit designers devised methods of providing temperature compensation which did not require esoteric resistors – usually based around differential pairs of matched transistors. Developments of these principles into custom synthesizer chips have effectively removed the need for additional temperature compensation. Unfortunately, the tuning problems had created a characteristic sound – which is one reason why the 'beating oscillator' sounds heard on vintage analogue synthesizers are emulated in fully digital instruments which have excellent temperature stability.

Tuning problems fall into four categories:

- overall tuning
- scaling
- high frequency tracking
- controllers

Because of the differences in the response of components to temperature, the tuning of an analogue synthesizer can change as it warms up to operating temperature. This can be compensated for manually by adjusting the frequency control voltage, or automatically using an 'auto-tune' circuit (see below). Some synthesizers used temperature controlled chips to try and provide constant temperature conditions for the most critical components: usually the transistors or diodes in the exponential converter circuits. These 'ovens' have been largely replaced in modern designs by careful compensation for temperature changes.

Temperature drift of the octave interval is the problem that most people mean when they say that analogue synthesizers go out of tune. Trying to match two exponential curves means that two interdependent parameters need to be changed: the offset and the scaling. The offset sets the lowest frequency that the VCO will produce, whilst the scaling sets the octave interval to get the doubling of frequency for each successive octave. On a monophonic instrument this is not so hard, and any slight errors only help to make it sound lively and interesting. For polyphonic analogue synthesizers this process can be very time consuming, and very tedious. With lots of VCOs to try and adjust, the problem can begin to approach piano-tuning in its complexity.

One method used to provide an 'automatic' tuning facility for polyphonic analogue synthesizers was introduced in the late 1970s. A microprocessor was used to measure the frequencies generated by each VCO at several points in its range, and then work out the offset and scaling correction control voltages. Because of the complexity of this type of tuning correction, and its dependence on a closed system, it has never been successfully applied to a modular synthesizer. (Auto-tuning is covered in more detail in Section 3.3 on DCOs.)

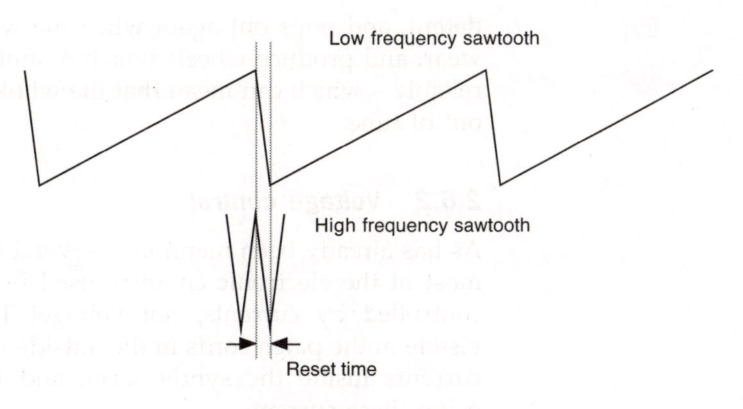

Figure 2.6.1 At low frequencies, the rising part of the sawtooth waveform is much longer than the fixed reset time. But at higher frequencies, the reset time becomes a significant proportion of the cycle time of the waveform and so the frequency is lower than it should be. This high frequency tracking problem needs to be compensated for in the control voltage circuitry of the VCO.

High frequency tracking is the tendency of analogue VCOs to go 'flat' in pitch at the upper end of their range. This is normally most noticeable when two or more VCOs are tuned several octaves apart, and although often present in single VCO synthesizers, is only apparent when they are used in conjunction with other instruments. Most VCOs use a constant current source and an integrator circuit to generate a rising voltage, and resetting the integrator when the output reaches a given voltage. This produces a 'sawtooth' waveform. The higher the current, the faster the voltage rises, and the sooner it will be reset: which produces a higher frequency sawtooth waveform. At low frequencies, the time it takes to reset the integrator is not significant in comparison with the time for the voltage to rise. But at high frequencies the reset time becomes more significant until eventually the waveform can become triangular in shape – which means that only one part of the waveform is actually controlled by the current source, and so the oscillator is not producing a high enough frequency. Some VCO designs generate a triangular waveform as the basic waveform and so do not suffer from this problem.

Controllers are another source of tuning instability. The stability of the pitch produced by a VCO is dependent on the control voltages that it receives. So anything mechanical that produces a control voltage can be a source of problems. Slider controls are one example of a mechanical control which can be prone to movement with vibration, whilst pitch-bend devices with poor detents can cause similar 'mechanical' tuning problems. The detent mechanism varies. One popular method involves using the pitch bend wheel itself – it has two of the finger notches opposite each other. One is used to help the user's fingers grip the wheel, whilst the other is used to provide the detent – a spring steel cam follower clicks into place when it is in the detent, and pops out again when the wheel is moved. This can wear, and produce wheels which do not click into position very reliably – which can mean that the whole instrument is then put out of tune.

2.6.2 Voltage control

As has already been mentioned several times, despite the name, most of the electronic circuitry used in synthesizers is actually controlled by currents, not voltage! The voltages which are visible in the patch cords in the outside world are converted into currents inside the synthesizers, and the control is achieved using these currents.

Two 'standards' are in common usage:

One volt/octave

The 1 volt/octave system uses a linear relationship between the control voltage and pitch – which in practice means that there is a logarithmic relationship between voltage and frequency. This means that small changes in voltage become more significant at higher frequencies – just where small changes in pitch might become significant and audible as tuning problems.

Exponential

The exponential system uses a linear relationship between the control voltage and the frequency. Because this method provides more resolution at high frequencies, it can be argued that it is a superior method to the 1 volt/octave system, since minor tuning errors at low frequencies are less objectionable.

Despite the apparent advantage of the exponential system, the most popular method was the 1 volt/octave system. Conversion boxes which enabled interworking between these two systems were available in the 1970s and 1980s – but they are very rare now.

2.6.3 Circuits

VCO

The basic oscillator circuit for a VCO uses a current to charge a capacitor. When the voltage across the capacitor reaches a preset limit, then it is discharged, and the charging process can start again. This 'relaxation' oscillator produces a crude sawtooth

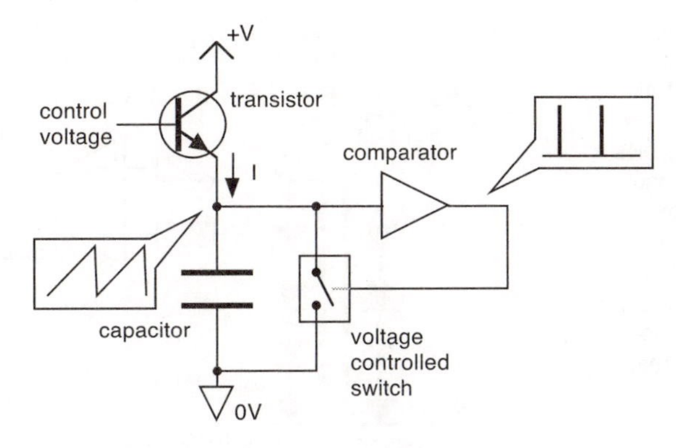

Figure 2.6.2 A relaxation oscillator circuit consists of a capacitor which is charged by a current, *I*, and discharged by a switch when the voltage across the capacitor reaches the point at which the comparator triggers. Two output waveforms are available: the sawtooth voltage from the capacitor; and the reset pulses from the comparator.

output, which can then be shaped to produce other waveforms. By varying the current which is used to charge the capacitor, then the time it takes to reach the limit changes, and so the frequency of the oscillator changes. By using a voltage to control the current, perhaps with a transistor, then the oscillator becomes voltage controlled. This type of circuit forms the basis of many VCOs.

VCF

Simple low-pass filters use RC networks to attenuate high frequencies. By making the resistor variable, it is possible to alter the cut-off frequency. This RC network forms a single pole filter, which has poor performance. Two or four-pole filters improve the performance, but require more resistors and capacitors. This requires separate buffer stages, and multiple variable resistors. One way to produce several variable resistors uses the variation in impedance of a transistor or diode as the current through it is varied. By arranging a cascade of RC networks, where the transistors or diodes have the 'voltage-controlled' current flowing through them, it is possible to make a low-pass filter whose cut-off frequency is controlled by the current which flows through the chain of transistors. This is the principle behind the 'ladder' filters used in Moog synthesizers.

The basic Moog-type filter uses two sets of transistors or diodes in a 'ladder' arrangement. The important parts of the filter are the base-emitter junction resistance, and the capacitors which connect the two sides of the ladder. Current flows down the ladder, and the input signal is injected into one side of the ladder. Since the resistance of the junctions is determined by the current which is flowing, the RC network thus formed changes its cut-off frequency as the current changes. This gives the voltage (actually current) control over the filter.

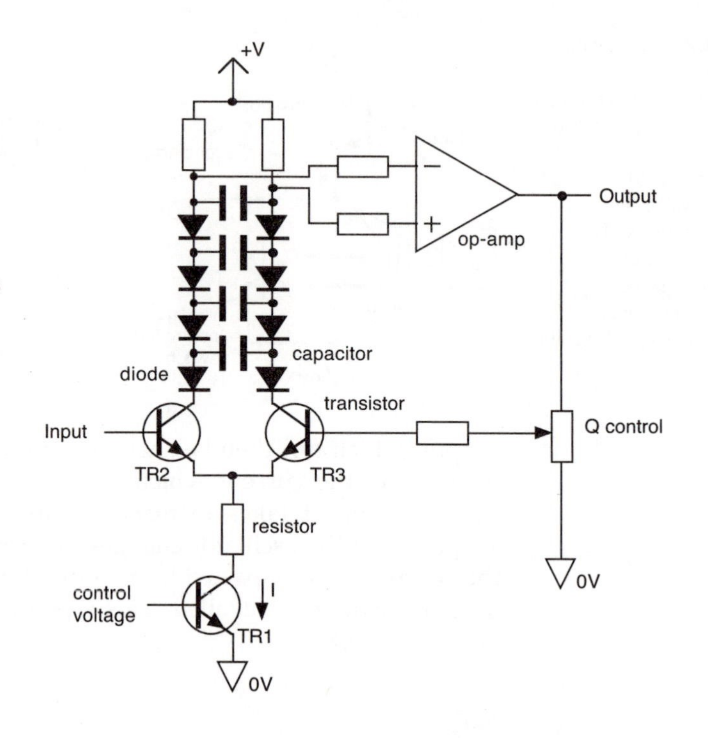

Figure 2.6.3 A typical 'ladder' filter. The current flows through the control voltage transistor, TR1, and then through the two chains of diodes. The diodes and the connecting capacitors form RC networks which produce the filtering effect, but with the diodes acting as variable resistors. The op-amp amplifies the difference between the two chains of diodes and feeds back this signal, thus producing a resonance or 'Q' control.

Another type of filter which is found in analogue synthesizers is the 'state variable' filter. This configuration had been used in analogue computers since valve days to solve differential equations. Once operational amplifiers were developed, making a state variable filter was considerably easier, and by using FETs or transconductance amplifiers the cut-off frequency of the filter could easily be changed by a control voltage.

A typical state variable filter is made in the form of a loop of three op-amps. It is a constant-Q filter. Three outputs are available: low-pass, high-pass, and band-pass (a band-reject can be produced by adding a fourth op-amp). Other types of multiple op-amp filters can be made: the bi-quad is one example whose circuit looks similar to a state variable, but the minor changes make it a constant-bandwidth filter and it only has low-pass and band-pass outputs.

2.6.4 Envelopes

It has been said that the more complex the envelope, the better the creative possibilities. The history of 'the envelope' is one of continuous evolution. The beginnings lie with organ technology, where RC networks were used to try and damp out the clicks caused by keying sine waves on and off – and then generated

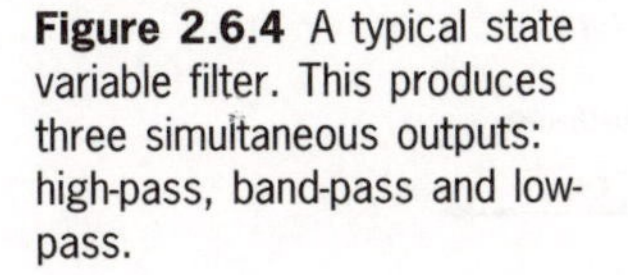

Figure 2.6.4 A typical state variable filter. This produces three simultaneous outputs: high-pass, band-pass and low-pass.

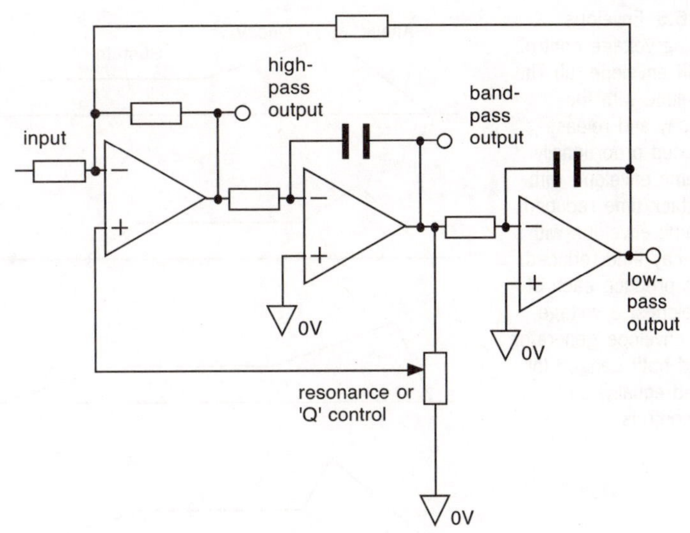

deliberately so that they could be added back in as 'key click'. Trapezoidal waveform generators followed, which provided control over the start and finish of the envelope. ADSR-type envelopes, and their many variants, were used for the majority of the analogue synthesizers of the 1970s and 1980s. The advent of digital synthesizers with complex multi-segment envelope generators has made the ADSR appear unsophisticated, and analogue synthesizers designed in the 1990s have tended to emulate the multi-segment envelopes by adding additional break-points to ADSR envelopes.

The suitability of an envelope has very little to do with the number of segments, rates, times or levels. Instead, it has to do with the way that things happen in the real world. There are several things to consider:

- Many instruments have envelopes with exponential attacks rather than the much easier to produce linear slopes which many analogue synthesizers use. One solution to this is to add in two or more attack segments and so produce a rough approximation of an exponential envelope. This is much easier to achieve in a digital instrument than in analogue circuitry.
- Envelopes often change their shape and their timing in ways that are related to the note's pitch and the velocity with which it was played. Most modular and monophonic analogue synthesizers are not velocity sensitive, and so instruments which depend on this sort of performance technique tend to suffer – pianos, for example. Changing the

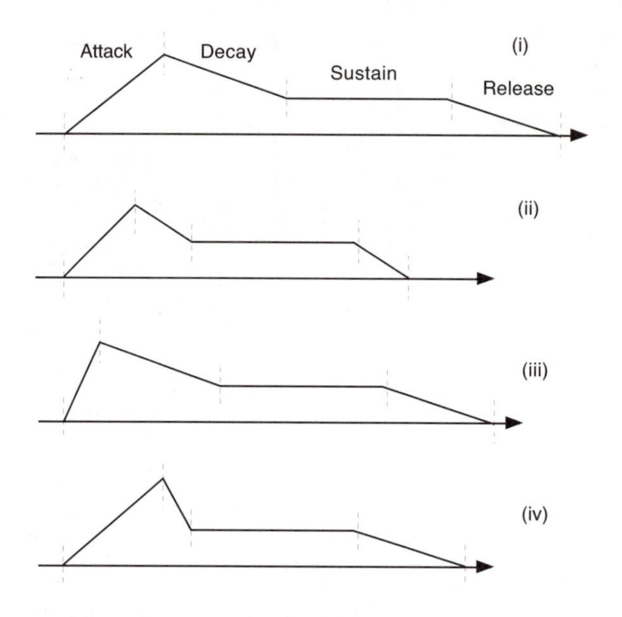

Figure 2.6.5 Envelope scaling using voltage control. (i) An ADSR envelope. (ii) The same envelope with the attack, decay and release times reduced proortionally. (iii) The same envelope with just the attack time reduced. (iv) The same envelope with just the decay time reduced. In order to produce each of these envelopes, a voltage-controlled envelope generator would need both ganged (all time altered equally) and individual controls.

attack times with pitch can be quite complex in an analogue synthesizer – you need an envelope generator with voltage controlled time parameters, and this can require a large number of additional patch-cords and control knobs.

Sophisticated multi-segment envelopes suffer from being harder for the user to visualize the shape of the envelope being produced. Probably the best compromise is an ADSR with a couple of attack, decay and release segments, and control over the slopes: 'function' generators meeting this sort of design criteria are beginning to appear. Envelope generator design research is still ongoing.

2.6.5 Discrete versus integration

Early analogue synthesizers used individual transistors to build up their circuits. This 'discrete' method of construction was gradually replaced by integrated circuits – usually op-amps for the majority of the analogue processing. Custom chips began to integrate large blocks of circuitry into single chips: a VCO or VCF, for example. Finally, by the mid 1980s, complete VCO, VCF, VCA, LFO and EG circuits could be placed on a single 'voice' chip intended for use in polyphonic analogue synthesizers.

2.6.6 Pre- & post-MIDI

The Musical Instrument Digital Interface (MIDI) signalled a major change in synthesizer technology (Rumsey, 1994). At a stroke, many of the incompatibility problems of analogue synthesizers were solved. Control voltages, gates and trigger pulses were replaced by digital data. The note control, control parameters, sound data; pitch bend and modulation controls were now standardized, and instruments could be easily interconnected.

Before MIDI, manufacturers were relatively free to use any method to provide interconnections between the instruments they produced, if at all. Commercial interests dictated that if you used a different control voltage, gate and trigger pulse system, then purchasers would only be able to easily interconnect to other products in your range. As a result, with a few exceptions, any interfacing between synthesizers from different manufacturers would require the conversion of voltages or currents. In addition, the performance controls were not fixed. Some manufacturers provided pitch bend controls and multiple modulation controls, whilst others only had switched modulation: on or off. If an instrument was programmable, then the sound data was normally stored on data cassettes – again in proprietary formats.

MIDI was intended to enable the interchange and control of musical events with and by electronic musical instruments. It replaced the analogue voltages, currents and pulses with digital numbers, and so provided a simple way to assemble simple instruments into a larger unit. The layering of one sound with another changed from requiring two tracks on a multi-track tape recorder, to being a simple case of connecting two instruments together with a MIDI cable.

The introduction of MIDI had a profound and lasting effect on synthesizer design. Because the MIDI specification included a standard set of performance controllers, it effectively froze the pitch bend and modulation wheel into the specification of a synthesizer. MIDI is also biased towards a keyboard-oriented way of providing control: monophonic pressure is one example of this. MIDI also provided a standardized way of saving sound data by using system exclusive messages, and the possibility of editing front panel controls remotely.

The uniformity of synthesizer design post-MIDI has meant that the emphasis has been placed onto the method of sound generation, rather than the functional design of the instrument. Although this has provided a wide variety of sounds, it has also meant that alternative controllers for synthesizers have tended to be ignored: the guitar synthesizer being one example.

2.6.7 Before and after microprocessors

The adoption of MIDI was also accompanied by a consolidation in the use of microprocessors. Microprocessors had begun to be used in polyphonic synthesizers to provide memory functions for storing sounds, but MIDI made the use of a microprocessor almost obligatory.

Before microprocessors, analogue synthesizers did not have auto-tune facilities, or memories for sounds. Interfacing was via analogue voltages and the complexity meant that only two or three instruments would be connected together. Front panel controls actually produced the control voltages that controlled the synthesizer sound circuitry.

Once microprocessors were incorporated in synthesizer designs, then auto-tuning was introduced for polyphonic synthesizers. Memories for sounds, and storage on floppy disk, data cassette or via MIDI system exclusive messages was possible. MIDI cables could be used to connect many instruments together. Front panel controls were scanned by the microprocessor to determine their position and thus produce a control voltage – or the front panel controls were replaced by a parameter system using buttons and a single control to select a parameter and edit it.

The changes in synthesizer design post MIDI and post microprocessors are most evident in rack-mounting synthesizer modules, which have very little in common with the exterior appearance of analogue synthesizers of the late 1970s: no keyboard, no control knobs, no data cassette, no CV sockets, and no performance controls.

2.7 Example instruments

Moog Modular (1965)

The Modular Moog synthesizers comprise a number of modules which are placed in a frame which provides their power. Connections between modules are made using quarter inch front panel jack connectors. Models were available where the number and choice of modules was pre-determined, or the user could make their own selection. The system shown here (Figure 2.7.1) provides enough facilities for a powerful monophonic instrument, although producing polyphonic sounds does require a large number of modules, and can be very awkward to control. Further information on the modules can be found in Martin Newcomb's guide to the Museum of Synthesizer Technology (Newcomb, 1994).

Figure 2.7.1 Moog Modular.

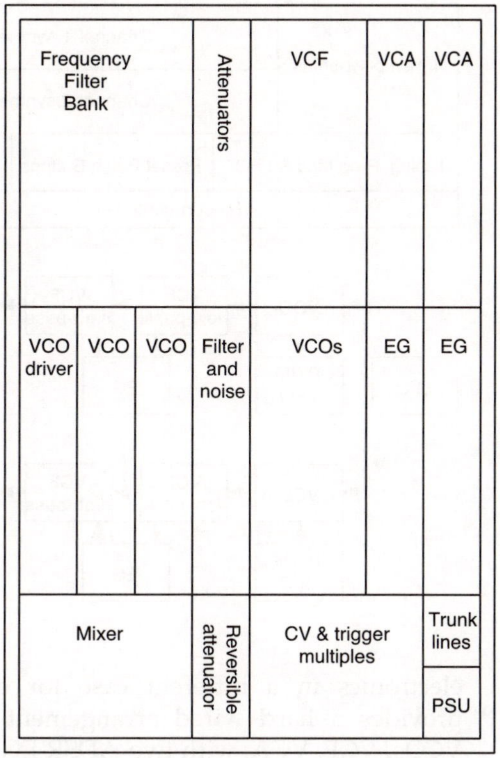

Figure 2.7.2 Mini Moog.

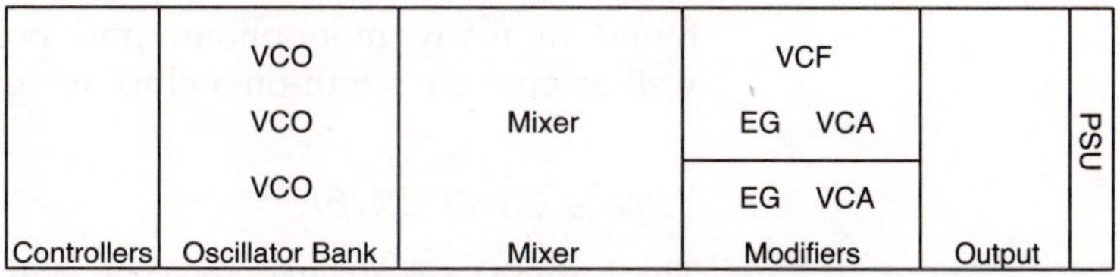

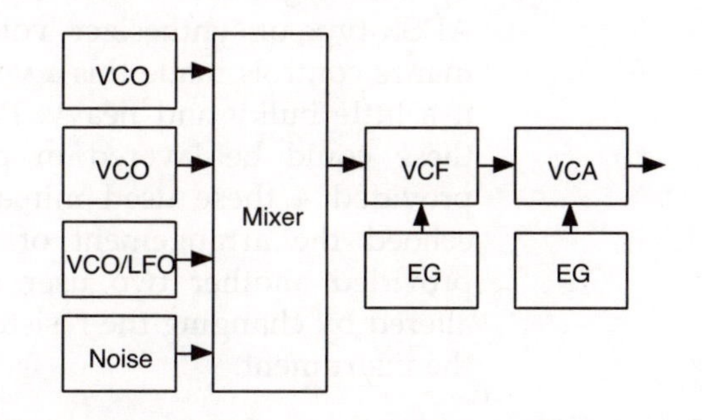

Mini Moog (1969)

The Mini Moog was intended to provide a portable monophonic performance instrument (the Sonic Six repackaged similar

Figure 2.7.3 CS-80.

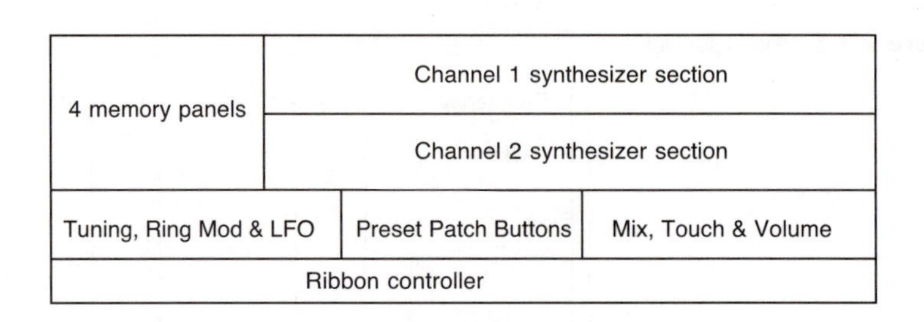

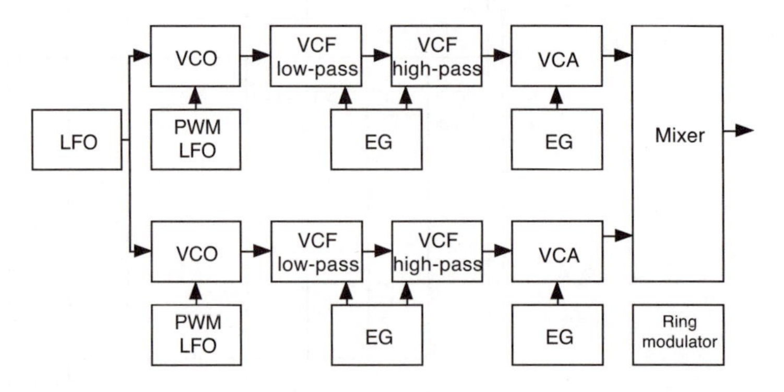

electronics in a different case for educational purposes). It provides a hard wired arrangement of synthesizer modules: VCOs, VCF, VCA, with two ADSR EGs. This topology has since become the de facto 'basic' synthesizer 'voice' circuit, and can be found in many monophonic and polyphonic synthesizers, as well as custom 'synth-on-a-chip' integrated circuits.

Yamaha CS-80 (1978)

The Yamaha CS-80 was an early polyphonic synthesizer made up from eight cards, each containing a dual VCO/VCF/VCA/ADSR type of synthesizer 'voice' circuit. Comprehensive performance controls made this a versatile and expressive instrument, if a little bulky and heavy. Preset sounds were provided, and these could be layered in pairs. Four user memories were provided – these used miniature sliders and switches which echoed the arrangement of the front panel controls, which provided another two user memories. The presets could be altered by changing the resistor values on a circuit board inside the instrument.

Sequential Prophet 5 (1979)

The Prophet 5 was essentially five 'Mini Moog' type synthesizer voice cards connected to a polyphonic keyboard controller. The

Figure 2.7.4 Prophet 5.

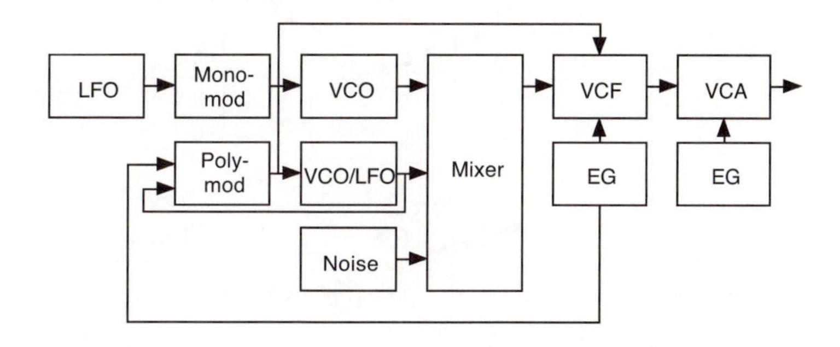

major innovation was the provision of digital storage for sounds, although the ability to use one VCO to modulate the other, called 'Poly-mod' by Sequential, allowed the production of unusual FM sounds.

Roland SH-101 (1982)

The SH-101 was intended for live performance, and contained a simplified basic synthesizer 'voice' circuit. The instrument casing was designed so that it could be adapted for on-stage use by slinging it over the shoulder of the performer – a special hand grip add-on provided pitch-bend and modulation controls.

Oberheim Matrix 12 (1984)

The Matrix-12 (and the smaller Matrix-6) was a modular synthesizer in a case which was more typical of a performance synthesizer. The front panel extends the use of displays which was

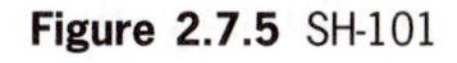

Figure 2.7.5 SH-101

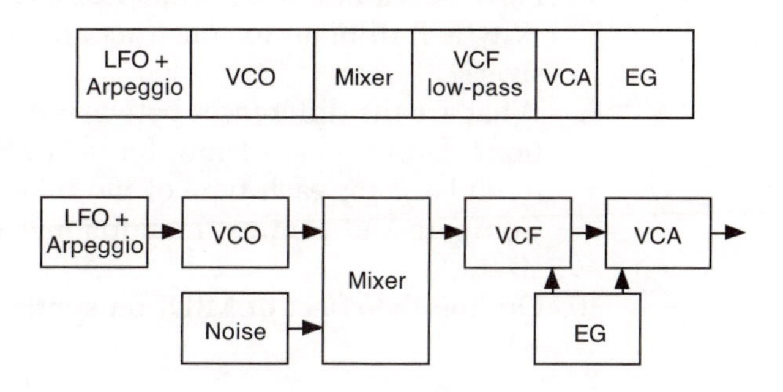

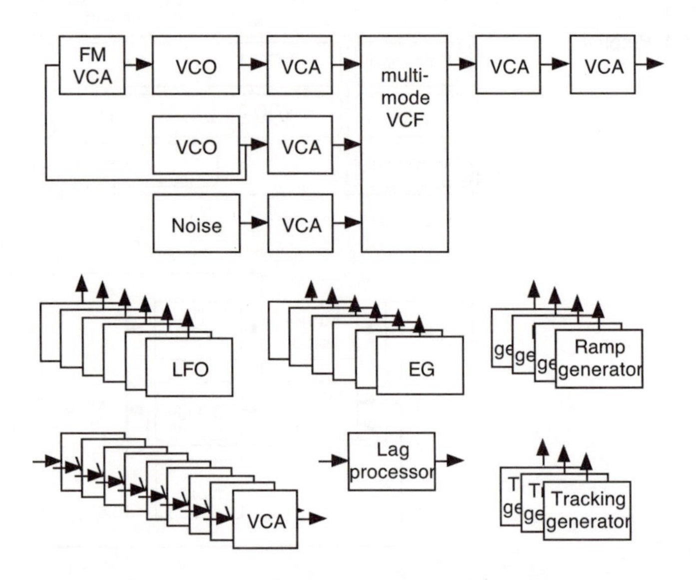

Figure 2.7.6 Matrix-12.

pioneered in earlier OB-X models – this time using green cold-cathode displays to provide re-assignable front panel controls. The wide range of processing modules made this a versatile and powerful instrument. Only one voice from the 12 available is shown in this diagram.

2.8 Questions

1. Name three ways of producing sound electronically using analogue synthesis, and briefly outline how they work.
2. Describe the 'source and modifier' model for sound synthesis.
3. What are the basic analogue synthesizer source waveforms and their harmonic contents?
4. What are the four major types of filter response curve? What effect do they have on an audio signal?
5. What are the main parts of an envelope? Include examples of the envelopes of real instruments.
6. How do vibrato and tremolo differ?
7. Why is it difficult to construct an analogue additive synthesizer?
8. What are the differences between AM, FM and ring modulation? Draw a spectrum for a 1 kHz carrier and 100 Hz modulator for each type of modulation.
9. Compare and contrast monophonic and polyphonic synthesizers.
10. Outline the effect of MIDI on synthesizer design.

Time line

Date	Name	Event	Notes
1768–1830	Jean Baptiste Fourier	French mathematician who showed that any waveform could be expressed as a sum of sine waves	Basis of Fourier (additive) synthesis and FFT (Fast Fourier Transform)
1807	Jean Baptiste Joseph Fourier	Fourier published details of his Theorem, which describes how any periodic waveform can be produced by using a series of sine waves	The basis of additive synthesis
1870s	Gavioli	Fairground organs from Gavioli begin to use real instruments to provide percussion sounds	
1928	Maurice Martinot & Ondes	Invented the Ondes Martenot – an early synthesizer	Controlled by a ring on a wire – finger operated
1930s	Ondes	Ondioline – an early synthesizer	Uses a relaxation oscillator as a sound source
1963	Don Buchla	Simple VCO, VCF and VCA-based modular synthesizer: 'The Black Box'	Not well publicised
1963	Herb Deutsch	First meeting with Robert Moog. Initial discussions about voltage controlled synthesizers	
1965	Paul Ketoff	Built the 'Synket', a live performance analogue synthesizer for composer John Eaton	Commercial examples like the Mini-Moog and Arp Odyssey, soon followed
1965	Robert Moog	First Moog Synthesizer is hand-built	Only limited interest at first
1966	Don Buchla	Launched the Buchla Modular Electronic Music System – a solid-state, modular, analogue synthesizer	Result of collaboration with Morton Subotnick and Ramon Sender
1969	Peter Zinovief?	EMS (Electronic Music Studios) produces the VCS3, the UK's first affordable synthesizer	The unmodified VCS3 is notable for its tuning instability
1969	Robert Moog	Minimoog is launched. Simple, compact monophonic synthesizer intended for live performance use	Hugely successful, although the learning curve is very steep for many musicians
1970	ARP Instruments	ARP 2600 'Blue Meanie' modular-in-a-box released	
1970	ARP Instruments, Alan Richard Pearlman	ARP 2500. Very large modular studio synthesizer	Uses slider switches – a good idea, but suffer from crosstalk problems
1971	ARP Instruments	The 2600, a performance-oriented modular monosynth in a distinctive wedge shaped box	
1978	Electronic Dream Plant	Wasp Synthesizer launched. Monophonic, all-plastic casing, very low-cost, touch keyboard – but it sounded much more expensive	Designed by Chris Hugget & Adrian Wagner

Date	Name	Event	Notes
1978	Sequential Circuits	Sequential Circuits Prophet 5 synthesizer – essentially five Mini-Moog-type synthesizers in a box	A runaway best seller
1978	Yamaha	Yamaha CS series of synthesizers (50, 60 and 80), the first mass-produced successful polyphonic synthesizers	Korg, Oberheim and others also produced polyphonic synthesizers at about the same time
1980	Roland	Jupiter-8 polyphonic synthesizer	8 note polyphonic, programmable poly-synth
1981	Moog	Robert Moog is presented with the last Minimoog at NAMM in Chicago	The end of an era
1981	Roland	Roland Jupiter-8. Analogue 8-note polyphonic synthesizer	
1982	Moog	Memorymoog – six note polyphonic synthesizer with 100 user memories	Cassette storage! Six Minimoogs in a box!
1982	Roland	Jupiter 6 launched – first Japanese MIDI synthesizer	Very limited MIDI specification. 6 note polyphonic analogue synth
1982	Sequential	Prophet 600 launched – first US MIDI synthesizer	6 note polyphonic analogue synth – marred by a membrane numeric keypad
1984	Sequential	Sequential launch the Max, an early attempt at mixing home computers and synthesizers	A complete failure – too early for the market
1984	Sequential Circuits	SixTrak. A multi-timbral synthesizer with a simple sequencer	The first 'workstation'?

3 Hybrid synthesis

Hybrid synthesis is the name usually associated with methods of synthesis which are not completely analogue or digital. These borderline methods were most important during the changeover from analogue to digital sound generation in the early 1980s, but the underlying techniques have also become part of the all-digital synthesis methods. With the increase of interest in 'analogue' synthesis in the 1990s, it is intriguing to note that very few of the instrument which are now being designed are truly analogue – in many ways they are actually hybrids!

Synthesis methods which combine more than one technique or method of synthesis to produce a composite sound are described as 'layered' or 'stacked', and are covered in Chapter 6. Although these methods are sometimes called 'hybrid' methods, I prefer the word 'composite' synthesis.

It is possible to divide hybrid synthesizers into different classes. One possible division is based on the roles of the digital and analogue parts:

- Digital control of the parameters of analogue synthesis – as used in many programmable analogue monophonic synthesizers.
- Digital control of the oscillator (DCO) with the remainder of the instrument analogue, perhaps with digital control of the parameters.

- Digital oscillator with analogue modifiers, and with digital control of the analogue parameters. This is the form which many of the mid-1990s' 'retro' analogue synthesizers used.

Another classification might be made on the method used to produce the sound. This section divides hybrid synthesizers using this method. Wavecycle, wavetable and DCO technologies are all discussed.

The predominant method of hybrid synthesis uses digital sound generation and control of parameters with analogue filtering and enveloping. This uses the technology which is most appropriate for the task. Hybrid synthesizers are often characterized by a more sophisticated raw sound as the output from the 'source' part, which is due to the use of digital technology – especially in wavetable-based synthesizers. In fact, this availability of additional waveshapes, beyond the 'traditional' analogue set of sine, triangle, sawtooth and rectangular waveforms, could be seen as the major differentiator between analogue and hybrid instruments.

Although the idea of mixing analogue with digital was pioneered in the late 1970s, most notably with the Wasp synthesizer, it still forms the basis of many of the most successful hybrid (and hybrid masquerading as analogue) instruments. In fact, the recent trend of using digital circuitry to replace traditionally analogue functions like filters or oscillators follows on from hybrid synthesis. Also, the hybrid design philosophy of using a complicated oscillator and conventional modifiers also forms the basis of all S&S (samples and synthesis) instruments.

3.1 Wavecycle

A wavecycle is another word for waveform, although it emphasizes the word cycle, which is very significant in this context. It is used here to emphasize the difference between the 'static, sample-based' replay-oriented wavecycle oscillators, and the 'dynamic, loop-based' wavetable oscillators.

Analogue synthesizers incorporate VCOs or oscillators which can typically produce a small number of different waveforms with fixed waveshapes, where each cycle is identical to those before and after it. The one exception to this is a PWM waveform, where the shape of the pulse can be changed using a control voltage – often from an LFO for cyclic changes of timbre.

Hybrid synthesizers which use wavecycle-based sound generation can use this single-cycle mode, but they can also produce additional waveshapes, and use more complex schemes where more than one cycle of the waveform is used before the shape is repeated. The logical conclusion to this is for there to be a large number of cycles, each different, and with no repetition at all – and the result is then called a sample. The only really important differences between these examples are the length of the audio sample, and the amount of repetition. Each method has its own strengths and weaknesses.

3.1.1 Single-cycle

Single-cycle oscillators produce fixed waveforms, somewhat like an analogue synthesizer – although the selection of waveshapes is often much larger. The method of producing the waveform is often a mixture of analogue and digital circuitry, although as digital technology has fallen in price, so the use of predominantly digital circuit design has increased.

Possibly the simplest method of controlling a waveshape is the pulse width control, which is sometimes found on analogue synthesisers. With a single control, a variety of timbres can be produced: from the 'hollow'-sounding square wave with the missing second harmonic, through to narrow pulses with a rich harmonic content and a thin, 'reedy' sound.

Pulse waveforms usually only have two levels, although there are variants which have three, where the pulses are positive and negative with respect to a central zero value. Multiple levels were used in one of the first 'user-programmable' waveforms, called 'slider scanning'. In this method, the oscillator runs at several times the required frequency, and is used to drive a counter circuit, which then controls a electronic switch called a multiplexer. The multiplexer 'scans' across several slider controls, and thus creates a single waveform cycle where the voltage output for each or the stages is equivalent to the positions of the relevant slider. By setting half of the sliders to the maximum voltage position, and the remainder to the minimum, then a square waveform is produced, but a large number of other waveforms are also possible (Figure 3.1.1).

Slider scanning oscillators are limited by the number of sliders which they provide. Eight or 16 sliders are often used, and this means that the oscillator is running at eight or 16 times the frequency of a VCO producing the same note conventionally. The counter is normally arranged so that it switches in each slider in turn, and when all of the sliders have been scanned, it returns to the first slider again. The sliders thus represent one

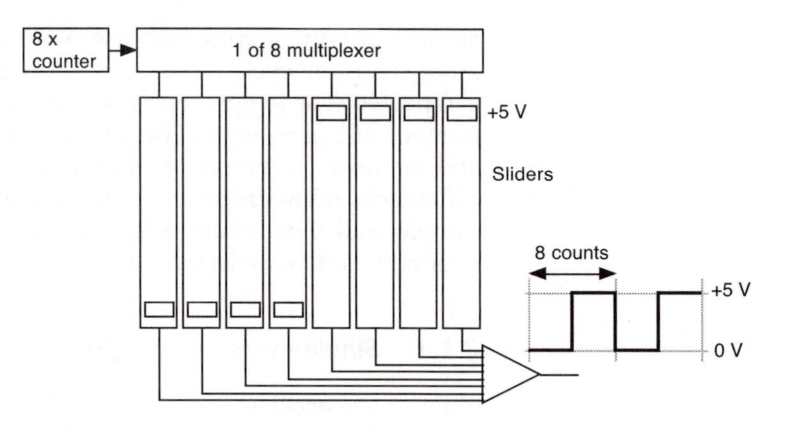

Figure 3.1.1 This eight slider scanner circuit runs a counter at eight times the required frequency. The counter causes a '1 of 8' multiplexer to sequentially activate each of the eight slider controls, which produce a voltage dependent on their position. The slider outputs are summed together to produce the output waveform.

cycle of the waveform. This type of counter is called a Johnson counter, although it is possible to use counters which scan back and forth along the sliders, in which case the relationship between the sliders and the cycle of the waveform is less obvious: although it appears to be only half of a cycle, the reversal of the scan direction merely adds in a time reversed version of the sliders, and this sounds like a second cycle of a two-cycle waveform. The pitch is thus unaltered. Slider scanners thus provide single-cycle (or two-cycle) waveforms where the shape of the waveform is static (unless the slider positions are changed) and repeated continuously.

For detailed control over the waveform, the obvious solution is to add more sliders. Providing many more than 16 separate sliders quickly becomes very cumbersome, and it makes rapid selection of different waveforms almost impossible. One alternative is to provide pre-stored values for the slider positions in a memory chip (ROM or RAM), and to use these values to produce the output waveform. In this way many more 'sliders' can be used, and the waveform can be formed from a large number of separate values, instead of just eight or 16. The difficulty lies in producing the values to put in the memory – preset values can be provided, but sliders or another means of user control of the values is preferable. A minimalist approach might be to provide two displays: one for the 'slider' number, and another for the value at that slider position.

Drawing the values on a computer display screen has been used as one method of providing a more sophisticated user interface to large numbers of sliders, but this can be very tedious to use and difficult to achieve the desired results. Trying to set the positions of several hundred sliders on a screen to produce a particular timbre is also hampered by the relationship between the shape drawn and the timbre produced – which is not intuitive for most users, and requires considerable practice before a specific timbre can be quickly set up. The simpler oscillators which scan through values in a memory chip are very economical in their usage of memory, and a large number of waveforms can be made available in this way. As with analogue synthesizers, selecting a waveform is easier if they are arranged in some sort of order: either a gradually increasing harmonic content, or else in groups with variations of specific timbres: pulses, multiple sine waves added together, etc.

Scanning across a series of voltages, which are set by slider positions, using a multiplexer is straightforward. Replacing the slider positions with numbers then requires some way of converting from a number to a voltage. This is achieved using a circuit called a 'digital-to-analogue converter', normally abbreviated to DAC, which converts a digital number into a voltage. By sequentially presenting a series of numbers at the input to the DAC, a corresponding set of output voltages will be produced (Figure 3.1.2).

By storing the numbers in ROM, a large number of preset waveforms can be provided, merely by sending different sets of numbers from the ROM to the DAC. This is easily accomplished by having additional control signals which set the area of the memory which is being used. For user-definable waveforms, RAM memory chips are used, where values can also be stored in the chip instead of merely recalled. Often a mixture of ROM and RAM is used to provide fixed and user-programmable waveforms. But there are alternatives to using memory chips. By

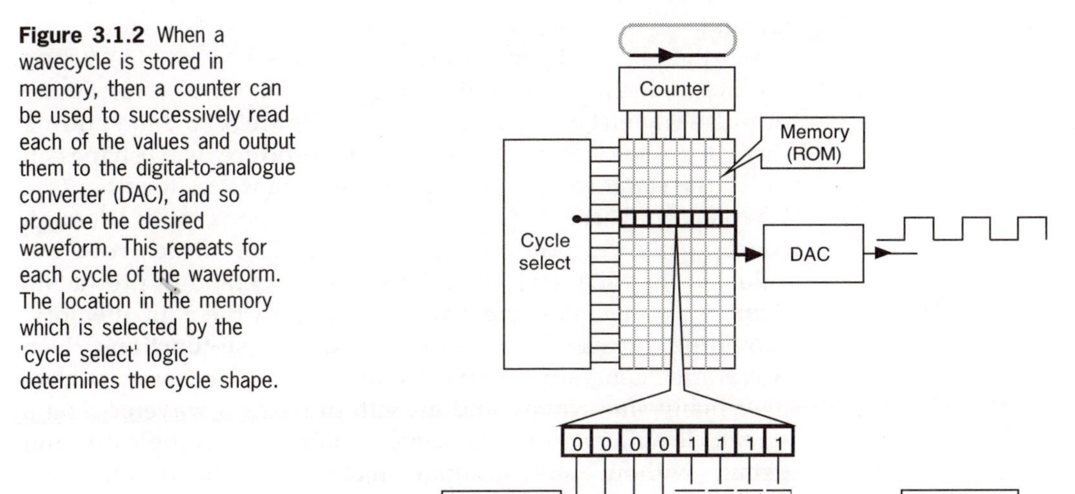

Figure 3.1.2 When a wavecycle is stored in memory, then a counter can be used to successively read each of the values and output them to the digital-to-analogue converter (DAC), and so produce the desired waveform. This repeats for each cycle of the waveform. The location in the memory which is selected by the 'cycle select' logic determines the cycle shape.

using the output of a counter as the input to the DAC, then a number of different waveforms can be produced: it depends on the way that the counter operates. Assuming that the DAC converts a simple binary number representation, then a simple binary counter would produce a sawtooth-like staircase waveform. There are a large number of types of counter that could be used for this purpose, although dynamically changing the type of counter is not straightforward. Some other types which might be used include the up-down and Johnson counters mentioned earlier, and Gray-code counters.

By using digital feedback between the stages of a counter, it is possible to produce a counter which does not just produce a short sequence of numbers in sequence, but a very much longer sequence of numbers in a fixed but relatively unpredictable order. These are called pseudo-random-sequence generators, and they can be used to produce noise-like waveforms from a DAC. Actually this is a multi-cycle waveform (see below) rather than a single cycle. But by deliberately using the wrong feedback paths (or by resetting the count), it is possible to shorten the length of the sequence so that it produces sounds with a definite pitch, where the length of the sequence is related to the basic pitch that is produced. In effect, the length of the sequence becomes the length of one cycle of the waveform. Slight changes in the feedback paths or the initial conditions of the counter can produce a wide variety of waveforms from a relatively small amount of circuitry with simple (if unintuitive) controls, and no memory is required (unless the paths and initial conditions need to be stored, of course!).

Since digital circuitry is concerned with only two values: on and off, or one and zero, the square or pulse waveform is a basic digital waveform, in much the same way as sine waves are the underlying basis of analogue. By taking several square or pulse waveforms at different rates (Figure 3.1.3), and adding them together, they can be used to produce waveshapes in much the same way as adding several sine waves together (see Chapter 4). These are called Walsh functions, and although conceptually very different from the systems which read out values from a memory chip, the simplest method of producing them is to calculate the values and store them in memory. Providing a user interface to a Walsh-function-driven waveform generator would require comprehensive control over many sine waves, and as with drawing a waveform on a screen, it suffers from the same problems of complexity and detail, without any intuitive method of determining the settings.

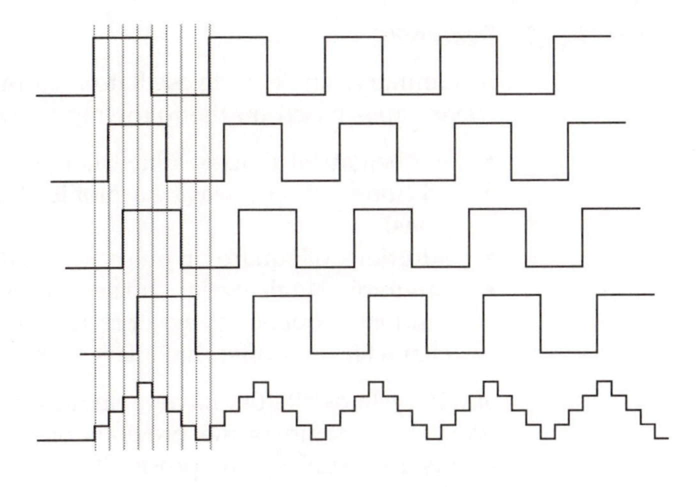

Figure 3.1.3 Walsh functions combine square or pulse waveforms to produce more complex waveforms. In this example, four square waves are added together to produce a crude 'triangle' type of waveform. Each position in the output is produced by adding together the level of each of the component waveforms – the final output waveform is shown at one quarter of the actual size that would be expected if the four waveforms were added together.

Filtering of outputs

All of the methods of producing arbitrary waveshapes using digital circuitry described above produce outputs which tend to have flat segments connected by sudden transitions. Since these rapid changes produce additional (often unwanted) harmonics, the outputs need to be filtered so that the final output is 'smooth' – usually with a low-pass filter whose cut-off frequency is set to be at the highest required frequency in the output (Figure 3.1.4). Because the frequency of the output waveform from an oscillator can change, then the filter needs to track the changes in frequency, which means that the VCO needs to be coupled to a VCF, and set up so that the cut-off frequency follows the oscillator frequency, usually by connecting the same control voltage to the VCO and VCF. In some circumstances, the additional frequencies are deliberately allowed to pass through the filter. Because these frequencies are linked to the oscillator frequency, they are actually harmonics of it, albeit high harmonics. Removing the filter means that a waveform which might appear to be a sine wave from the slider positions, is actually a sine wave with extra harmonics. The ability to switch the filter in and out, together with knowledge of how the waveforms are being produced, is very useful if the most is to be made of the potential of single-cycle oscillators.

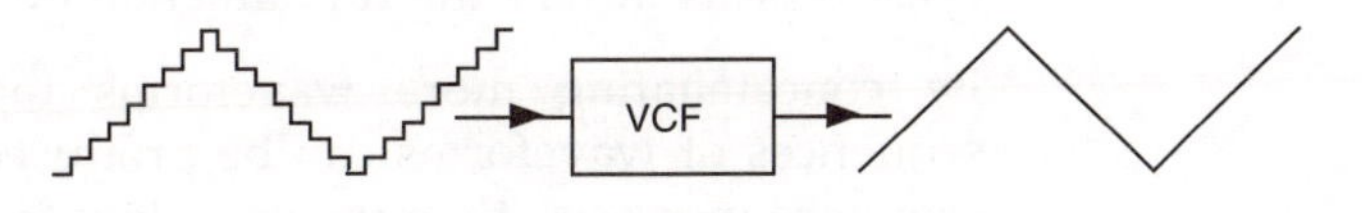

Figure 3.1.4 Filtering the output of a generated wavecycle waveform can smooth out the abrupt transitions and produce the required shape.

Waveshapes

In summary, single-cycle oscillators normally have a selection of waveshapes based on the following types:

- mathematical shapes: sine, triangle, square, sawtooth
- additions of sine wave harmonics (organ 'drawbar' emulations)
- additions of square or pulse waves (Walsh functions)
- random: single-cycle 'noise' waveforms from pseudo-random sequence generators tend to be very non-white in character, and often have large amounts of high harmonics.

Single-cycle oscillators have a characteristic fixed timbre – each cycle is the same as the previous one, and the next one. This means that subsequent processing through modifiers is often used to make the sound more interesting to the ear.

3.1.2 Multi-cycle

The important thing about multi-cycle oscillators is that although they can have many cycles of waveform which they output in sequence, the same set of cycles repeats continuously. This is different to a sample, where the sample may only be played through in its entirety once.

The technology for producing multi-cycle waveforms is very similar to single-cycle oscillators, although the user interface is often restricted to merely choosing the specific waveforms for each cycle, rather than providing large numbers of slider or other controls. The basic method uses a memory chip and DAC, just as with single-cycle oscillators. The difference is that the area of memory which is being cycled can also be controlled dynamically. A simple example might have two areas of ROM memory: one containing a square waveform; and the other a sawtooth waveform. By setting the ROM to output first the square, then the sawtooth, and then repeating the process, the output will be a series of interspersed square and sawtooth cycles. This two-cycle waveshape has a harmonic structure which incorporates some elements from each of the two types of waveform, but also has additional lower frequency harmonics which are related to half of the basic cycle frequency. This is because the complete cycle repeats at half the fundamental cycle rate.

By concatenating more waveforms together, more complex sequences of waveforms can be produced. As the length of the sequence increases, then the extra low frequency harmonics also drop in frequency. To take an extreme example, imagine one cycle of a square waveform followed by three cycles of a silence waveform. The equivalent is a pulse waveform with a frequency of a quarter of the square wave cycle: two octaves below the pitch which was intended. With eight cycles in the sequence, then the lowest harmonic component will be three octaves down, and with 16 cycles, the frequency will be four octaves down. If the length of the sequence is not a square of two, then the frequency that is produced may not be related to the cycle frequency with intervals of octaves.

For example, if a square wave cycle is followed by two cycles of silence, then the effective frequency of the pulse waveform is a third of the basic cycle frequency, which means that the lowest frequency will be an octave and a fifth down – and the ear will interpret this as the fundamental frequency, and so the oscillator has apparently been pitch shifted by an octave and a fifth. With longer sequences of single-cycle waveforms, this pitch change can be harmonically unrelated to the basic pitch of the oscillator. With more than one cycle of non-silence, the resulting harmonic structures can be very complicated. This can produce sounds which have complex and often unrelated sets of harmonics, which gives a bell-like or clangorous timbre. The pseudo-random-sequence generators mentioned earlier are one alternative method of producing sequences of single cycle waveforms, and exactly the same length-related pitch-shifting effect happens. Chapter 4 looks at these effects in more detail.

For very long sequences of cycles, the pitch-shift can become so large that the frequency becomes too low to be heard, and it is then only the individual cycles which are heard. This means that by concatenating a series of pulse waveforms which gradually change their pulse width, it is possible to produce a repeated multi-cycle waveform which sounds like a single-cycle PWM waveform. (By altering the number of repeated cycles of each different pulse width waveform, it is possible to change the effective speed of pulse width modulation. In fact, this is exactly how wavetable oscillators work – see Section 3.2.)

There are two methods for reading out the values in a wave-cycle memory. When the values are accessed by a rising number, then the shape is merely repeated, whilst by accessing the same values using a counter which counts up and down, then each alternate cycle is reversed in time. This can be a powerful technique for producing additional multi-cycle waveforms from a small wavecycle memory. If repeats of the cycles can be inverted as well, then even more possibilities are available. All of these variations on a single cycle can produce changes in the

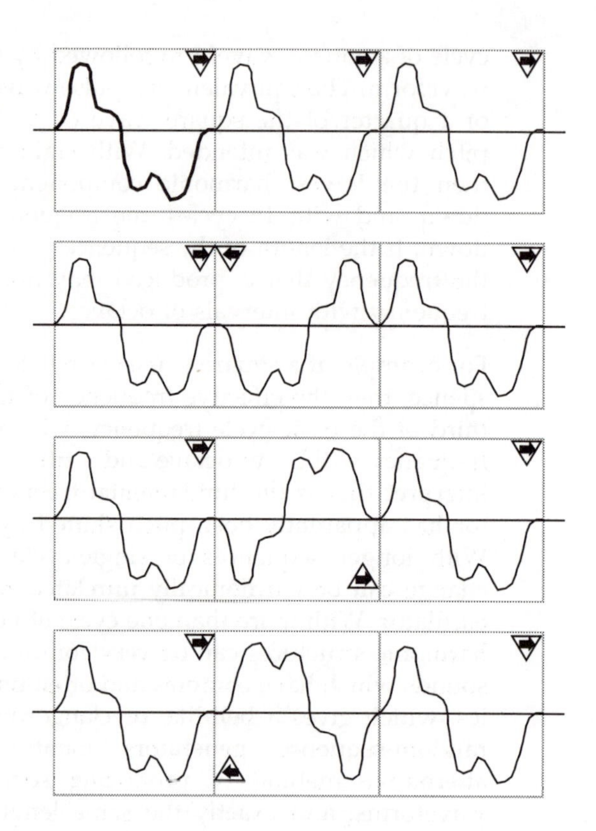

Figure 3.1.5 Pairs of wavecycles can be arranged in different ways by exploiting the symmetry (or lack of) the waveshapes. These four examples show a wavecycle followed by the four possibilities of reversing or inversion. The transitions between the wavecycles can be smooth or abrupt depending on the shape of the wavecycle.

spectrum of the sound: from minor detail through to major additional harmonics where the transition between the repeats is not smooth.

Fixed sequences of single cycles can be thought of as short samples – the PWM waveform is one example where a complete 'cycle' of pulse width modulation is repeated to give the same audible effect as a pulse waveform which is being modulated. Many other dynamically changing multi-cycle waveforms can be produced.

Multi-cycle oscillators can also be likened to granular synthesis, since they both concatenate cycles of waveforms, although granular synthesis normally works on groups of cycles rather than individual cycles (see Chapter 5). Roland's 'RS-PCM' and many other soundcards and computer sound generators all often use loops of multi-cycle 'samples' to provide the sustain and release portions of an enveloped sound – and sometimes even the attack portion. This technique is equivalent to changing the waveform of a multi-cycle oscillator dynamically, which is an advanced form of wavetable synthesis – see Section 3.2.

Multi-cycle oscillators normally have a selection of single-cycle waveshapes plus the following additional types:

- concatenations of mathematical shapes: sine, triangle, square, and sawtooth cycles in sequences
- symmetry variations of mathematical and other shapes
- PWM waveshapes which change their harmonic content with time
- waveshapes which change their harmonic content with time, but not in a regular sequence (i.e. not progressively as in PWM)
- shapes with additional non-harmonic frequencies (clangs, chimes and vocal sounds)
- noise: more cycles means that the noise produced can be more 'white' in character than that from single-cycle oscillators.

Interpolation is a method of producing gradual changes from one wavecycle shape to another, rather than the abrupt changes which occur when wavecycles are concatenated. Section 3.2 deals with this in more detail.

Sample

For very long sequences of single cycles, the complete sequence may not repeat whilst a note is being played, and it then becomes a sample rather than a multi-cycle waveform. Samples are usually held in either ROM or RAM memory, and because of the length, the amount of memory used can be quite large. For example, for 16-bit values, where there are 44 100 values output per second (the same as a single channel of a CD player) then 705 600 bits (just over 86 KBytes) are required to store just one second of audio. This means that it requires a megabyte of memory to store just under 12 seconds of audio sample. Obviously long samples are going to require large quantities of memory. Because of this, most hybrid synthesizers of the 1970s and 1980s used very short samples, and it is only with the availability of low-cost memory in the 1990s that sampling techniques became more widespread, and this was in an all-digital form.

Trying to reduce the amount of memory which is required to store cycles affects the quality of the audio. Hybrid wavecycle synthesizers suffer from the resolution limitations of their storage. At low frequencies there aren't enough sample points to adequately define the waveshape, whilst at high frequencies the circuitry may not run fast enough. For example, suppose that a single cycle of a waveform is represented by 1024 values. At

100 Hz, this means that the VCO needs to run at 1024 times 100 Hz, which is 102.4 kHz. But at 1000 Hz, the VCO needs to run at 1.204 MHz, and at 10 KHz, the VCO is oscillating at 10.24 MHz. Accurate VCOs with wide ranges, good temperature stability and excellent linearity at these frequencies are more normally found in very high quality radio receivers. More importantly, affordable late-1970s memory technology was beginning to run out of speed at a few megahertz. Reducing the number of bits which are used to represent the waveform cycles also reduces the quality by introducing noise and distortion. The whole of the sampling process is covered in more detail in Chapters 1, 3 and 4.

It is a common fallacy that the ability to produce complex waveshapes is all that is required to recreate any sound. In practice, most methods of producing a waveshape do not have the required resolution to provide enough control over the sound. This is covered in more detail in Section 2.3 on additive synthesis.

Modifiers

When the oscillator has produced the raw source waveform, then most hybrid synthesizers pass it through a modifier section which is typical of those found on analogue synthesizers. This is usually a VCF and VCA, with associated envelope generator control. Curiously, whereas most hybrid synthesizers attempt to improve on the selection of available waveforms for the oscillator, the VCF is often still just a simple resonant low-pass filter. This puts a great deal of emphasis on the oscillator as the prime source of the timbre, and means that the possibilities for the changing of the timbre by the modifiers is just the same as an analogue synthesizer. This means that the resonant filter sweep sound remains an audio cliché for both analogue and hybrid synthesizers. Only on some all-digital S&S synthesizers are the filtering capabilities enhanced significantly.

In a historical perspective, hybrid synthesizers reached a peak of popularity in the early 1980s, after polyphonic analogue synthesizers and just before the all-digital synthesizers. The many 'analogue' synthesizers of the mid-1990s' 'retro' revival of analogue technology, are often not truly analogue, but are actually modern hybrids, where the 'VCOs' are actually sophisticated wholly-digital DCOs which use the methods described above to produce their waveforms, but coupled with a conventional analogue modifier section in a standard hybrid synthesis way. This revival of hybrid synthesis is, in its turn, being incorporated into all-digital instruments which use a mixture of synthesis methods.

3.2 Wavetable

Initially, wavetable synthesis might appear to be very similar to multi-cycle wavecycle synthesis. Both methods use sequences of cycles to produce complex waveshapes. The major difference lies in the way that the cycles are controlled. In multi-cycle wavecycle synthesis, the chosen sequence of cycles is repeated continuously, whereas in wavetable synthesis, the actual waveform which will be used can be chosen on a cycle-by-cycle basis.

3.2.1 Memory

Wavetable synthesis is based around memory – even more strongly than wavecycle synthesis where there are a few methods which do not use large quantities of memory, pseudo-random sequence based waveform generators, for example. But wavetable synthesis uses the memory as an integral part of the synthesis process, since the cycle being used is dynamically selected by controlling the memory.

Just as with single-cycle wavecycle synthesis, a cycle of a waveform is stored in a memory chip, and successive values are retrieved from the memory and sent to a DAC where they produce the output waveform (Figure 3.2.1). The values are retrieved in order by using a counter which steps through the memory in an ascending sequence of memory locations.

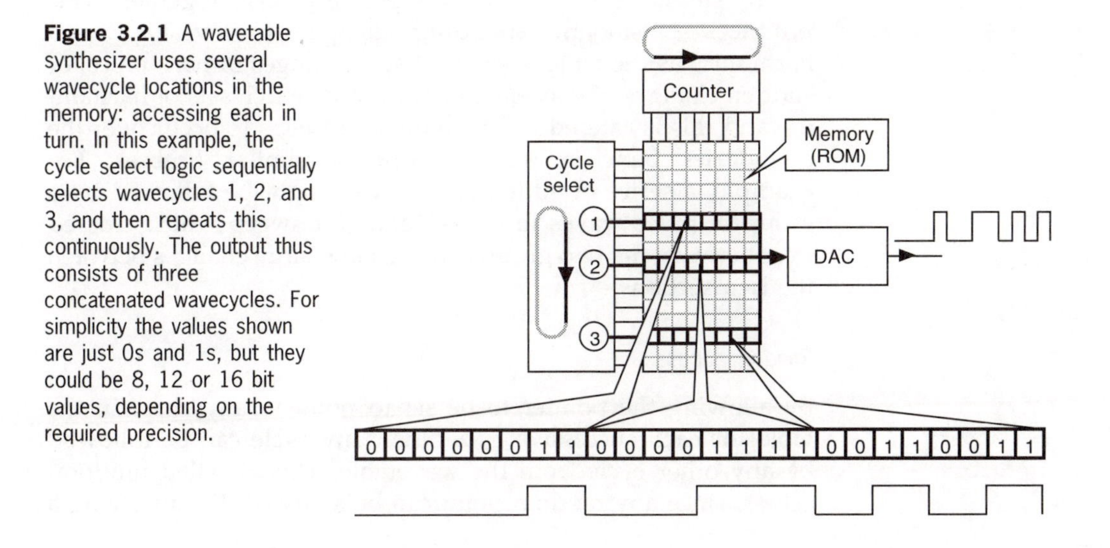

Figure 3.2.1 A wavetable synthesizer uses several wavecycle locations in the memory: accessing each in turn. In this example, the cycle select logic sequentially selects wavecycles 1, 2, and 3, and then repeats this continuously. The output thus consists of three concatenated wavecycles. For simplicity the values shown are just 0s and 1s, but they could be 8, 12 or 16 bit values, depending on the required precision.

Determining which values the counter steps through is set by controlling which part of the memory is used. In a single-cycle wavecycle oscillator, the control signals are set to point to the specific single-cycle waveform, and the oscillator then outputs that waveshape continuously.

In a wavetable oscillator, the control signals which determine where the waveform information is stored can be changed dynamically as the oscillator is outputting the waveshape. Normally the changes are made as one cycle ends and another begins, so that the waveshape does not change mid-way through a cycle. The control signals which set where the cycle is retrieved from can be thought of as modulating the shape of the output waveform, although they are really just pointing to different parts of the memory. The name 'wavetable' comes from the way that the memory can be thought of as being a table of values, and so the control signals just point to cycles within that wavetable. There are two basic ways that the cycle being used in the table can be changed: swept and random-access.

Swept

By incrementing the pointer so that it points to successive cycles in the wavetable, the control signals effectively 'sweep' the resulting waveshape through a series of waveforms. The fastest rate at which this can happen is when only one cycle of each waveform is used before moving to the next waveform, although by omitting waveforms the sweep speed can be increased. Wavetables which are intended to be swept in this fashion are normally arranged so that the waveforms are stored in an order where similar sounding waveforms are close together. This produces a 'smooth' sounding change of waveshape and harmonics as the table is swept. Large changes of harmonics, or sudden changes of waveshape, can produce rich sets of harmonics, and this is catered for by allowing sweeps to occur over the boundaries between these groups of similar timbres. For example, a series of added sine waves might be followed by a group of pulse waves in the table, and a sweep which crossed over between the two groups would have large changes between the two sections.

Random-access

By allowing the pointer to be set to point to anywhere in the table for each successive cycle, then any cycle can be followed by any other cycle from the wavetable. This is called random-access, since any random point can be accessed. By supplying a

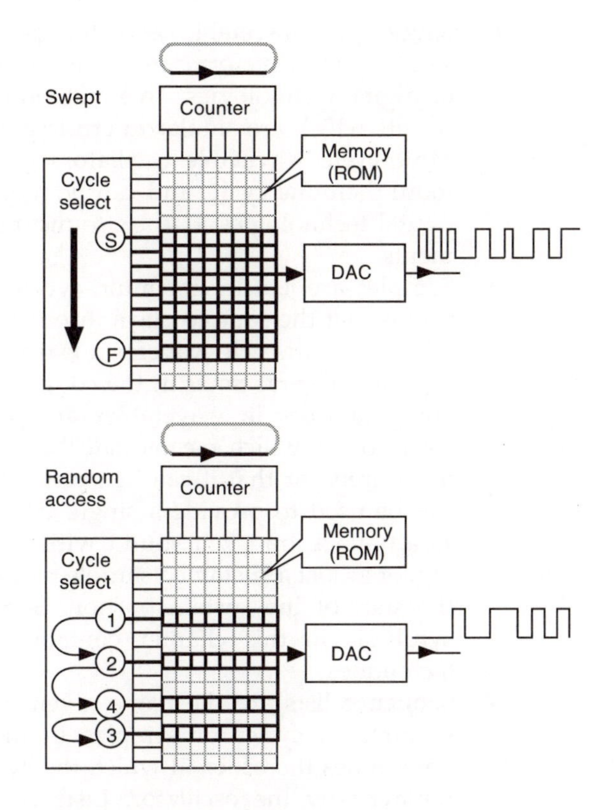

Figure 3.2.2 A swept wavetable outputs each of the cycles between the start and finish points in the memory. A random access wavetable only outputs the specific cycles which have been chosen.

series of pointers, the waveform can be swept – so a sweep is in fact a special case of random-access (Figure 3.2.2). More normally, a series of values are used to make the pointer access a sequence of waveform cycles. This can be a fixed sequence, in which case the wavetable behaves as a multi-cycle wavecycle oscillator, or a dynamically changing sequence of pointer locations, in which case the modulation of the waveform is characteristically that of a wavetable.

3.2.2 Table storage

The actual storage of the waveforms inside the wavetable can be of several forms. Some hybrid oscillators only have one of these types. Others provide two types or all three. The naming conventions differ with each manufacturer: wavesamples and hyperwaves are just two examples of names used for samples and multi-cycle waves respectively. Some types are found in analogue/digital hybrids as well as digital instruments which emulate analogue hybrids, whilst the more complex types are only found in digital instruments. The major types of table storage are:

- Single-cycle wavetable oscillators provide large numbers of single-cycle waveforms, and can be implemented in hybrid or digital technologies. This method can be used to provide results which are similar to crude granular synthesis.
- Multi-cycle wavetable oscillators contain waveforms with more then one cycle, and can be implemented in hybrid or digital technologies, but are found mostly in digital instruments.
- Samples are just longer multi-cycles, although the implication is that the sample plays through once or only partially, whilst a multi-cycle waveform is usually short enough to be repeated several times in the course of a note being played. Some samples in wavetables are provided with multiple start points, which means that the sample can be played in its entirety, or that it can be started mid-way through. This can be used to provide a single sample which can be used as an attack transient sound with a sustain section following, or as just a sustained sample by playing the sample from the start of the sustain portion. Section 3.4 on using S&S synthesis in this chapter contains more detail on these techniques.
- Sequence lists are the name given to the sequential set of pointers to cycles or samples in the wavetable. This list determines the order in which the cycles or samples will be replayed by the oscillator. Lists can automatically repeat when they reach the end, reverse the order when they reach the end, or merely loop the last cycle or sample. Some sequences allow looping from the end of the list to an arbitrary point inside the sequence list, which allows a set of cycles or samples to be used for the attack portion of the sound, whilst a second set of cycles or samples is used for the sustain and release portions of the sound. This interdependence of the oscillator and envelope is common in sample based instruments, whereas in analogue synthesis the VCO and EG are normally independent.
- Mixed modes. Some hybrid wavetable oscillators allow mixtures of single-cycle, multi-cycle and sample waveforms to be used in the same sequence list. Additional controls like repetitions of single or multi-cycle waveforms, or even the length of time that a sample plays, may also be provided.

Multi-samples

The term 'multi-samples' can be applied to the result of a sequence list which causes samples to be played back in a different order to that in which they were recorded. These samples

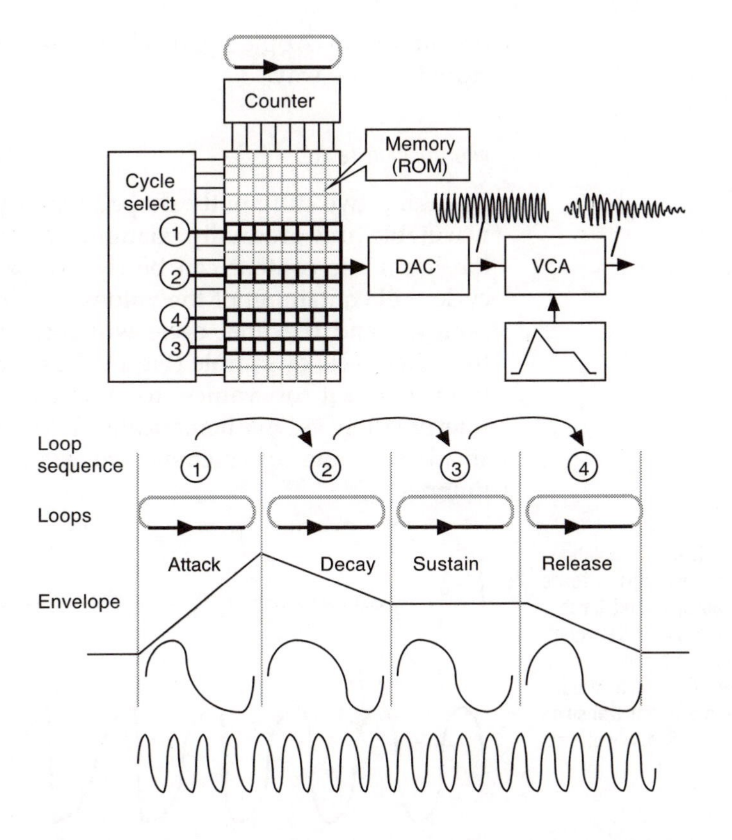

Figure 3.2.3 Loop sequences control the order in which looped wavecycles are replayed. In this example, the loop sequence is controlled by the envelope. Wavecycle 1 loops during the attack part of the envelope, followed by wavecycle 2 during the decay segment. The sustain segment is produced by looping wavecycle 3, and wavecycle 4 loops during the release part of the envelope.

may be looped, in which case some interaction with an envelope generator is usually used to control the transitions between the individual looped samples. Roland's 'RS-PCM' and many other soundcards and computer sound generators all often use loops of multi-cycle 'samples' to provide the sustain and release portions of an enveloped sound – and sometimes even the attack portion. This technique is equivalent to changing the waveform of a multi-cycle oscillator dynamically, which is an advanced form of wavetable synthesis.

'Multi-samples' is also used to mean the use of several different samples of the same sound, but taken at different pitches.

Loop sequences

A loop sequence is the name for the sequence of samples which are used in a multi-sample. It provides the mapping between the envelope segments and the samples which are looped in that segment (Figure 3.2.3). Loop sequences are sometimes part of a complete definition which includes the multi-sample and loop sequence information: for example, the musical instrument definitions in Apple's QuickTime and Roland's 'RS-PCM' are stored in this way.

Interpolation tables

By taking two differently shaped wavecycles or samples from a wavetable and gradually changing from one shape to another, the harmonic content can be dynamically changed. The initial cycle will contain all of the values from one of the two cycles or samples, and the final cycle will contain only the values from the other cycle or sample (Figure 3.2.4). The process of changing from one set of values to another is called interpolation. Interpolation is mathematically intensive, but requires only small amounts of memory to produce complex changing timbres.

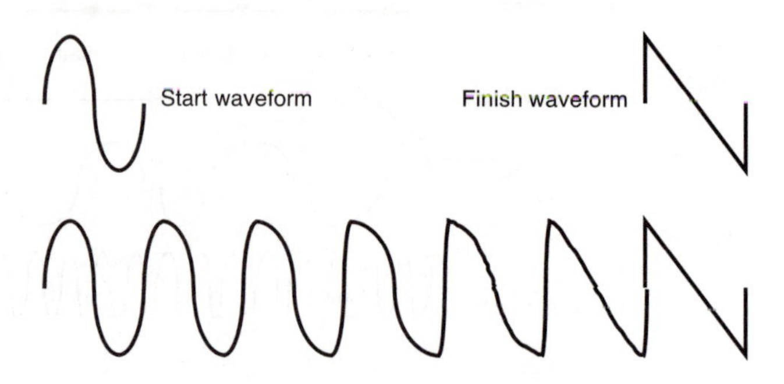

Figure 3.2.4 Interpolation allows two waveforms to be defined as start and finish points, and the 'in-between' wavecycles are then calculated (or interpolated) to produce a smooth transition between the two waveforms.

Although a good idea in principle, interpolating between two waveshapes does not always produce a musically useful transition because the changes in harmonic content may not be smooth. The relationship between the shape of a waveform and its harmonic content is not a simple one, and minor changes in the shape of a waveform can produce large changes in the harmonic content. Interpolation can emphasize this effect by producing timbres which change from one sound to another, but which pass through many other timbres in the process, rather than the smooth 'evolution' which might be expected.

Rather than using interpolation as a mathematical transformation of information about a waveform, a much more satisfactory method is to use interpolation to produce the changes between one spectrum and another. This method does produce smooth changes of timbre (or very un-smooth changes, depending on the wishes of the user!). This sort of spectral transformation is described in Section 5.6 on analysis-synthesis techniques.

3.2.3 Additional notes

- Wavetable synthesis is a term used to cover a wide range of techniques, and as a result, there are as many definitions of wavetable synthesis as there are techniques.
- The differences between single-cycle wavetable synthesis and sampling are actually greater than the differences between multi-cycle wavecycle and sampling. Very few samplers have the facility to alter the order in which the parts of a sample are played back!
- Soundcard manufacturers tend to describe almost any hybrid technique as 'wavetable' synthesis.
- By loading a wavetable oscillator with a set of multi-cycle waveforms which have been generated from the addition of sine waves, then an additive synthesis engine can be produced using hybrid digital and analogue techniques.

3.3 Digitally controlled oscillators

Digitally controlled oscillators (DCOs) are the digital equivalent of the analogue VCO. DCOs have much in common with wavecycle synthesis. In fact, they can be considered to be a special case of the most basic wavecycle oscillator: one which only produces the 'classic' synthesizer waveforms (sine, square, pulse and sawtooth).

DCOs are combinations of analogue and digital circuitry and design philosophies – they are literally hybrids of the two technologies. They were originally developed in order to replace the VCOs used in analogue synthesizers with something which had better pitch stability. The simple exponential generator circuitry used in many early analogue VCOs was not compensated for changes in temperature, and so the VCOs were not very stable and would go out of tune. Replacing the VCOs with digitally controlled versions solved the tuning stability problems, but changed some of the characteristics of the oscillators because of the new technology. The DCO is also notable because it marked the final entry of the era of three character acronyms: VCO, VCF, LFO and VCA. Subsequent digital synthesizers moved away from acronyms towards more accessible terms like oscillator, filter, function generator and amplifier.

It is worth noting that although early VCO designs did suffer from poor pitch stability, the designs of the late 1970s included good temperature compensation, and the custom chips developed in the early 1980s had excellent stability. But the damage to the reputation of analogue VCOs had been done, and DCOs have now replaced the VCO permanently for all but the most purist of analogue users. DCO based synthesizers were, in turn, eclipsed by FM in the mid-1980s, although the foundations of FM actually lie in DCO technology.

3.3.1 Digitally-tuned VCOs

The simplest DCOs are merely digitally-tuned VCOs. A microprocessor is used to monitor the tuning of the VCOs and retune it when necessary. This process was usually carried out when the instrument was initially powered up, although it could be manually started from the front panel. The technique isolates the VCO from the keyboard circuitry (see Chapter 7), and then uses a timer to measure the frequency by counting the number of pulses in a given time period. This measurement process is carried out near the upper and lower frequency limits of the VCO by switching in reference voltages – with additional measurement points as required to check the tracking of the VCOs (see Chapter 2). From this information, the two main adjustments can be calculated: an 'offset' voltage can then be generated to bring the VCO back into 'tune'; and a 'tracking' voltage can be used to set the tracking. The offset and tracking voltages for all the VCOs in a synthesizer would be stored in

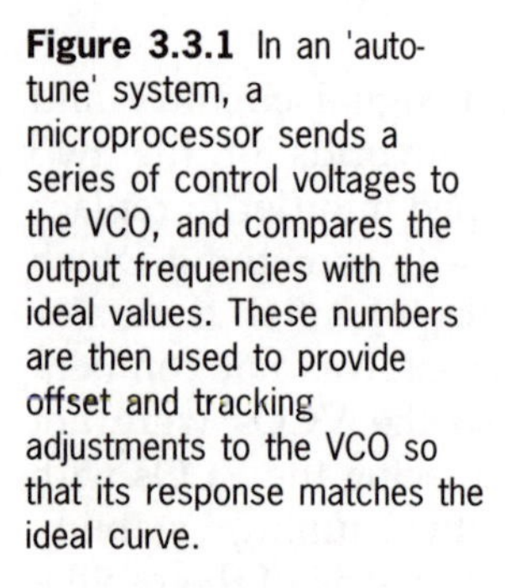

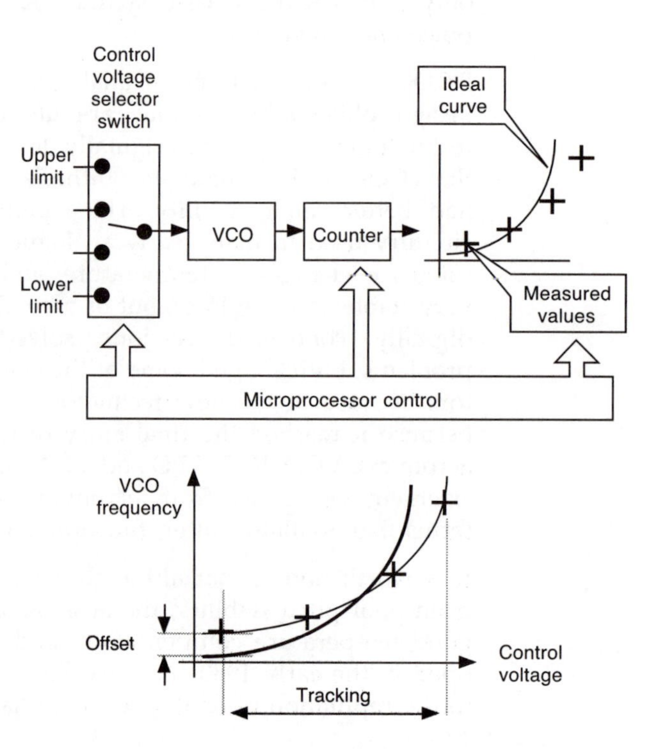

Figure 3.3.1 In an 'auto-tune' system, a microprocessor sends a series of control voltages to the VCO, and compares the output frequencies with the ideal values. These numbers are then used to provide offset and tracking adjustments to the VCO so that its response matches the ideal curve.

battery-backed memory. This technique is often known as 'auto-tune' (Figure 3.3.1).

A variation of this technique is currently used in some analogue-to-digital converter chips (ADCs), where the circuit monitors its own performance and recalibrates itself continuously, for each sample. In early auto-tune synthesizers, the time required to measure the frequencies at the various points meant that it was not possible to continuously tune the VCOs – it could take several minutes to retune all of the VCOs in a polyphonic synthesizer. It would be possible to arrange the voice allocation scheme so that a VCO being tuned could be removed from the 'pool' of voices, but this would effectively reduce the polyphony by one. Some synthesizers allowed voices to be disabled in exactly this way if they could not be tuned correctly by the auto-tune circuitry – which reflects the poor reliability record of the VCOs in some early polyphonic synthesizers.

Another combination of digital microprocessor technology with analogue VCOs occurs in synthesizers where the keyboard is scanned using a microprocessor and the resulting key codes are turned into analogue voltages using a digital-to-analogue converter (DAC) and then connected to the VCOs. Although suited to polyphonic instruments, this technique has been used in some monophonic synthesizers, particularly where simple sequencer functions are also provided by the microprocessor. The Sequential Pro-One is one example of a monophonic instrument which uses 'digital' storage of note voltages, and the use of a microcontroller chip allows it to also provide two short sequences with a total of up to 32 notes clocked by the LFO.

3.3.2 Master oscillator plus dividers

A nearer approach to a 'true' DCO uses ideas taken from master oscillator organ chips. A quartz crystal controlled master oscillator provides a high frequency clock, which is then divided down to provide lower rate clocks through a series of divider chips (Figure 3.3.2). By using a high rate of master clock and correspondingly large divider ratios, it is possible to generate all the frequencies which are required for all the notes for a synthesizer from just a few chips. These notes can then be gated from the keyboard circuitry, with the end result being a polyphonic oscillator which is derived from a stable crystal-controlled master oscillator.

Rather than have separate stages of dividers for each note, an obvious design simplification is to have 12 dividers which

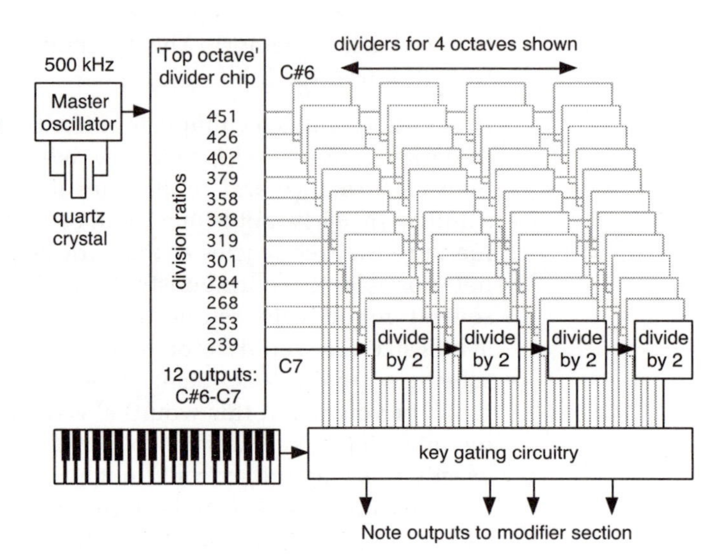

Figure 3.3.2 A 'top octave' divider system uses a high-frequency master oscillator and dividers to provide all the required frequencies for all the notes on a keyboard. In this example, the master oscillator frequency is 500 kHz, and the division values required to produce the 12 top notes in an octave are shown. Each of these frequencies then needs to be further sub-divided down to produce the lower octave notes.

produce the highest required frequencies, and then divide each of these outputs successively by two to produce the lower octaves (this presupposes that the scale used will be fixed: usually equal temperament, and that each octave has identical ratios between the notes). This is called a 'top octave' method, and with the right division values and a high enough clock, it gives very good results. For example, with a master clock frequency of 500 kHz, the division required to produce a C#6 at 1108.73 Hz is 450.96, which is almost exactly 451, whilst for the next note, the D6 at 1174.66 Hz, the division is 425.65, and so an integer value of 426 will produce an output frequency which is slightly too low. In a real world design, you might expect that the clock frequency and division values would be chosen to minimize the errors by setting real division values which are as near to integers as possible. In practice, a real-world custom top octave synthesizer chip, the General Instruments AY1-0212A, used exactly these division values, as shown in Table 3.3.1.

As the table shows, using integer dividers only makes a slight difference in the output frequency. Using individual separate division stages only improves the accuracy slightly. Taking the values for the C1, C5 and C7 division values given above, if the 239 value for the C7 'top octave' division is then divided down by successive dividers, this is equivalent to doubling the effective division value: which would thus have the values of 956 for the C5 frequency and 15 296 for the C1 frequency. The 956 division value is identical to the one used, but the 15 296 is slightly too large, which means that output frequency will be too

Table 3.3.1 DCO dividers 2

Clock (Hz)	Note	Note frequency (Hz)	True divider	Integer divider	Actual note frequency (Hz)	Frequency error (%)	Frequency difference (Hz)
500000	C#0	17.32	28861.84	28862	17.32	–0.0006	0.0001
500000	C#0	17.32	28861.84	28862	17.32	–0.0006	0.0001
500000	D0	18.35	27241.95	27242	18.35	–0.0002	0.0000
500000	D#0	19.45	25712.97	25713	19.45	–0.0001	0.0000
500000	E0	20.60	24269.82	24270	20.60	–0.0008	0.0002
500000	F0	21.83	22907.66	22908	21.83	–0.0015	0.0003
500000	F#0	23.12	21621.95	21622	23.12	–0.0002	0.0001
500000	G0	24.50	28861.84	28862	17.32	–0.0006	0.0001
500000	C#0	17.32	20408.40	20408	24.50	0.0020	–0.0005
500000	G#0	25.96	19262.97	19263	25.96	–0.0002	0.0000
500000	A0	27.50	18181.82	18182	27.50	–0.0010	0.0003
500000	A#0	29.14	17161.35	17161	29.14	0.0020	–0.0006
500000	B0	30.87	16198.16	16198	30.87	0.0010	–0.0003
500000	C1	32.70	15289.03	15289	32.70	0.0002	–0.0001

Clock (Hz)	Note	Note frequency (Hz)	True divider	Integer divider	Actual note frequency (Hz)	Frequency error (%)	Frequency difference (Hz)
500000	C#4	277.18	1803.86	1804	277.16	–0.0075	2.0769
500000	D4	293.66	1702.62	1703	293.60	–0.0222	0.0652
500000	D#4	311.13	1607.06	1607	311.14	0.0038	–0.0118
500000	E4	329.63	1516.86	1517	329.60	–0.0090	0.0297
500000	F4	349.23	1431.73	1432	349.16	–0.0190	0.0662
500000	F#4	369.99	1351.37	1351	370.10	0.0275	–0.1018
500000	G4	392.00	1275.53	1276	391.85	–0.0372	0.1459
500000	G#4	415.30	1203.94	1204	415.28	–0.0054	0.0223
500000	A4	440.0	1136.36	1136	440.14	0.0320	–0.1408
500000	A#4	466.16	1072.59	1073	465.98	–0.0379	0.1768
500000	B4	493.88	1012.39	1012	494.07	0.0387	–0.1911
500000	C5	523.25	955.57	956	523.01	–0.0454	0.2374

Clock (Hz)	Note	Note frequency (Hz)	True divider	Integer divider	Actual note frequency (Hz)	Frequency error (%)	Frequency difference (Hz)
500000	C#6	1108.73	450.97	451	1108.65	–0.0074	0.0825
500000	C#6	1108.73	450.97	451	1108.65	–0.0074	0.0825
500000	D6	1174.66	425.66	426	1173.71	–0.0810	0.9511
500000	D#6	1244.51	401.76	402	1243.78	–0.0586	0.7289
500000	E6	1318.51	379.22	379	1319.26	0.0569	–0.7512
500000	F6	1369.91	357.93	358	1396.65	–0.0188	0.2620
500000	F#6	1567.98	318.88	319	1567.40	–0.0371	0.5819
500000	G6	1108.73	450.97	451	1108.65	–0.0074	0.0825
500000	G#6	1661.22	300.98	301	1661.12	–0.0054	0.0904
500000	A6	1760.00	284.09	284	1760.56	0.0320	–0.5634
500000	A#6	1864.66	268.15	268	1865.67	0.0542	–1.0116
500000	B6	1975.53	253.10	253	1976.28	0.0382	–0.7546
500000	C7	2093.00	238.89	239	2092.05	–0.0454	0.9498

Based on a GI AY1-0212A TOS chip

Figure 3.3.3 Dividers can be used as filters. In this example, a single pulse is missing from the 4.096 MHz clock. Subsequent divide-by-two stages reduce the effect of the missing clock by 'averaging' out the frequency.

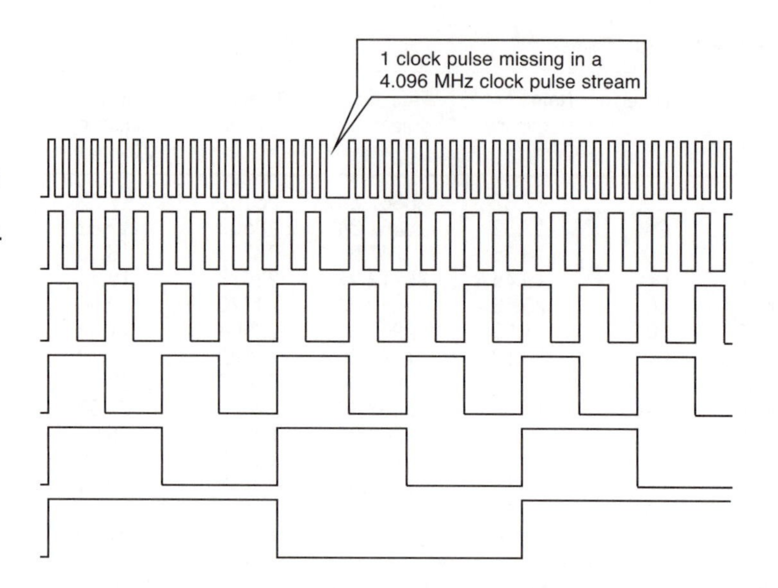

low – by about 0.0149 Hz, or 0.00045 cents. The difference in division values thus produces only very slight differences in the output frequencies. For comparison purposes, the human ear can detect pitch changes of a minimum of about 5 cents, whilst the Emu Morpheus has fine tuning steps of 1.5625 cents, the Yamaha SY99 has micro-tuning steps of 1.171875 cents, and the MIDI Tuning Standard has steps of 0.0061 cents.

Changes in the pitch of these types of 'master oscillator plus divider' DCO might be achieved by using a voltage-controlled crystal oscillator to make minor changes in pitch for pitch bend or vibrato effects, but it is difficult to change the frequency of a crystal oscillator enough for satisfactory pitch control. A more satisfactory method uses a rate adapter, which is a counter-based circuit which removes just one clock occasionally from a continuous clock signal and so reduces the effective clock rate. This 'gapped' clock needs to be followed by the equivalent of a low-pass filter to remove the effects of the jitter in the clock pulses, but this type of DCO has just such filters in the form of divider circuits.

For a 4.096 MHz clock, removing just one clock pulse with a rate adapter can be thought of as changing the effective frequency to 2.048 MHz whilst the clock pulse is missing, but then the next clock pulse restores the frequency to 4.096 MHz again (Figure 3.3.3). The actual number of pulses per second (which is what frequency measures) varies depending on when and how the measurement is taken. If the frequency is measured by timing from the start of one clock pulse to the start of the next, then the frequencies of 4.096 MHz and 2.048 MHz are correct. The brief change in frequency is large when measured in this way, but by measuring more than one clock pulse, the change in frequency reduces as the number of clock pulses used for the measurement increases. This process of averaging the frequency over several clock cycles is just what happens when a divider circuit is used to divide down the output of a rate adapter.

The missing clocks are an extreme form of the random variation in the time between successive clock edges in a digital system: called jitter. Caused by unstable clock sources, or noise affecting the switching point of gates, jitter usually varies randomly around the true edge position: some clocks are slightly closer together, whilst others are slightly further apart. More complicated circuits like accumulator/divider circuits can provide very small frequency changes without the need for large numbers of divider stages to filter out the jitter.

The main indicator of the use of this type of 'master oscillator plus dividers' DCO is global pitch control: particularly for portamento. Glissando can be produced by altering the way that the keyboard circuitry controls the gating of the note frequencies, and can be carried out on a 'per voice' basis – so the glissando can be polyphonic, with held notes being unaffected. If portamento is provided, then it can be implemented by producing a series of pitch bends from note to note, and this may well produce audible glitches in the transitions – but more importantly, because the pitch bend rate adaption is done before the divider circuitry, it will affect all of the notes which are being played. This type of portamento is thus called 'monophonic' portamento. Because of this 'global' pitch change, synthesizers which have this type of DCO do not usually provide pitch envelopes which change the pitch of a note, since any new notes which are played would pitch bend any pre-existing held notes, which is not very useful musically. The pitch bend and vibrato normally affect all notes that are being played, and individual pitch control for pitch bend or vibrato on a 'per voice' basis is more unusual.

The upper frequency limit of many top octave synthesizer chips was limited – for a 500 kHz input, the chip mentioned above can only produce a C7 at 2093 Hz, which is an octave below the top note of an 88-note piano keyboard. Also, having lots of simultaneous frequencies produced by a large number of divider chips can induce a characteristic buzzing sound in the audio output if care is not taken with wiring layout and circuit board design – the onomatopoeic commonly used term for this problem is 'beehive' noise. 'Electronic pianos' and 'string machines' of the mid-1970s which used top octave synthesis were notorious for this extraneous noise.

Another useful hint that this type of DCO is being used is the lack of any 'detune' facility if there is more than one DCO provided in the voice. Since the only way to provide fine resolution pitch changes is by using the rate adapter (of which there is usually only one), which produces global pitch changes, then it is not possible to achieve the slight 'detuning' effects of two VCOs. By using two rate adapters and two sets of divider circuits, then it is possible to produce detune, but this almost doubles the required circuitry. Many 'master oscillator plus divider' synthesizers provide 'sub-oscillators', which are merely the output of the gated notes divided by 2 or 4 to give extra outputs which are one or two octaves down in pitch from the main output. Chorus is often also provided to try and reproduce the effect of detuned VCOs (see Chapter 6).

3.3.3 Wave shaping

The basic output of most simple DCOs is a pulse or square wave at the required frequency. In order to emulate a conventional VCO, this needs to be converted into the 'classic' analogue waveforms: sine, square, sawtooth, triangle and pulse. This can be done using analogue electronics, but a much more flexible system can be achieved by using wavecycle/wavetable techniques. By setting the DCO to produce an output which is much higher in frequency, a look-up table can be used to store the values for each point in the waveform and a DAC can be used to produce the required waveforms. The purity of the waveforms produced is limited only by the number of bits used to represent each point on the waveform, and the highest output frequency of the DCO, which sets the number of points which can be used.

Since several of the 'classic' analogue waveforms have lots of symmetry, the number of points which need to be stored in order to produce a single cycle can be minimized. For a square wave, it can be argued that only two values need to be stored, but this is inadequate for the remaining waveforms. Whilst the sine and triangle waveform can be perfectly described with only a quarter of a cycle, the sawtooth and pulse waveforms require at least half a cycle. By using 256 points to define a half-cycle of the waveform, this is thus the equivalent of having 512 separate stored points – which means that the DCO needs to run at 512 times the cycle frequency. By exploiting the symmetry or

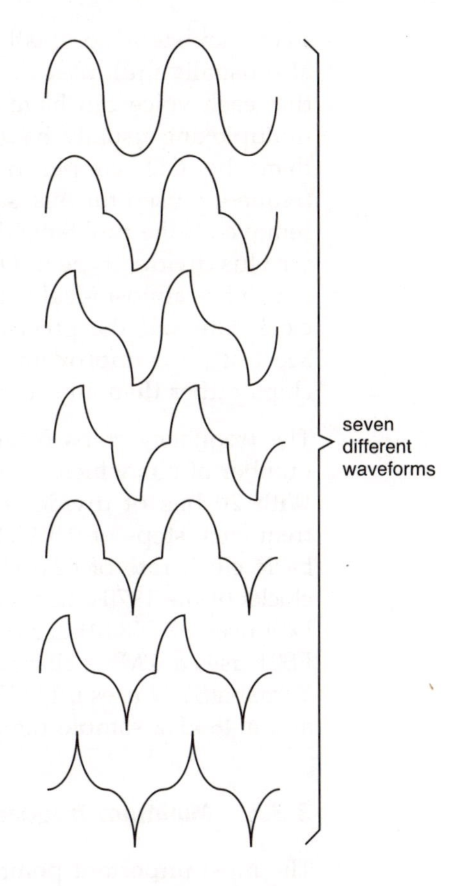

Figure 3.3.4 Symmetry can be used in a wavetable synthesizer to produce many waveforms from a small segment of a complete waveform. In this example, a quarter cycle of a sine wave is used to generate a sine waveform, plus six other waveshapes.

asymmetry of waveforms, a number of waveforms can be produced by using the same set of points.

Sometimes 8-bit values are used to store the waveform point values, but for only a doubling of the memory requirements, 16-bit values give a huge increase in the perceived quality (the doubling of the number of bits produces a disproportionately large increase in the audio quality – see Chapter 4). For high pitched notes, the whole of the wavetable need not be used, since only one or two harmonics will be audible, and so less points are required in the table – this can be achieved by only using every other value, or perhaps even missing out three points, and only using every fourth value.

3.3.4 High resolution DCOs

The sophisticated DCOs of the mid-1990s use higher frequency oscillators and similar division techniques to those of the mid-1970s, but with much finer resolution: sufficient to provide frequency steps so small that they are almost inaudible. They also usually multiplex the rate adapter and division circuitry so that each voice can have an effectively independent DCO. The multiplexing usually happens at a very high rate, often higher than the CD sample rate of 44.1 kHz: 48 or 62.5 kHz are frequently used for this 'sample rate' clock. These enhancements remove all the problems described above for the 'master oscillator plus divider' type of DCO, and give a tone generation source which has almost ideal performance – limited only by the master clock rate and the precision of the dividers and rate adapters. Most of these improvements are due to the availability of faster chips rather than any major changes in design.

The frequency steps from realizable oscillators depend on the number of bits which are used to control the frequency changes. With 20 bits of divider resolution, then it is possible to have frequency steps of 0.3% at 20 Hz, and 0.005% at 1 kHz, using a basic clock rate of 62.5 kHz. For comparison with the 500 kHz clocks of the 1970s, here are some mid-1990s figures: the Roland D50 uses 32.768 MHz for its tone generator ASIC, the Yamaha FB01 uses a 4 MHz clock and a 62.5 kHz sample rate, whilst the Yamaha SY99 uses 6.144 MHz clocks for its tone generator chips and a 48 kHz sample rate.

3.3.5 Minimum frequency steps

The most important pointer to a good DCO design (apart from temperature stability) is the minimum step in frequency which can be made. This is most apparent when the pitch bend control is used. Some DCOs have audible jumps or steps in pitch, which shows that insufficient frequency resolution is available. A more rigorous method of verifying the size of the frequency steps can be achieved by detuning two DCOs so that they beat together. The pitch differences required for slow beating are quite small. For example, if two frequencies are 1 Hz apart, then they will beat once every second. If they are 0.1 Hz apart, then the beat will cycle once every 10 seconds. For a pitch difference of 0.01 Hz, then the beat will take one 100 seconds to complete one cycle. So, to measure the minimum frequency step, you leave one DCO unchanged, and apply the smallest pitch change that you can produce to the other – this is probably not going to be audible, but by listening to the beats you can hear when the two DCOs go from the same frequency (no beats) to slightly different in frequency, when the beating will start. By timing the length of one cycle of the beat, you can work out the difference in frequency.

3.4 Sample and synthesis

Sample and synthesis (S&S) is a generic term for the many methods of sound synthesis which use variations on a sample playback oscillator as the raw sound source for a VCF/VCA synthesis modifier section. The samples are normally stored in ROM memory using pulse code modulation, normally abbreviated to PCM. This is just a technical term for the conversion of analogue values to digital form by converting each sample into a number, but the acronym has become widely used in manufacturer's advertising literature. The source sample playback is much the same as for a DCO driving a large wavetable, whilst the modifier sections are usually based on the VCF/VCA structure of analogue synthesizers.

Although the use of the term 'S&S' has been introduced for instruments where the modifiers are digital emulations of the VCF and VCA section of an analogue synthesizer, S&S is not necessarily restricted to digital instruments. It can also be produced with analogue equipment, and in fact, instruments like the Mellotron and Birotron could be considered to be S&S synthesizers which use magnetic tape instead of solid state memory. Many of the larger wavetable instruments and early samplers replay digital samples and then process them through analogue modifiers.

The availability of low-cost, high-capacity ROM memory is one of the major factors in the change from simple wavecycle DCOs to sample replay instruments with hundreds of sampled sounds. In the same way, advances in digital technology have allowed a gradual changeover from analogue modifiers to digital emulations. So S&S synthesizers start out as hybrid instruments with a DCO driving a sample replay, processed by analogue filters, but end up as completely digital instruments. A typical S&S synthesizer of the mid-1990s mixes many of the features of a sampler with a modifier section which has the processing capability of an analogue synthesizer of 20 years ago – complete with detailed emulations of resonant VCFs.

3.4.1 Samples

Unlike dynamic wavetable synthesizers, the samples that are provided with S&S instruments are normally replayed singly rather than being sequenced into an order. The only available source of the raw sound material for subsequent modification is thus a collection of preset sounds or timbres. S&S instruments then allow the processing of this raw sample 'source' of sound through one or more 'modifiers', and so allow different sounds to be synthesized. The modifiers are usually just some sort of filtering and enveloping control. The complexity of the processing varies a great deal – some have just low-pass filtering and simple envelopes, whilst others have complicated filtering that can changed in real-time, and loopable or programmable function generators instead of envelopes. In general, the most creative possibilities for making interesting sounds are provided by the most elaborate processing functions.

Most S&S instruments have their sample sets held in ROM memory, which means that there is a fixed and limited set of available source sounds. Many General MIDI and low-cost 'home' keyboards use S&S technology to produce their sounds – replaying the sounds is relatively straightforward and can provide high-quality sounds. In a typical GM instrument, a large proportion of the memory is taken up with a multi-sampled piano sound, and the rest is almost entirely devoted to other orchestral or band instruments. These instrument samples are chosen because they have the correct characteristics for the instruments that they are intended to sound like – if they do not sound correct, then they fail to sound convincing. Unfortunately, because these audio fingerprints are so effective at identifying a sound as being of a particular type, it is not easy to make any meaningful modifications to the sample – a violin sample still tends to sound like a violin, regardless of most changes to the envelope and the filtering. The sample sets in most S&S instruments thus represent a pre-prepared set of clichés: all readily identifiable, and all very difficult to disguise – rather like the audio equivalent of a 'fingerprint', in fact.

This fingerprint analogy can also be extended to the modifiers of the source sounds as well. If the only filtering available is a low-pass filter, then there will be a characteristic change of harmonics as the filter frequency is changed – and this can be just as distinctive as a specific sample. Filters with alternative 'shapes' like high-pass, notch, band-pass and comb filters can help to give extra creative opportunities for sound making. Again, the creative potential is reflected in the complexity of the available processing.

There are many synthesizers and expander modules which use the S&S technique to produce sounds, and it has been very successful commercially for a number of reasons. It is comparatively easy to design an S&S instrument which incorporates sounds like the General MIDI (GM) set, and it will have a broad range of applications, from professional through to home use.

Because S&S instruments use pre-defined and fixed samples in ROM form, there is also considerable scope for selling add-ons like extra sample ROMs. Despite this, because many samplers also have the same sort of synthesizer processing and modification stages but their samples are held in RAM instead of ROM, the creative possibilities of a sampler are much wider!

3.4.2 Topology

Because S&S instruments have a 'pre-packaged' set of samples, they are sometimes described as merely sample-replay instruments, and not true synthesizers. For the case of a single sample being replayed by an S&S instrument, the only changes that can be made to the sample are restricted to the modifier section, which allows changes to the filtering and envelope of the sound. But almost all S&S instruments provide rather more than this 'basic' mode of replay: normally either two independent sets of 'sound source and modifier', or two separate sound sources processed by a single modifier. In addition, some instruments also allow more than one sound to be triggered from the same note event, and so several samples can be combined.

The ability to trigger the playback of several different samples from one event opens up considerably more synthesis possibilities. Some early S&S instruments used an 'attack and sustain' model, where one sample was used to produce the attack portion of a sound, whilst a simplified 'subtractive synthesizer'-type section was used to produce the sustained portion, with a cross-fade between the two portions of the sound. As the technology developed, two sample replay sound sources could be used to produce the sound, and this allowed a more flexible

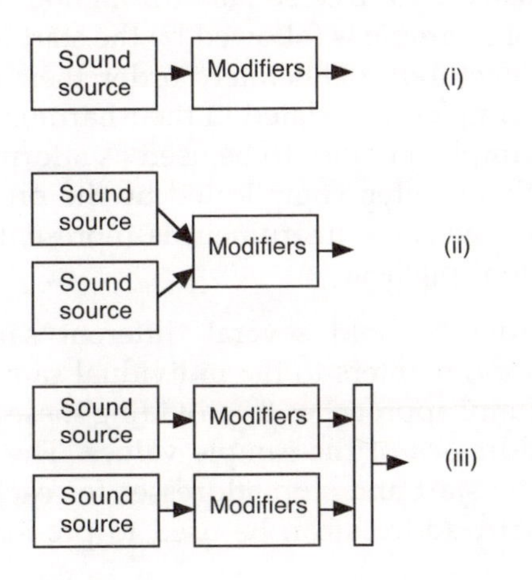

Figure 3.4.1 The basic S&S topology is a single sound source followed by one modifier section, as shown in (i). But most S&S synthesizers have either two sound sources which share a single modifier section (ii), or two separate sets of sound source and modifier, as in (iii).

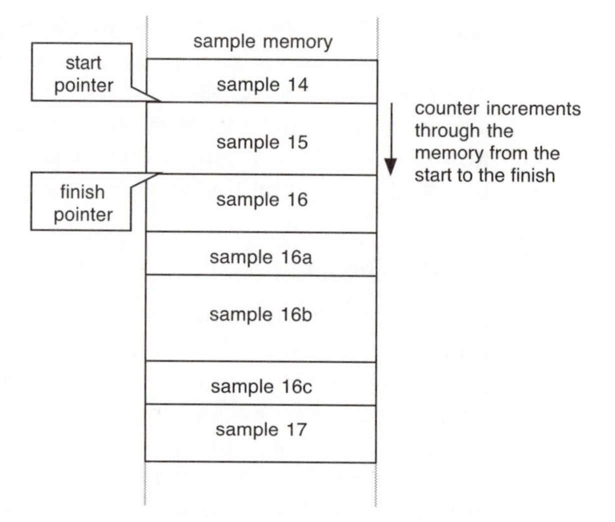

Figure 3.4.2 Sample memory is often arranged as a single contiguous block of ROM or RAM (or a mixture of the two). Sample replay consists of setting pointers to the beginning and end of the required sample, and then loading a counter with the start location and incrementing a count until the finish location is reached. Often it is possible to set the start and finish pointers to encompass several samples – in the example shown only sample 15 will be replayed, but by moving the finish pointer then the multi-samples of sample 16 could also be included.

division of their roles. Chapter 6 describes some of the ways in which two or more separate sound sources can be combined to produce composite sounds.

3.4.3 Counters and memory

The basic process for reproducing a sample from memory involves using a digital counter to sequentially access each sample value in the sample memory. The first sample is pointed to by the counter, which then increments to point to the next value, and this repeats until the entire sample has been read out. In practice, these retrieved values may be used as the input to an interpolation process, but the counter and memory structure remains the foundation of the replay technique. Samples are normally organized serially throughout the memory device – the end of a sample is followed by the start of the next sample. Some manufacturers deliberately order their samples so that successive samples are related in their harmonic content, which allows the sample memory to be used as a form of dynamic wavetable. But this is often complicated by the provision of multi-samples where the same instrument is represented by samples taken at different pitches.

In order to hold several different samples in one block of memory, pointers to the individual samples are required. There are many approaches to providing these pointers to the locations or addresses of the sample values. The simplest method specifies the start and stop addresses for each of the samples, where the start address can be used to pre-load the counter, and the stop address can be used to stop the counter when the end of the sample is reached. Alternatively, start and length parameters can be used when the counter merely adds an offset to the start address, since then the length parameter stops the counter when the count equals the length. By changing the length parameter, the playback time of the sample can be controlled.

If it is required to commence playback of the sample after the true start of the sample, then an offset parameter may be used to add an offset value to the start address which is loaded into the counter. Some instruments allow the length parameter to be set to longer than the sample, in which case the playback will continue into the following sample. Offsets can sometimes be used to provide similar control over the start address, and can cause the replay to commence from a different sample altogether. E-mu's Morpheus is one example of an S&S instrument which implements start, offset, and length parameters, as well as an ordered sample ROM to allow wavetable-like usage.

When offsets are applied to the start and end of sample replay, then the sample values at those points may be numerically large, which can produce clicks in the audio signal, especially if the sample is looped or concatenated with another sample. Some S&S instruments only allow start and stop addresses to be selected when the sample values at those addresses are close to zero – although this is useful to prevent clicks in the output, it can be a problem when trying to loop the sample. Most samples normally start and finish with values which are close to zero.

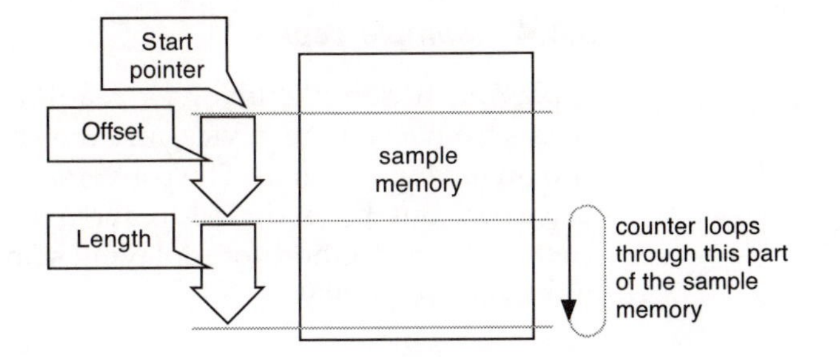

Figure 3.4.3 Sample replay parameters provide additional control over how the counter starts and loops whilst replaying a sample. An offset parameter allows the start of the sample to be later (or earlier) in the sample memory, whilst a length parameter allows the offset or start to be changed dynamically without altering the replay time of the sample.

Looping samples can involve either the entire sample between the start and the stop addresses, or any portion in between. Some instruments allow loops to extend beyond a single sample – sometimes even through the whole of the sample memory. The obvious loop is to play the sample through to the end of the loop, then to return to the start of the loop, and then repeat the section between the start and end of the loop. Loops can be set to occur a number of times, or for a specified time period, or they may be controlled by the envelope generator. Loops can be forwards, where the end of the loop is immediately followed by the start of the loop, or can alternately move forwards and backwards through the looped section of the sample. Alternate sample looping can help to prevent audible clicks when the sample values at the loop addresses are not at zero crossing points. Another possibility is to invert the sample playback for each alternate repetition, again so that clicks are minimized.

The predominant use of the loop is to provide a continuous sound when the envelope generator is in the sustain portion of the envelope – these are called sustain loops. But it is also possible to have attack or release loops, where the start or end of notes can be extended without requiring long samples. Loops are a way of minimizing the storage requirements for sounds which are required to have long attack, sustain or release envelopes. Roland's RS-PCM technology is an example of an S&S technique where each sample is closely connected to the envelope generator, and so has separate attack, decay, sustain and release loops.

Storing the parameters required for playing back a sample requires two separate storage areas. A look-up table is required to map the samples to their addresses in the sample memory, and this can also contain details of the length of the sample, zero-crossing points for potential start and offset addresses, as well as default loop addresses. These default control parameters can often be replaced by values held as part of the complete definition of a sound.

3.4.4 Sample replay

Replaying a sample involves reading the individual sample values from a storage device, and then converting these numbers into an analogue signal. The conversion from digital to analogue is carried out by a digital-to-analogue converter (DAC) chip. There are two methods of replaying samples: variable frequency, and fixed frequency.

Variable-frequency playback

The easiest method of replaying a wavecycle or wavetable would be to output the sample values at a rate controlled by an oscillator or DCO. This is called variable-frequency playback. The oscillator steps through the values which specify the waveform, and this is converted into an audio signal by a DAC. Although simple to understand conceptually, this technique has several major limitations:

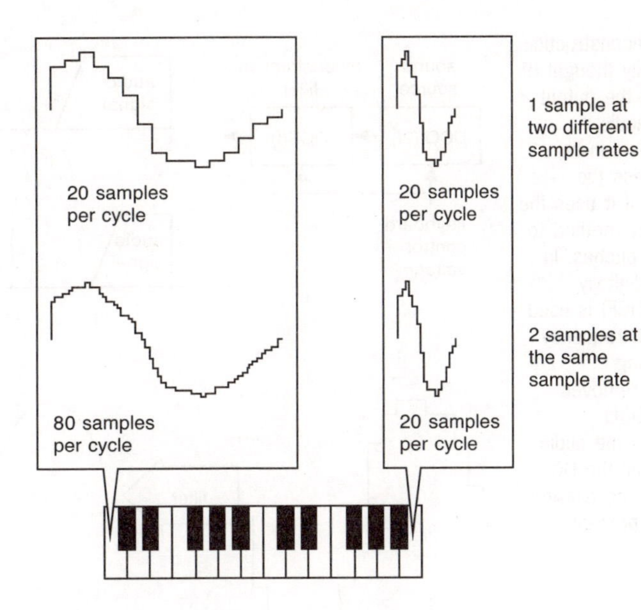

Figure 3.4.4 Multi-sampling is often used to provide several different samples mapped onto a keyboard, but it can also be used to provide different degrees of detail in a given sample. This diagram compares two methods of shifting down by two octaves in pitch: using one sample and slowing down the replay rate provides the same number of sample points regardless of the output pitch, whilst the use of two multi-samples enables the same sample rate to be used for each sample, thus increasing the amount of detail which is available compared to the single sample method.

- The same number of sample values are replayed regardless of frequency: Because the oscillator is merely stepping through a fixed series of values, the detail contained within the waveform is constant, but the same is not true for the spectrum: at low pitches all of the harmonics may be below the half-sampling frequency, whilst for high pitches then only one or two harmonics may be below the half-sampling frequency. The sample should thus ideally have more detail when it is used to produce low-pitched sounds, because this is where the harmonic content is most important, whilst for higher pitched sounds less detail is required because less harmonics will be heard. One technique which can be used to provide the required detail in samples is to use different sample rates for different pitches.
- The half-sampling frequency changes as the pitch changes: Because using a DCO to control the replay rate means that the half-sampling rate tracks the playback pitch, then the reconstruction filter also needs to track the half-sampling rate. (Note that early sample playback devices did not always do this. Instead they set the reconstruction filter so that it filtered correctly for the highest playback pitch, which meant that for lower pitches aliasing was present in the output signal.) Tracking means that a low-pass VCF is required to follow the changes in playback pitch, so that frequencies above the half-sampling rate are not heard in the output audio signal. Such VCFs have much more stringent

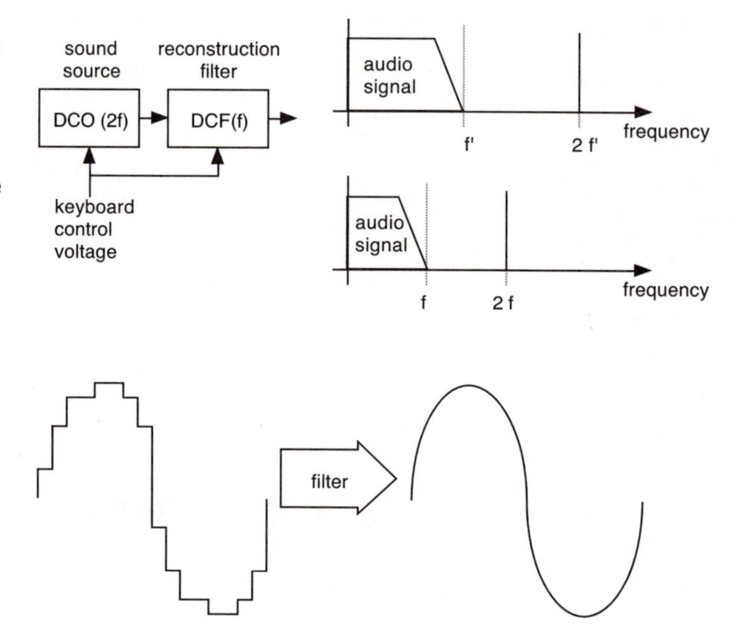

Figure 3.4.5 Reconstruction filters are normally thought of as being used in the output stage of digital audio systems, but they can be required to process the output of a DCO if it uses the variable frequency method to provide different pitches. In this example, a digitally controlled filter (DCF) is used to track the DCO frequency so that any aliasing components are removed before any post-DCO modifiers process the audio. The DCF 'smooths' the DCO waveform so that no aliasing components are present.

design criteria than the VCFs found in analogue synthesizers: the 24 dB/octave roll-off slope of a typical analogue synthesizer VCF is not adequate for preventing aliasing, and slopes of 90 dB/octave or more are often required, with 90 dB or more of stop-band attenuation.

- The playback is monophonic: Because the sample replay rate is set by an oscillator or DCO, variable frequency playback requires a separate sample replay circuit for each individual pitch that is playing. Each DCO in a polyphonic synthesizer can produce a differently pitched monophonic audio sample, and these analogue outputs are then processed by analogue filters in the modifier section.

Fixed-frequency playback

Fixed-frequency playback uses just one frequency for the sample rate, but changes the effective number of sample values which are used to represent the different pitches by calculating the missing values. It has the advantage that only one sample rate is used, and so a fixed frequency sample playback circuit can be polyphonic – and can be connected to a digital modifier section, which allows a completely digital synthesizer to be produced where the digital-to-analogue conversion happens after all the digital processing. Fixed-frequency playback is now almost exclusively used in digital synthesizers and samplers.

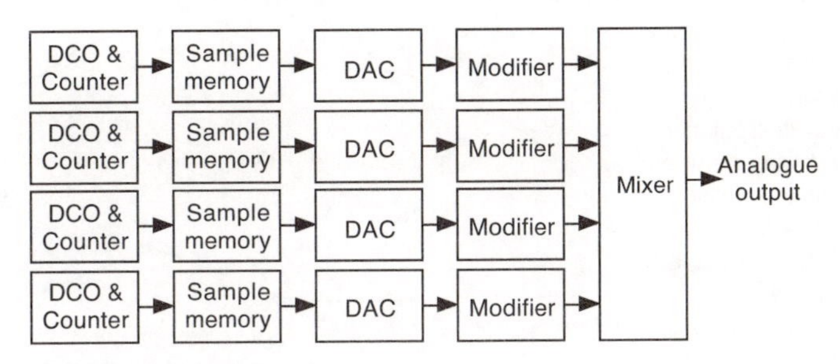

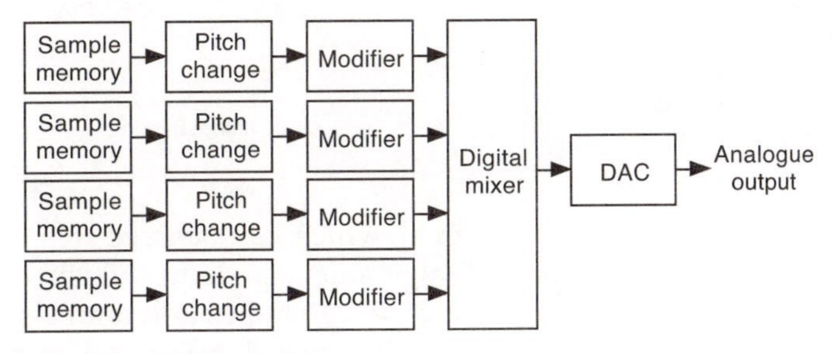

Figure 3.4.6 Variable frequency playback requires a separate DCO and counter to access the sample memory, followed by a DAC to convert the sample into an analogue signal for processing by the modifier section. Fixed frequency playback changes the pitch of the sample and allows the use of a digital modifier section – with a single DAC to convert the output to analogue.

3.4.5 Interpolation and pitch shifting

Changing the playback sample rate is not the only way of changing the playback pitch of a sampled sound. Consider a sample of a sound: a single cycle of a given pitch will contain a number of sample points, where the number is related to the cycle time for the waveform, and the sample rate. So if 256 sample values represent a single cycle of a waveform at one pitch, then for a lower playback pitch more sample values would be required, whilst for a higher playback pitch less sample values are required. But it is possible to take the existing sample values and work out what the missing values are by a process called interpolation. This is used in fixed-frequency sample playback.

Interpolation attempts to represent the waveform by a mathematical formula. If the sample values are thought of as points on a graph, then interpolation tries to join up those points. Once the points are joined up, then any sample values in between the available points can be calculated. The simplest method of interpolating merely joins the sample points with straight lines – this is strictly called linear interpolation, although it is often erroneously shortened to just interpolation. Although this is easy to do, real-world waveforms which consist of lots of straight lines joined together are rare! A better approach is to try and produce a curve which passes through the sample points.

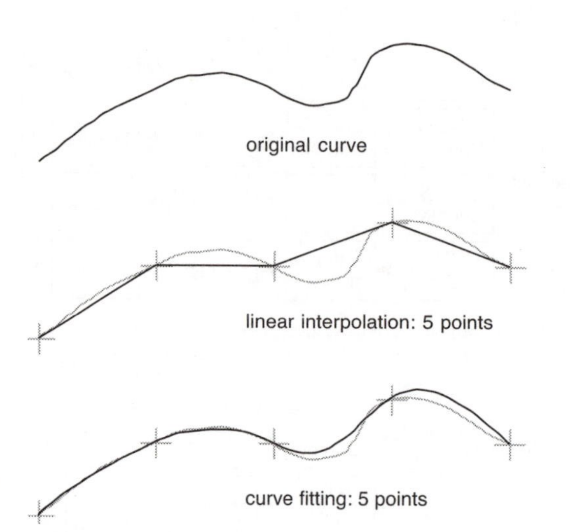

Figure 3.4.7 Interpolation is used to calculate missing or intermediate values in a sample. Linear interpolation draws straight lines between the sample points, whilst polynomial curve fitting attempts to match a curve to the sample points. In this example, a sample curve is shown, together with a linear interpolation based on five sample points, and a curve fitted interpolation. The linear interpolation misses some of the major features, whilst the curve fitting produces a much better fit.

One method which can achieve this uses polynomials: general purpose algebraic equations which can be used to represent almost any curve shape. Polynomials are categorized by their degree, and in general, n points can be matched by an (n–1)th degree polynomial. So for two sample values, a first degree polynomial is used – which turns out to be the formula for a straight line. If more sample values are used to work out the shape of the curve, then higher order polynomials can be used: three points can be fitted by a quadratic (or second degree) equation, whilst four points require a cubic equation. Manufacturers rarely reveal how they do their interpolation – in general, the lower cost the implementation, the lower the degree of polynomial which is used, and the poorer the resultant audio quality. Interpolation using polynomials with degrees higher than 1 is sometimes called differential interpolation to distinguish it from linear interpolation. It is also possible to design filters which can interpolate, and these are used in many digital systems.

An alternative technique which can reduce the number of sample values at high frequencies is literally to remove samples, or conversely to add in extra sample values at low frequencies. The simplest way to do this for an octave shift up or down is to either miss every other sample, or to repeat each sample. This is called decimation, and it is crude but effective. Since there are no calculations involved, it is easier to implement than interpolation, but it can produce distortion in the output. Because of the relatively low cost of ROM, sampling at a higher rate than is required can be used. Known as 'over-sampling', the idea is to provide more points for the interpolation processing. Since the points are closer together, and there are more available for the calculations, the interpolation quality improves. Over-sampling can be at anything from twice the required rate to 64 times or more. The performance of the memory and the interpolation processing requirements limit the over-sampling rate.

3.4.6 Quality

Sample reproduction quality is determined by:

- sample rate in kHz (affects the bandwidth)
- sample size in bits (affects the SNR)
- interpolation technique (linear or polynomial: affects the distortion when pitch transposing)
- the anti-aliasing and reconstruction filters (affect the distortion and SNR)

The CD sample rate of 44.1 kHz has become widely used in electronic musical instruments, with some instruments using even higher rates of 48 kHz or more. Samplers often have a range of available sampling frequencies, so that their memory usage can be maximized – sampling at 32 kHz or 22.05 kHz can reduce the amount of storage which is required for sounds which have restricted bandwidths.

Sixteen bit sample size has become the norm. Internal processing is often higher, but conversion chips designed for CD players and DAT recorders (which are based on 16 bits) are widely used in synthesizers and samplers. Some 18-bit converters are being used, and as 20-bit and higher resolution converters become available, they are likely to be incorporated.

Interpolation techniques depend on the processing power which is available. Microprocessors and digital signal processing (DSP) chips continue to increase their performance, and so more sophisticated interpolation techniques will become possible, which should improve the quality of sample replay and transposition.

Analogue filter technology is almost at the theoretical limits, and so any improvements are likely to take place by adding in digital filtering. By increasing the sample rate inside the conversion chips, it is possible to augment the anti-aliasing and reconstruction filters with additional digital filtering using DSP chips. This allows enhanced performance, and yet outside the conversion chips, the samples can still be at at sample rate of 44.1 or 48 kHz.

Synthesizers and samplers will continue to follow developments in audio technology. Future developments are likely to include more digital processing, and less analogue electronics.

3.5 Early versus modern implementations

Section 3.3 discussed the technology of DCOs, and mentioned the difference between the design of instruments in the mid-1970s, and those of the mid 1990s. This section summarizes the differences between hybrid synthesizers since the 1970s.

The 1970s

In the 1970s, hybrid instruments were just developing. VCOs were gradually being enhanced by the addition of auto-tune and digital control features, as well as programmability of the complete synthesizer with the change in emphasis from 'live' user programming to instant access via large numbers of memories. There were two distinct types of keyboard synthesizers: versatile monophonic or polyphonic synthesizers with rather more limited functionality – all based on mixtures of analogue and digital circuitry, and fully polyphonic 'electronic pianos'/'string machines' and multi-instruments (string and brass) – all based on top octave chips plus dividers followed by simple filtering and enveloping circuits. The second category was already in decline: the polyphonic digital instruments of the 1980s would cause their complete disappearance.

The synthesizers had limited wavecycle waveforms, and if any controls were provided for the levels in the wavecycle, they would be on a 'one control per function' basis. The display would use LEDs, or perhaps a discharge tube/fluorescent display. Waveform samples would be in 8 bits, and the sample rate would be between 20 and 30 kHz, giving an upper limit for frequency output of between 10 and 15 kHz. Control would be via 4- or 8-bit microcontrollers, adapted from chips intended for simple industrial control applications. The interfacing would be via analogue control voltages, gates and trigger pulses – or perhaps from a proprietary digital bus format.

The 1980s

The release of the Yamaha all-digital FM synthesizers in the early 1980s, saw all the other manufacturers trying to catch up and releasing hybrids whilst their development teams worked on the digital instruments which would begin to appear in the late 1980s. These hybrids used digital enhancements to make the most of analogue oscillators, and eventually replaced the VCO completely with a digital equivalent. Portamento was the first casualty of this conversion, but by the end of the decade it had re-appeared as the clock speed of chips made more sophisticated

DCOs possible. Early designs used medium and large scale integrated circuits containing tens, hundreds or thousands of digital gates. Wavecycle was joined by wavetable, usually with either 8-bit or 12-bit waveform samples.

The display gradually replaced the front panel knobs as the centre of attention during the programming process, although a 2 row by 16 character LCD display (which might be back-lit) was not ideal. Individual controls were replaced by 'parameter access', where a single slider or knob was used to change the value of a parameter which was selected by individual buttons. Eight-bit and 16-bit microprocessors were used to control the increasingly complicated functionality, especially once MIDI became established. Interfacing polarized rapidly from proprietary interface busses to MIDI within a couple of years of the launch of MIDI in 1983.

The 1990s

The 1990s opened with a preponderance of all-digital instruments, and a consolidation of sampling. But this was quickly followed by a resurgence of interest in analogue technology, and some manufacturers began to rework older designs or even design completely new instruments from scratch. Although often labelled 'analogue', many of these instruments are actually hybrids – most often they use DCOs rather than VCOs. Even the 'pure' analogue instruments will have considerable amounts of digital circuitry used for control and programming purposes.

DCOs now use multiplexed circuitry to provide independent 'oscillators', and these use sophisticated accumulator/divider type techniques to provide very fine resolution frequency control – typically on custom chips made for the individual manufacturer (application-specific integrated circuits, ASICs). With plenty of processing power available, the wavecycle and wavetable generation techniques were joined by sampling, and wave sequencing, normally with 16-bit waveform samples and better than CD sampling rates (greater than 44.1 kHz). Displays have increased in size, with 4 row by 40 character back-lit LCDs (and larger) in common usage, some with dot-addressable graphics modes instead of just character-based displays. Allied to this is the increasing importance of a graphical user interface (GUI), perhaps with a mouse used as a pointing device, but almost certainly with softkeys or assignable buttons. Computer-based editing software has helped to make the front panel display almost superfluous on some rack-mounting instruments. Control functions are provided by 16- or 32-bit microprocessors,

Table 3.5.1 Comparisons

	Early designs (1970s)	Later designs (1980s)	Current designs (1990s)
DCO	Digitally controlled VCOs, top octave synthesizers	Master oscillators, rate adapters and dividers	Multiplex, accumulator/dividers
Technology	Analogue/digital	MSI/LSI digital logic	ASICs
Waveform	Wavecycle	Wavecycle, wavetable	Wavecycle, wavetable, sampling
Display	LEDs	16 × 2 LCD	dot-matirx LCD (4 × 40)
Parameter entry	Individual sliders	Slider and button selector	GUI, mouse, softkeys
Sample bits	8	12	16
Control	4 and 8-bit microcontrollers	8 and 16-bit microcontrollers	16 and 32-bit microcontrollers
Interfacing	Analogue: CVs, gates	MIDI	MIDI (ZIPI?)

perhaps with a digital signal processor (DSP) for handling the more complex signal processing functions. Interfacing is via MIDI, although by the end of the decade, a faster alternative like ZIPI is likely to also be available. The conversion from analogue to digital is almost complete – often only the VCFs and enveloping is analogue in modern hybrid instruments.

Table 3.5.1 summarizes these points.

The future

The mid-1990s saw the release of all-digital instruments which replaced even the VCFs with digital 'software-based' equivalents, and the era of 'emulation' began. With software now capable of producing complex imitations of entire analogue instruments and even models of real instruments on DSPs, the current hybrid designs may well be the last: software emulations will almost certainly price analogue designs out of the market by the end of the decade.

3.6 Example instruments

Fairlight CMI Series I (1979)

The Computer Musical Instrument came from Australia, and combined computer technology with sampling technology, using voice cards which were a hybrid mix of analogue and digital technology on the earlier models. The first models offered plug-in 8-bit wavecycle and wavetable synthesis cards that had evolved into 16-bit sample replay cards by the time that the Series III model came out in 1985. Additive synthesis, 'draw your own waveform', step-time rhythm programming, and many other innovations made this a very popular instrument with those who could afford the high purchase price.

PPG Wave 2.2 and Waveterm (1982)

The PPG Wave 2.2 combines wavetable oscillators with analogue filtering and enveloping, whilst the Waveterm added sampling capability and sequencing facilities. The wavetable memory offered 1800 basic waveforms, whilst the samples were only 8-bit. Later models like the Wave 2.3 and EVU were 12-bit.

Roland Juno-60 (1982)

The Roland Juno-60 (and its memory-less version, the Juno-6) both had DCOs, and provided low-cost polyphonic synthesis (albeit with no velocity sensing on the keyboard, arpeggios instead of portamento, Roland's proprietary digital communication bus (DCB) instead of MIDI, and only one DCO per voice).

Roland D50 (1987)

The Roland D50 was arguably the first commercial synthesizer to use S&S synthesis – although it uses the confusing term 'linear arithmetic (LA)' to describe the technique, and the implementation is only partial in comparison to later instruments. (The first full S&S implementation was probably the Korg M1, although it did not have resonant filters.) The D50 provides a combination of analogue synthesizer emulation plus a simplified S&S. The analogue synthesizer provides the classic synthesizer waveforms as the source material for a resonant filter and VCA modifier

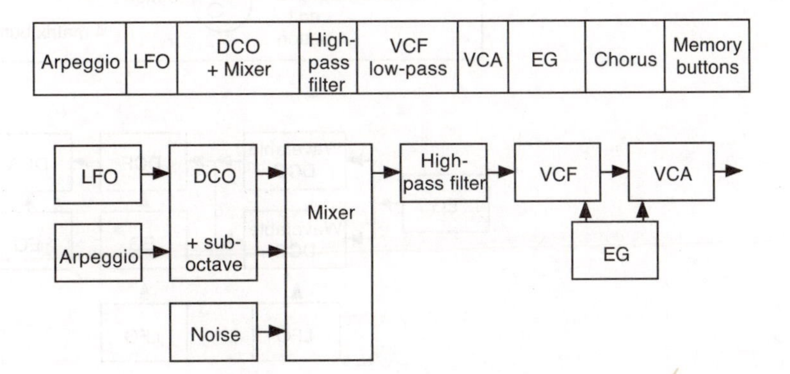

Figure 3.6.1 Juno-60.

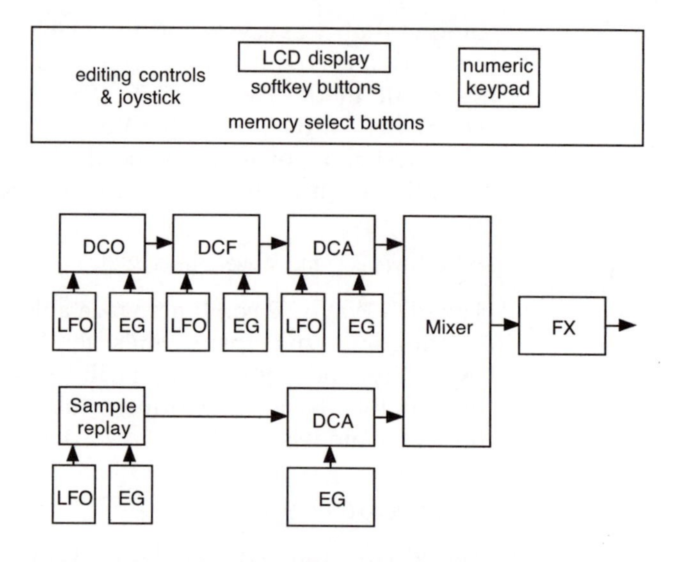

Figure 3.6.2 The Roland D50 mixes simple sample replay technology with a basic DCO/DCF analogue synthesis emulation.

section. The sample replay is more primitive, with just a VCA and no filtering. The normal mode of operation is to use the sample part to provide the attack for a sound, whilst the sustained sound is provided by the synthesizer part. When it was released, this combination of sample realism plus analogue familiarity proved to be a strong contender against the ubiquitous FM of the time.

The D50 was one of the first commercial polyphonic synthesizers to incorporate comprehensive built-in effects: EQ, chorus and reverb. It also marks the end of the front panel as a guide to the operation of the synthesis method. Instead, the front panel clearly shows the influences at the time: diagrams from FM synthesizers, joystick from vector synthesizers, and a large softkey-driven display to simplify the editing.

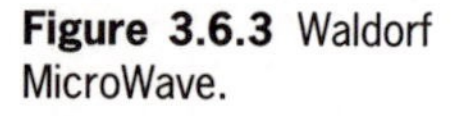

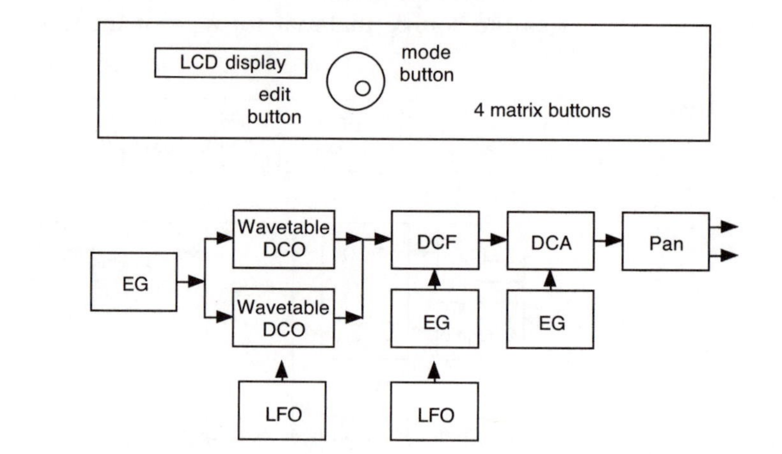

Figure 3.6.3 Waldorf MicroWave.

Waldorf MicroWave (1989)

The Waldorf MicroWave is essentially a 'PPG Wave'-type of wavetable synthesizer, but redesigned to take advantage of the available electronics of the late 1980s.

3.7 Questions

1. What is a hybrid synthesis technique?
2. What are the differences between cycles, multi-cycles and samples?
3. Give four examples of single cycle waveshapes.
4. What is the difference between multi-cycle wavecycle and wavetable synthesis?
5. How does a 'top octave' oscillator produce audio frequency outputs?
6. How does the frequency resolution of a DCO affect detuning and pitch bending?
7. Why are modern 'analogue' synthesizers really hybrids?
8. How can dividers reduce the effect of jitter?
9. How does 'auto-tuning' work?
10. How are the contents of a wavetable converted into an audio signal?

Time line

Date	Name	Event	Notes
1970s	Ralph Deutsch	Digital generators followed by tone-forming circuits	The popularisation of the electronic organ and piano
1975	Moog	Polymoog is released	More like a 'master oscillator and divider' organ with added monophonic synthesizer
1982	PPG	Wave 2.2, polyphonic hybrid synthesizer, is launched	German hybrid of digital wavetables with analogue filtering
1983	Oxford Synthesiser Company	Chris Huggett launches the Oscar, a sophisticated programmable monophonic synthesizer	One of the few monosynths to have MIDI as standard
1986	Sequential	Sequential launches the Prophet VS, a 'Vector' synth which uses a joystick to mix sounds in real-time	One of the last Sequential products before the demise of the company
1989	Waldorf	Microwave, a digital/analogue hybrid based on wavetable synthesis	Effectively a PPG Wave 2.3 brought up to date

4 Samplers

A sampler is the name given to a piece of electronic musical equipment which records a sound, stores it, and then replays it on demand. There are thus three important functions:

- record the sound
- store the recording on some sort of storage medium
- replay the stored sound.

A sampler combines all of these functions into one unit, and this makes it very different from almost all of the other examples of synthesizers described in this book. Most synthesizers can fulfil the last two functions – store and replay – but the distinguishing feature of a sampler is its ability to record sounds.

This definition of a sampler in terms of its functionality is important because it enables a wide range of equipment to be classified as being samplers, whereas the commonly used term is often restricted to merely electronic music equipment which stores sounds in RAM memory. Using the functional description, the following can all be described as samplers:

- tape recorder
- cassette recorder
- video recorders
- DAT recorder (DCC, MiniDisc, etc.)
- echo effects unit
- computers with sound input and output facilities.

All of these 'samplers' represent ways to record, store and subsequently replay sounds. In some of these cases, the sounds will probably be naturally occurring sounds that can be recorded with a microphone, but this does not prevent the process of collecting the sounds, storing and manipulating them, and then replaying them from being called 'sound synthesis'.

Within this wider context, any of the techniques that have already been described can become part of a larger synthesis system by utilizing sampling. The 'source and modifier' model can be used to describe the working of an analogue subtractive synthesizer, but it can also be used to describe the process of using a synthesizer merely as the source of sounds which are then recorded, stored, modified, and finally replayed using a sampler which acts as the modifier of those sounds.

Samplers thus form a bridge between the analogue and the digital synthesizer, since they span the two technologies with very similar instruments. Analogue sampling can be tape-based or chip based, although analogue sound storage chips have been largely ignored since digital technology became available. Digital sampling has increasingly used the technology and approach of synthesis, and this has led to the convergence of sampling and synthesis.

(One popular usage of the word 'sampler' which is not covered by this type of definition is the recorded collections of material from more than one source – which are also called samplers.)

4.1 Tape-based

Samplers based on tape recording techniques have a long history. The first 'tape' recorders did not use tape at all, but used wire instead. Plastic tape covered with a thin layer of iron oxide is much easier and safer to handle than reels of wire, and far easier to cut and splice!

4.1.1 Tape recording

The underlying idea behind how a tape recorder works is very simple. The sound signal is converted into an electrical signal in a microphone, and this signal is then amplified, converted into a changing magnetic field and stored onto tape. By passing this magnetized tape past a replay head, the changes in the magnetic field are picked up, amplified and converted back into sound again.

Magnetic tape is made up from two parts:

- a plastic material which is chosen for its strength, wear-resistance, temperature stability
- magnetic coating which is chosen for its magnetic properties.

It is actually possible to record and replay sounds using a fine layer of iron rust placed onto the sticky side of adhesive tape – although this is not recommended as a practical demonstration. The commercial versions of recording tape are just more sophisticated versions of this 'rust on tape' idea.

A tape recorder is a mixture of mechanical and electronic engineering. The mechanical system has to handle long lengths of fragile tape, pulling it across the record and replay heads at a constant speed, and ensuring that the tape is then wound onto the spool neatly. This requires a complex mixture of motors, clutches and brakes to achieve. The pulling of the tape across the heads is achieved by pressing the tape against a small rotating rod called the capstan. The tape is held onto the capstan with a rubber wheel called the pinch roller. The spool which is supplying the tape is arranged so that it provides enough friction to provide sufficient tension in the tape to press it against the record and replay heads as it is pulled past. Once past the capstan and roller, the tape is then wound onto the other tape spool. When the tape is wound forwards or backwards, the pinch roller is moved away so that the tape no longer presses against the capstan or the heads, and the spools can then be moved at speed.

The electronic part of a tape recorder has two sections: record and replay. The record part amplifies the incoming audio signal, and then drives the record head with the amplified signal plus a high frequency 'bias' signal. The combination of the two signals allows the response of the magnetic tape to be 'linearized'. Without the bias, the tape recorder would produce large amounts of distortion. The replay section merely amplifies the signal from the replay head (no bias is required for replay).

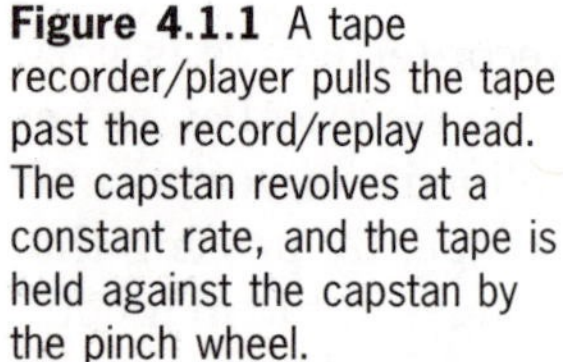

Figure 4.1.1 A tape recorder/player pulls the tape past the record/replay head. The capstan revolves at a constant rate, and the tape is held against the capstan by the pinch wheel.

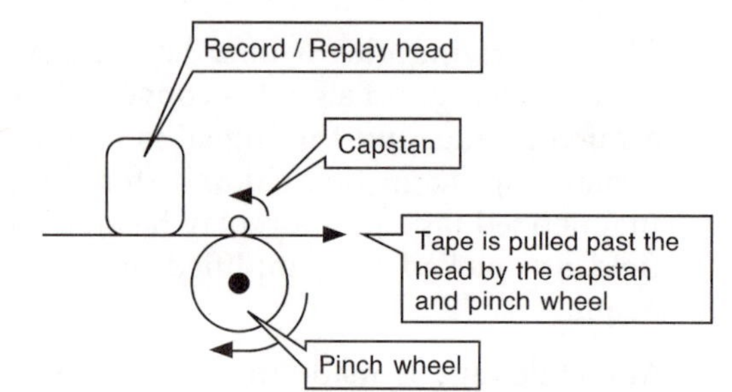

4.1.2 Mellotrons

The word 'Mellotron' is a trademarked name for one type of sample replay musical instrument which uses short lengths of magnetic tape. The concept is simple – the practicalities are rather more involved. The basic idea is to have a tape replayer for each key on the keyboard. A long capstan stretches across the whole of the keyboard. Pressing a key pushes the tape down onto the capstan and pulls it across the replay head. The tape is held in a bin with a spring and pulley arrangement to pull it back when the key is released. The length of tape is thus fixed, and so the key can only be held down for a limited time. Loops of tape cannot be used because the start of the sound would not be synchronized with the pressing of the key – by arranging for the tape to be pulled back into the bin each time the key is released, it automatically goes back to the start point of the sound, ready for the next press of the key. There have been several other variants on the same idea from other manufacturers, but the Mellotron is the best-known.

Because the capstan is the same size for each key, the tape for each key needs to be recorded separately, with each tape producing just one note – although several tracks are available on each tape, with a different sound on each track. The tapes are thus multi-sampled at one note intervals. Recording user

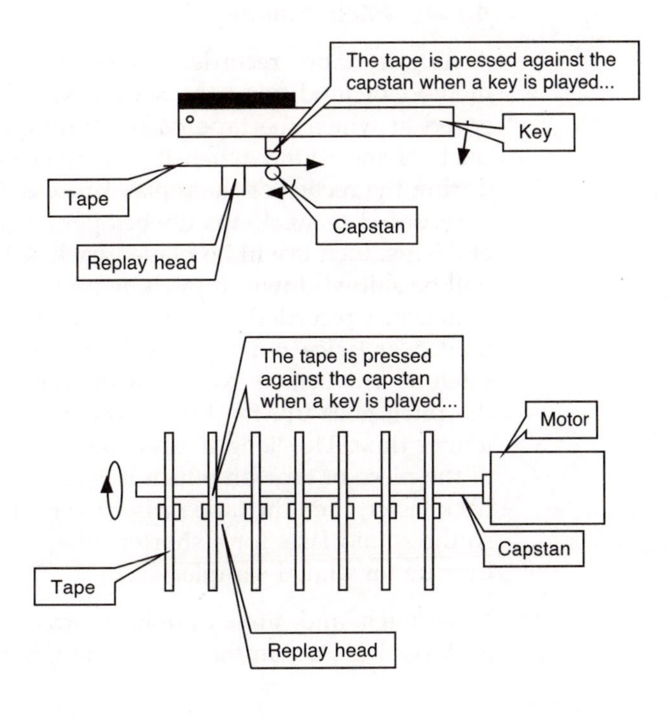

Figure 4.1.2 A side view and top view of a tape sample replay instrument. The capstan spans the whole of the keyboard, and revolves continuously. When a key is pressed down, this presses the tape against the capstan, which pulls the tape across the replay head.

samples for such a machine requires time, patience and attention to detail: the levels of the sounds must be consistent across all the tapes, for example. The 'frames' which contain complete key-sets of the tape bins can be changed, but this is not a quick operation. Because of the difficulty of recording your own sounds onto a tape, these tape samplers can almost be regarded as being sample replay instruments rather than true samplers.

4.3.1 Tape loops

By looping a piece of tape around and joining the end to the beginning with splicing tape, it is possible to create a continuous loop of tape which will play the same piece or recorded material repeatedly. The only limitation on the size of the tape is physical: short loops may not fit around the tape recorder head and capstan, whilst long loops can be difficult to handle as they can easily become tangled.

The repetition of a sequence of sounds produces a characteristic rhythmic sound, which can be used as the basis of a composition. As with the Mellotron tape player, synchronizing the playback of the start of a loop is difficult, and synchronizing two loops requires them to be exactly the same length. Tape loops are thus used for asynchronous purposes.

4.1.4 Pitch changes

Analogue tape recorders have one fundamental 'built-in' method of modifying the sound: speed control. Changing the speed at which the tape passes through the machine alters the pitch of the sound when it is played back. This can be either during the record or the replay process. For example, if a sound is recorded using 15 ips (inches per second), and then replayed at 7.5 ips, then it will be played back at half the speed, and thus will be shifted down in pitch by one octave. Conversely, sounds which are recorded at 7.5 ips and replayed at 15 ips will be played at twice normal speed, and will thus be shifted up in pitch by one octave. Note that the pitch and time are linked: as the pitch goes up, the time shortens, whilst lower pitch means longer time. The 'length' of a sound is exactly that – the length of the piece of tape on which it is recorded. If the tape is played back faster, then the tape passes over the replay head faster, and so the sound lasts for a shorter time. (The same is not necessarily true for digital samplers. . .)

This 'pitch and time doubling' was used to great effect by guitarist Les Paul in the 1950s. Using the technique of recording at a slow tape speed, and then replaying at a faster tape speed, he was able to achieve astonishingly fast and complex performances on guitar. The same technique is still a powerful way of changing the pitch of sounds, or for enabling virtuoso performances at slow tempos.

4.2 Analogue sampling

Analogue sampling covers any method which does not use tape or digital methods to store the audio signals.

The most common technology which meets these requirements is the 'bucket-brigade' delay line, or analogue delay line. This uses the charge on a series of capacitors to represent the audio signal, rather than the magnetic field used in tape systems or the numbers used in digital systems. The sampling process is merely the opening of an electronic switch to charge up the first capacitor in the delay line. The size of the voltage determines the amount of charge which is transferred to the capacitor: the higher the voltage which is being sampled, then the more charge which is stored in the capacitor. Effectively, the capacitor acts as a store for the voltage, since the presence of the charge in the capacitor is shown by the voltage across the capacitor. The switch then opens and the charge is held in the capacitor since there is no significant leakage path. Another switch is then used to transfer the charge to the next capacitor in the delay line, where it again produces a voltage. The original capacitor is then available to sample the next point on the incoming audio signal. This process continues, with the sample voltages moving along the delay line formed by the capacitors – hence the term 'bucket-brigade' delay lines.

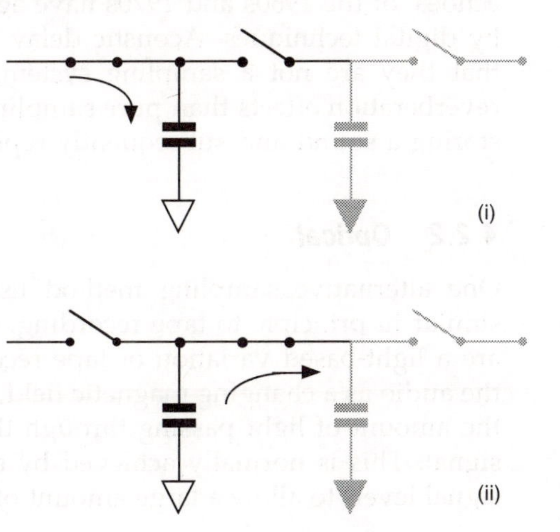

Figure 4.2.1 An analogue delay line moves charge along a series of capacitors connected by switches. The input voltage is stored on the first capacitor (i). The charge is then transferred to the next capacitor (ii). This repeats for the entire chain, and so the input voltages move along the capacitors.

Because each section of the delay line is just a capacitor and some electronic switches, it is easy to fabricate, and so several thousand can be placed on a single integrated circuit chip. The sampling and transfer of charges requires a high frequency clock signal, but the control circuitry is straightforward. This simplicity of control and application made analogue 'bucket brigade' delay lines popular in the 1970s and early 1980s for producing echo, chorus and reverberation effects. At least one monophonic sampler was produced using analogue delay lines in the early 1980s, but it was rapidly superseded by digital versions.

The limitations of the analogue delay line technique are twofold: firstly, the capacitors are not perfect, so some of the charge leaks away causing signal loss, distortion and noise; but more importantly, the high frequency clock signals tend to become superimposed on the output audio signals and this degrades the usable dynamic range of the delay line. Because of these problems, digital sampling technology has replaced analogue delay lines in the majority of cases – although recent advances in the technology mean that some analogue delay lines continue to be used in specialist applications like audio storage for telephone answering machines.

4.2.1 Delay lines

An alternative to bucket brigade delay lines moving charge around is to use metal springs or metal plates to carry the sound signals acoustically. Sounds are transferred to the metal using modified loudspeaker drivers, and the delayed sound signals are recovered with contact microphones. The physical size of these acoustic delay lines is large, and the 'spring lines' and 'plate echoes' of the 1960s and 1970s have again been largely replaced by digital techniques. Acoustic delay lines have the advantage that they are not a sampling system, but are more suited to reverberation effects than pure sampling – they are not suited to storing a sound and subsequently replaying it.

4.2.2 Optical

One alternative sampling method uses a technique which is similar in principle to tape recording. Optical film sound tracks are a light-based variation of tape recording. Instead of storing the audio as a changing magnetic field, the film sound track uses the amount of light passing through the film to store the audio signal. This is normally achieved by arranging for large audio signal levels to allow a large amount of light to pass through the film, whilst small signals allow less light through. A photodetector and lamp are used to convert the transmission of light into an audio signal.

This modulation of light by an audio signal is normally achieved by using the audio waveform to control the width of a slot, and so the amount of light which passes through the film. Variable density (opacity) film can also be used, but this is rare for film use, although it has been used for experimental systems where film is used to produce sound by literally painting onto it to control the amount of light which passes through it at any given instant. By passing the resulting film through a lamp and photodetector, the optical version of the audio can be converted into sound. Although flexible, the complexity of producing the required degree of detail is enormous and very time-consuming. At least one manufacturer produced an optical sample replay machine in the 1980s, but as with all analogue methods, this was not a success against the digital competitors.

4.3 Digital sampling

Digital sampling is based around three electronic devices:

- Analogue-to-digital converters (ADCs)
- Memory devices (RAM, flash EPROM, etc.)
- Digital-to-analogue converters (DACs).

Coincidentally, these three devices carry out the three major sampling functions:

- The ADC records the sound
- The memory devices store the recording
- The DAC replays the stored sound.

A sampler works in three modes: record, edit/store and replay. The record mode is used to convert signals from a continuous analogue form into a numeric digital representation. The digital data which represents the sound is then held in RAM memory inside the sampler, and this is edited in the second mode: edit/store. When an audio signal is recorded by a sampler, the start of recording is normally set to before the actual start of the sound, so that the initial attack part of the sound is captured. Once in RAM memory inside the sampler, the sample data needs to be edited so that the start of the sound is at the start of the sample data. This ensures that when the sample is replayed, it will start playing without any time delay. Once edited, the sample data is then stored in some sort of permanent storage like hard disk. This sample data can be reloaded into the sampler's RAM memory when it is required. The replay mode takes the sample data in the sampler memory and converts it back to an analogue audio signal.

The major division between an S&S instrument and a sampler can be considered to be the type of memory: S&S instruments use fixed ROM memory and so the samples cannot be edited, whilst samplers use volatile RAM memory where the samples can be edited.

4.3.1 Editing

The most important editing function which is required by a sampler is the trimming of the unwanted portions of the raw samples – 'before' and 'after' the wanted sample. This trimming or 'topping and tailing' process allows the sampler user to set the start and the end of the sound. This can be especially important if the raw sample is noisy, because then the start of the sound may not be apparent, and a compromise may been to be made between finding the true start of the sample and hearing noise. Some samplers provide automatic functions which will trim a sample using criteria which can be adjusted to suit the user. Although this trimming function is of great importance to a user who produces their own samples, the majority of sampler users merely use the sampler to replay pre-prepared samples, and so the trimming function is not as important as might be supposed. But the ability to manipulate segments of audio is essential for the user who wishes to use a sampler as a synthesizer rather than merely a sample replay device.

Most S&S synthesizers have only limited sample manipulation facilities and no sample editing facilities – the samples are in ROM memory, and so the only manipulations which are possible are changes in direction, start point and loop points. S&S instrument thus rely almost entirely on their synthesizer modifier section to make changes to the timbre of the samples. In contrast, samplers normally have a powerful sample manipulation and editing section as well as a synthesizer type modifier section. The synthesis modifier facilities are thus less critical to the operation of the sampler – and in fact, many samples of filter sweeps and related modifier sounds are available so that the modifier section is less important. By having the sample editing facilities available, changes can be made to individual samples rather than the global filtering which is available in a modifier section.

Three of the most powerful of these sample editing techniques are looping, stretching and resampling.

Looping

Looping consists of implementing the equivalent of a loop of tape, but in the digital domain instead of with a physical loop of magnetic tape. In the simplest case, this is merely a repetition of the same portion of the sample, but it may also be controlled by an envelope generator so that the loop does not stay at a fixed volume. The transition from the end of the loop to the beginning of the loop is the equivalent of the splice in a physical tape loop, but the control of this transition can be much more sophisticated because the sample is stored in a digital form.

The basic method of joining the end of the loop to the beginning is to splice the two points together. If the end and start of the loop do not have the same level, then the resulting 'glitch' will be audible as a click in the loop. There are several approaches to avoiding this problem. It is possible to arrange for the splice to be made only when the two levels meet one or more criteria:

- same level
- same slope
- same rate of change of slope.

The 'same level' criteria is often refined to when the audio waveform crosses the zero axis – this is called zero crossing. Since the zero axis is normally the effective 'silence' level of the sample, splicing at zero crossings can produce splices without clicks, although this is not guaranteed. Better techniques take into account the shape of the waveform at the transition.

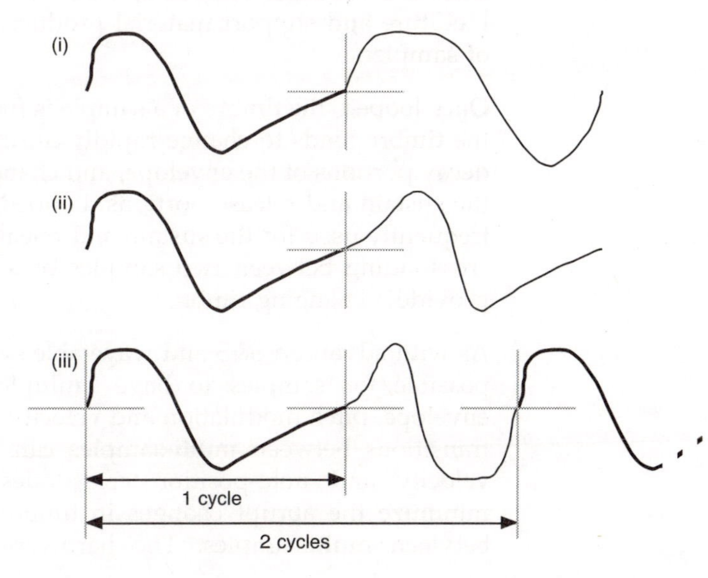

Figure 4.3.1 Splicing loops. (i) Choosing the same level does not guarantee a good splice. (ii) Matching the level and the slope can give a good splice. (iii) Even a good splice can alter the frequency of the loop if the same cycle time (or an integer multiple of it) is not maintained.

Matching the slopes of the two portions of the waveforms can reduce the level of any click, whilst matching the two waveforms so that the splice point occurs at similar points on both can minimize the click – although this restricts the available splicing points. By reversing the direction of playback at the splice point, some of the problems of matching the levels or slopes can be avoided, but this works only for short samples where the backwards and forwards playback of the loop will not be heard – long loops can sound very unusual if they are looped using this technique, although it is useful as a special effect. The length of the loop also affects the perceived pitch of the sound as it is replayed. In extreme cases, the looped section can shift its pitch markedly from the original pitch before looping. This is most obvious for short loops, especially where a single cycle of the waveform is being looped.

Even if clicks are not produced by the splice in a loop, the start and end of the loop may not have the same timbre. This produces a sudden change in the timbre which can be almost as noticeable as a click. The abrupt change in spectrum of the sound is often interpreted by the listener as a click even though examination of the waveform at the spice point shows no obvious mismatch of level or slope. Cross-fading between the start and end of the loop can be used to reduce the effect of a sudden change into a smoother transition between the two contrasting timbres. Unfortunately, even a cross-faded loop can still produce a cyclic variation in timbre which is apparent when the loop repeats. The subject of producing usable looped samples of sounds is an entire subject, and is often covered in detail in the literature and support material produced by the manufacturers of samplers.

Once looped, the timbre of a sample is fixed. In real instruments, the timbre tends to change rapidly during the initial attack and decay portions of the envelope, and changes more slowly during the sustain and release portions. Looped samples are thus most frequently used for the sustain and release parts of sounds, with cross-fading between two samples or a modifier filter used to provide a changing timbre.

As with advanced S&S and wavetable synthesis techniques, it is possible for samples to have multiple loops: each with an envelope, pitch modulation and velocity switching facilities. The transitions between multi-samples can also be modified with velocity and note-position cross-fades, which can help to minimize the abrupt changes in timbre which can be present between multi-samples. The hardware and software of the sampler itself determines exactly which methods can be applied to the samples. The sound manipulation possibilities which are opened by these techniques should not be underestimated – most samplers provide powerful synthesis capability even without using the 'synthesis' modifier section.

Looping a sound can markedly reduce the amount of storage which is required. For a ten second 'CD quality' stereo sound without any looping, then nearly 2 Mbytes of storage is required. If seven seconds can be produced by looping part of the sample during the sustain segment, then this can be reduced to half a megabyte.

Stretching

Stretching is the name given to the process of independently adjusting the timing or pitch characteristics of a sample. Transposition changes both the pitch and the time of a sample – if the sample is shifted up by an octave, then it plays back twice as fast and so a 'one second' sample lasts only half a second. In contrast, stretching aims to change the time without changing the pitch, or the pitch without altering the timing.

Changing the timing involves analysing the existing sample and either removing or adding sections, depending on whether the sample is being lengthened or shortened. For pitch changes, then the sample is either lengthened or shortened by the pitch change, and so sections again need to be added or removed. The length of the sections which are repeated or removed is normally quite short: at least one cycle, but short enough that the repeated sections are not heard as repeats. The transitions between the original and new sections suffer from the same problems as loop splice points, except that there are many more of them. This makes time and pitch stretching prone to audio quality problems – although this can be used as an effect on rapidly changing material to give a result which is similar to granular synthesis.

Re-sampling

Re-sampling is the name for using the sample record facility of a sampler to record the output of a sample replay. In a digital sampler it is the digital signals which are used, and so the loss in quality is not dependent on the analogue-to-digital or digital-to-analogue conversions. Re-sampling allows the sample rate of a sample to be changed, or for an LFO modulated sound to be stored as a sample, or for a filter sweep to be stored as part of a sample. It enables 'snapshots' to be made of the output of the sampler, and then the re-use of these sample snapshots as the raw material for further sounds by reprocessing the sample snapshots through the sample manipulation and modifier sections.

4.3.2 Multi-sampling

Multi-sampling is normally used merely to provide changes in sound across the keyboard. It is typically used for instruments which have marked differences in their harmonic structure for high and low pitches, most notably the piano. Samples are taken of the source sound played at different pitches, normally all at the same sample rate. The limiting case of multi-sampling is when each note on a music keyboard is sampled separately – so each note will be reproduced using a different set of sample values. This uses large amounts of memory, but provides the potential for the most accurate reproduction. Most multi-sampling is less extravagant than this, with samples being transposed to provide spans of an octave or perhaps a fifth rather than individual notes.

For multi-sampling where two or more samples are used across the whole keyboard range, the transition between samples can be important. As an example, consider two samples made of a piano: one from each extreme of the keyboard. The low pitched sound would be rich in harmonics, whilst the high pitched sound would be a sine wave plus a 'plink' transient hammer noise from the hitting of the string. The changeover from one sample to the other in the middle of the keyboard is likely to be very noticeable to a listener! Most multi-sampling does not use two samples taken from extreme ends of the range of an instrument. Instead, the aim is to provide enough samples to capture the characteristic sound of the instrument whilst minimizing the unwanted effects of transposing samples. Extreme transpositions of samples produce an effect called 'munchkinization', where the changes of pitch and timing emphasize the pitch change and give it a comic effect. This is particularly apparent on the spoken word or singing, although many instruments change their character noticeably when they are transposed by a large amount.

Since most playing of an instrument concentrates on the middle portion of the range of the instrument, most multi-sampling schemes involve having the most detail in this area. The transpose range of the multi-samples is thus small where the detail and transitions between multi-samples are important, but increases at the extremes of the range. This can be observed in many piano multi-sample sets – the bass notes often use a single transposed sample, whilst the high notes use a single 'plink' sample, with the smallest multi-sample ranges being present in the middle area of the keyboard. The transitions between these two extreme samples and those used in the central area are often the most striking, since this is where the largest compromise is made between choosing a suitable sample and ensuring a smooth transition between adjacent samples.

For instruments with large changes in timbre across their range, producing multi-sample sets can be complex and exacting work. Instruments which have a restricted range can also be a problem because most samplers will enable the playback of a sample over the complete range of the sampler – even if the source instrument cannot! This often means that a single violin multi-sample set is utilized as a violin, viola, cello and even double-bass – which although useful for providing synthetic textures which have some of the characteristics of real instruments, it is not suitable as a means of emulating real instruments. The most extreme example of the failure of transposition occurs in percussion instruments, where the fixed parts of the spectrum are essential to the timbre. Transposing a sample of a triangle or a tambourine produces instruments which merely sound wrong!

4.3.3 Storage

There are two forms of storage used in a sampler. The short-term internal storage is usually inside the sampler itself, whilst the longer-term storage is often external to the sampler and is frequently removable.

The storage which is used to hold the samples as they are made or replayed is normally fast read-write memory called RAM. RAM is an acronym for random access memory, and the name refers to the ability to rapidly access any location in the memory device at random. In contrast, a tape recorder is much more restricted in its access: it either plays back the audio, or it needs to be wound or rewound to reach an alternative location on the tape. RAM storage does not have this problem: any location can be accessed as quickly as any other. RAM storage comes in two forms: static and dynamic. Static RAM chips will hold their contents for as long as they are powered up, which makes them ideal for short-term storage using battery backup. Dynamic RAM chips lose their contents if they are not continuously 'refreshed' by the host microprocessor chip. Dynamic RAM chips are considerably cheaper than the static version, and so low-cost samplers are more likely to have dynamic RAM which will require backing up to another more permanent type of storage before powering down the sampler.

Longer-term storage is often associated with magnetic or even optical media, although a variation of read-only memory (ROM) technology called 'flash' EPROM allows long-term storage of samples in memory chips which do not require a backup battery. Flash EPROM can be internal to the sampler, or on a plug-in memory card. Suitable magnetic and optical media include floppy disks, as well as hard disks and CD-ROMs – in either fixed or removable forms. Memory cards can be either RAM or flash EPROM based, or may include a miniature hard disk drive, and will typically use the PCMCIA/PC card format. Samplers typically use the SCSI bus to interface to external memory devices, and this allows additional protocols like the SCSI variant of MIDI (SMDI) to be used to transfer samples at high data rates. Networking of samplers together over a local area network, or LAN, allows samplers to share common storage devices. The use of large amounts of on-line storage forces the use of detailed management of the storage to enable specific samples to be located, and then loaded samples into memory for editing and playback, with the edited versions then being catalogued and stored again.

Digital audio signals require large amounts of storage. For a 44.1 kHz sample rate, stereo 16-bit samples produce just over 1.4 Mbits per second, or about 600 Mbytes per hour. Eight-bit resolution samples halve these figures, but with a significant loss in quality. Reducing the sample rate restricts the bandwidth, which is only useful with sounds which have limited bandwidths like some bass and drum sounds. Table 4.3.1 shows some examples of storage requirements for sampled audio.

Table 4.3.1 Storage requirements for sampled audio

Resolution (bits)	*Sample rate (kHz)*	*Mono/stereo*	*Time (seconds)*	*Storage (kilobits)*	*Storage (kilobytes)*	*Storage (megabytes)*
8	32	Mono	1	250	31.3	
12	32	Mono	1	375	46.9	
16	44.1	Stereo	1	1378.1	172.3	0.2
16	44.1	Stereo	10	13781.3	1722.7	1.7
16	44.1	Stereo	60	82687.5	10335.9	10.1
16	44.1	Stereo	3600	4961250	620156.3	605.6
24	48	Stereo	1	2250	281.3	0.3
24	48	Stereo	10	22500	2812.5	2.7
24	48	Stereo	60	135000	16875	16.5
24	48	Stereo	3600	8100000	1012500	988.8

4.3.4 Types

There are three main types of digital sampler:

- stand-alone
- keyboard
- computer-based.

Stand-alone

Stand-alone samplers are normally designed to fit into a 19-inch rack mount case. Control and editing functions are carried out using the MIDI protocol, although some samplers also have provision for an external monitor and keyboard to provide improved access to the editing functions. Samplers are often controlled from a master keyboard or synthesizer keyboard, but some samplers are designed to be controlled from the front panel – for adding sound effects or replaying drum sounds, for example.

Keyboard

Keyboard samplers are essentially a stand-alone sampler placed in a larger case and with an added keyboard. Whereas S&S instruments have seen considerable success with this format of instrument, keyboard samplers have been less successful commercially.

Computer-based

Computer-based samplers are normally manufactured in the form of plug-in 'sound cards' – although some take advantage of the audio capabilities of some computers and are then merely software. Some cards are merely converters where the audio storage uses either the computer's own RAM or an additional card which contains dedicated RAM memory. Other cards may provide special-purpose processors to carry out digital signal processing (DSP) functions: these are sometimes referred to as DSP 'farms'. It is also possible to find all of these separate parts on a single card.

The conversion from analogue audio signals to and from digital data is sometimes carried out in a separate box outside the computer in order to optimize the conversion accuracy – the interior of a computer case is not an ideal location for a sensitive conversion system. In some systems, the external box is merely used to provide a convenient way to house all the connection sockets – plug-in cards for computers normally provide only a very small area of panel in which to locate input and output sockets.

Direct-to-disk recording can be thought of as a variation on computer-based sampling, although it has a different set of design goals. Whereas a sampler will normally record into RAM memory, and this sets a time limit on the length of the sample which can be recorded, a direct-to-disk recording unit will store the converted audio data on the hard disk directly, which means that the length of the recorded audio is limited only by the available hard disk size. This process places considerable demands on the computer and the hard disk, and in fact, the number of tracks which can be recorded and/or replayed simultaneously is determined by the computer's processing power and the rate at which data can be transferred to the hard disk storage.

4.4 Convergence of sampling with S&S synthesis

The fundamental differences between an S&S synthesizer and a sampler are often described as being related to the sample memory and the sample processing.

- **Sample memory**: There is a popular misconception that S&S synthesizers have permanent ROM memory whilst samplers have volatile RAM memory. This view ignores the way that both types of instrument have evolved. S&S synthesizers have acquired user sample RAM, whilst CD-ROM drives in samplers virtually relegate them to replay-only status.
- **Sample processing**: S&S synthesizers normally have a restricted set of controls for the replay of samples, but this is usually compensated for by the provision of a sophisticated synthesis section with a resonant filter and VCA. Samplers often concentrate more on the sample replay controls, with multi-sampling, looping, sample stretching and interpolation between one sample and another, although their subsequent processing is often just as capable as many S&S instruments.

The differences are thus less apparent than is often supposed. There is an ongoing convergence of functionality in both instruments. S&S instruments can have user sample RAM memory, and external sampling units can provide samples, although CD-ROMs are more frequently used to provide additional 'off-the-shelf' sounds, much as with a sampler. Samplers now use CD-ROMs to provide rapid access to raw sounds in much the same way that S&S instruments provide sample replay.

Samplers are sometimes used merely as replay devices. This wastes the creative potential of the synthesis sections which can be used to great effect in processing the samples and providing new sounds. There is some evidence of a stigma being associated with samplers because of this 'replay only' reputation, with some people preferring to use S&S instruments with user sample RAM memory instead of a 'sampler'. The convergence between S&S and sampling should soon produce instruments which are so difficult to categorize into either of the two types that the bias against samplers may change.

4.5 Example equipment

Ensoniq Mirage (1985)

The Ensoniq Mirage (Figure 4.5.1) was the first affordable commercial sampler. Although only monophonic and a grainy 8-bits of sample resolution, with 8-note polyphony and a very restricted memory (2 seconds total sample time at 15 kHz), this instrument changed synthesis and ushered in S&S instruments and samplers. In contrast to the basic sample replay that might be expected from a first instrument from a new company, the Mirage has LFO modulation and separate filter/VCA envelopes, with a velocity sensitive keyboard. The user interface is minimalistic, with a 2-digit LED display and yet there are plenty of features. Up to 16 multi-samples can be assigned across the keyboard, the low-pass filters have a resonance control and keyboard tracking, and samples can be looped. There is also a simple sequencer too!

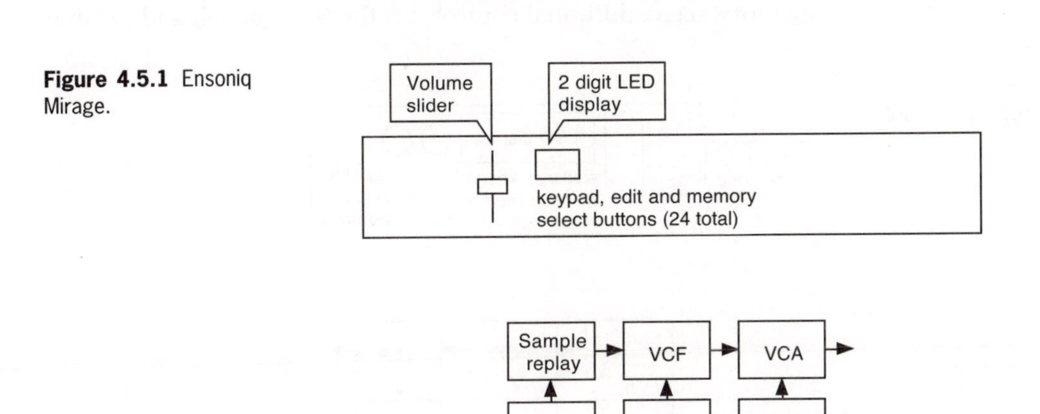

Figure 4.5.1 Ensoniq Mirage.

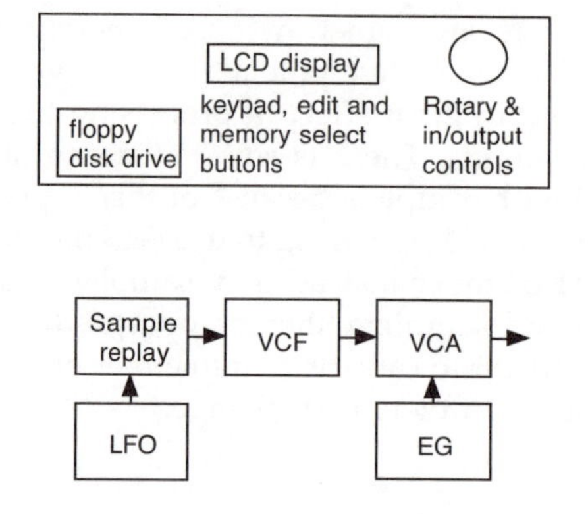

Figure 4.5.2 Akai S900.

Akai S900 (1986)

The Akai S900 (Figure 4.5.2) was probably the first serious rack-mount professional-quality sampler. Although only 12-bit, and 8-note polyphonic, had the facilities (like eight individual outputs) and software to make it almost a 'de facto' standard for sampling for several years, with the floppy disk format used for the samples also acquiring the status of a common exchange medium. The S900 had a maximum sample rate of 40 kHz giving 12 seconds of high quality monophonic sampling. Up to 32 samples could be assigned across the keyboard, and it provided facilities such as velocity switching/crossfading, and crossfade looping. Complete setups of the sampler can be saved onto disk, with 32 of these available at any one time.

Akai S1000 (1988)

The Akai S1000 (Figure 4.5.3) introduced 16-bit stereo sampling with 16-note polyphony, eight separate outputs, increased memory size, additional controls on the front panel, and sample

Figure 4.5.3 Akai S1000.

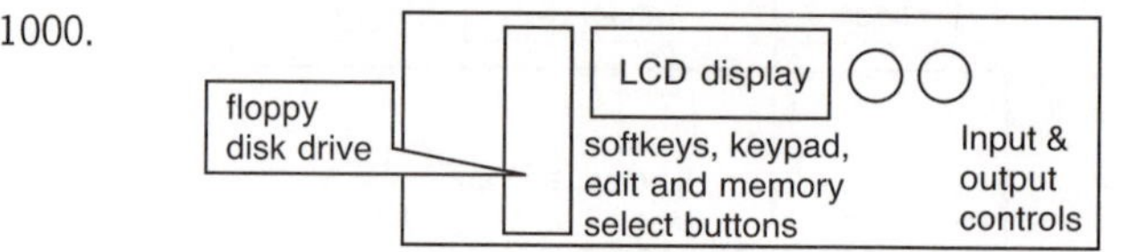

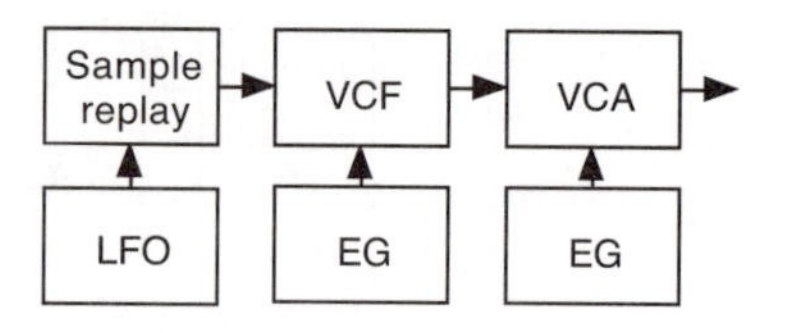

compatibility with the S900, as well as an optional SCSI hard disk interface. Sampling time on the standard model was almost 50 seconds at 22 kHz for monophonic samples. The modifier section looks much like an analogue synthesizer, with LFO modulation, and separate filter VCA envelopes. The sample playback control includes all the crossfading/velocity switching, looping points and forwards/backwards/alternating loop modes (and more) that you would expect on a second generation professional sampler.

4.6 Questions

1. What are the differences between a sampler and an S&S synthesizer?
2. What is the relationship between the speed of a tape and the playback pitch? What happens when the tape is played backwards?
3. What three electronic devices form the basis of a sampler?
4. What do anti-aliasing filters do?
5. What do reconstruction filters do?
6. What criteria need to be considered when looping a sample?
7. What editing functions would you expect to find in a sampler?
8. Outline the convergence of samplers with S&S synthesis.
9. Describe the limitations of analogue sampling techniques using optical and magnetic storage. Then show how these limitations were overcome by the use of digital techniques.
10. What electronic devices use analogue sampling? What audio applications have they been used for?

Time line

Date	Name	Event	Notes
1600s	Gottfried Leibnitz	Developed the mathematical theories of logic and binary numbers	
1642	Blaise Pascal	First mechanical calculator	Addition or subtraction only
1694	Gottfried Leibnitz	Devised a mechanical calculator which could multiply and divide	
1815–1862	George Boole	Father of Boolean algebra, which is used to describe the computations inside a computer	Symbolic logic-based algebra based on 'true' and 'false' values
1833	Charles Babbage	Invented the computer – intended for producing log tables	The electronic calculator eventually made log tables obsolete!

Date	Name	Event	Notes
1943	Colossus	The world's first electronic calculator	Built to crack codes and ciphers
1969	Max Mathews	Publishes *The Technology of Computer Music*	Includes manual for the 'Music V' sound synthesis program
1969	Philips	Digital master oscillator and divider system for organs	
1973	John Chowning	Publishes paper: 'The synthesis of complex audio spectra by means of frequency modulation', the definitive work of FM	FM introduced by Yamaha in the DX series of synthesizers 10 years later
1975	New England Digital	Synclavier is launched. Claimed to be the first 'portable' all-digital synthesizer	Expensive and bulky
1980	Electronic Dream Plant	Spider Sequencer for Wasp Synthesizer. One of the first low-cost digital sequencers	252 note memory, and used the Wasp DIN plug interface
1981	Casio	VL-Tone. Rhythm, drums, chords and monophonic synthesizer in a low-cost 'overgrown calculator'	Electronic music for the masses!
1983	Yamaha	Yamaha DX7 is released. First all-digital synthesizer to enjoy huge commercial success. Based on FM synthesis work of John Chowning	First public test of MIDI is Prophet 600 connected to DX7 at the NAMM show – and it works (partially!)
1984	Kurzweil	Kurzweil 250 provides 2 Megabytes of ROM sample playback	
1985	Yamaha	DX100 (4 operator mini-key) FM synthesizer launched	
1985	Yamaha	DX21 (4 operator full size keyboard) FM synthesizer launched	
1986	Yamaha	DX7II is revised DX7 (a mark II)	Optional floppy disk drive
1986	Yamaha	Electone HX series organ launched	Mixture of FM and AWM (sampling)
1987	Casio	Introduced the Casio CZ-101, probably the first low-cost multi-timbral digital synthesizer	Used phase distortion, a variant of waveshaping
1987	Kawai	K5 digital additive synthesizer launched	Powerful and not overly complex
1987	Roland	Roland D-50 combines sample technology with synthesis in a low-cost mass-produced instrument	S&S synthesis (Sample & Synthesis)
1987	Yamaha	Yamaha DX7II centennial model – second generation DX7, but with extended keyboard (88 notes) and gold plating everywhere	Limited edition
1988	Korg	Korg M1 is launched. Uses digital S&S techniques with a excellent set of ROM sounds	A runaway best seller. Filter has no resonance
1990	Korg	Wavestation is launched. An updated 'Vector' synth, using S&S, wavecycle and wavetable techniques	Powerful and under-rated

Date	Name	Event	Notes
1990	Technos	French-Canadian company Technos announces the Axcel – first resynthesizer	There was no follow up to the announcement
1990	Yamaha	SY77, a digital FM/AWM hybrid synthesizer, mixed FM and sampling technology	Followed in 1991 by the larger and more powerful SY99
1991	Roland	JD-800, a polyphonic digital S&S synthesizer	Notable for its front panel – controls for everything!
1993	E-mu	Morpheus synthesizer module is launched. Uses real-time interpolating filter morphs to change sounds	Sophisticated DSP
1995	Yamaha	VL1, world's first physical modelling instrument is launched	Duophonic, and very expensive

5 Digital synthesis

Digital synthesis of sound is the name given to any method which uses predominantly digital techniques for creating, manipulating and reproducing the sounds. Often the only 'analogue' part of a 'digital' instrument will be the audio signal that is produced by the digital-to-analogue converter chip (DAC) at the output of the instrument.

Most digital synthesis techniques are based very strongly on mathematics – even methods like digital S&S which often attempt to mimic, in software, the analogue filters found in subtractive synthesizers. The precision with which digital synthesizers operate has both good and bad aspects. Repeatability and consistency might seem to be a major advantage over the uncertainty which often occurs in analogue synthesizers, but this precision can also be a disadvantage. For example, whereas FM synthesis in an analogue synthesizer is difficult to control adequately because of the slight non-linearities of the frequency modulation inputs of many oscillators, in a digital synthesizer the precision of the calculations can mean that 'unwanted' effects like the cancellation of harmonics in a spectrum can happen. In an analogue synthesizer, the minor variations in tuning and phase would prevent this from happening – in a digital system, these may need to be artificially introduced.

This illustrates a very important point about digital sound synthesis. The degree of control which is possible is often seen as an advantage. But it also requires a considerable investment of time in order to be able to take advantage of the possibilities offered by the depth of detail which may be required – especially when there are potential problems if one does not fully understand the way that the synthesis works. This is very important in techniques like FOF, where forgetting to set some of the phase parameters can result in major changes to the sound which is produced.

In summary:

- Analogue synthesizers offer the rapid and often intuitive production of sounds, but they have intrinsic non-linearities, distortions and inconsistencies which can contribute to their characteristic 'sound'. If the speed of use and the available sounds are suitable, then the limitations may not matter.
- Digital synthesizers can provide a wider range of techniques, some of which are very powerful at the cost of complexity and difficulty of understanding. But they do not suffer from the built-in imperfections of analogue circuitry, and so these may need to be simulated – which adds to the task of controlling the synthesis and makes them less intuitive. The creative possibilities offered by digital synthesis are obtained at the expense of the detail required in setting up and controlling them.

Digital sounds

It has been said that digital sounds 'clean', whilst analogue sounds 'natural'. As a vague generalization, this is almost acceptable. It is possible to make very crisp, clear timbres using digital technology, but this is by no means the only tone colour which is available. In fact, digital technology often introduces its own distinctive 'dirt', 'grunge' and distortion into the signal.

Two of the commonest artifacts of using digits instead of analogue signals are quantization noise and aliasing.

- **Quantisation noise** is the grainy roughness that is typically found on the decay and release of pianos or reverbs. It happens because of the limited resolution of the numbers that digital systems use to represent audio signals – as the numbers get too small then errors get introduced and this appears as extra noise.
- **Aliasing** is a side effect of the process of sampling. It is caused by a combination of imperfect filtering and 'just good enough' sampling rates. Aliasing sounds like ring modulation, and is often heard as harmonically unrelated frequencies towards the top of the frequency spectrum.

Notice that both of these 'distortions' are due to imperfections in the way that digital works – and as such, are very similar to the limitations that are found in 'real' instruments, or even 'analogue' synthesizers. So the gross distortion which can be produced by overloading a filter in an analogue synthesizer is fundamentally no different to the aliasing in a digital sampler, or to the 'wolf' tones that can be obtained by careful blowing into wind instruments.

The important thing is that applying descriptions like 'natural' or 'clean' to a sound is a very personal and subjective thing. Some very 'natural-sounding' flutes and harps can be entirely synthetic in origin, whilst some 'clean' sounding clarinets might have high levels of distortion. Digital does not have any better claim on 'clean' sounds than any other method, nor does 'analogue' have a special reason to sound 'natural'. For example, there's no way that an analogue resonant filter-based 'Moog bass' sounds like anything in nature, because most real instruments don't have resonances that change in frequency quite to that extent!

5.1 FM

FM is an acronym for frequency modulation, an old idea which although possible on analogue synthesizers, was not really practical. Analogue synthesizer VCOs are subject to frequency drift, variation with temperature, non-linearities, high-frequency mis-tuning and other effects which lead to unrepeatable results when you try to use FM for generating melodic timbres. In fact analogue-based FM is very good for producing a variety of 'non-analogue' sounding special effects: bells, metallic chimes, ceramic sparkles and more. It was not until the advent of digital technology that FM became possible as a way of producing playable sounds.

FM essentially means taking the output of an oscillator and using it to control (modulate) the frequency of another oscillator. If you try this with two VCOs in a modular synthesizer then you are almost guaranteed to get some bell-like timbres at the output of the second, 'modulated' VCO, especially if the only control input to the VCO is the exponential control input (FM should really use the linear control input – often marked FM!). In synthesizers, FM is used as a synonym for 'audio FM', where the two oscillator frequencies are both in approximately the audio frequency range: about 20 Hz to 20 kHz. FM radio uses an audio signal to modulate a very much higher frequency which is then used to carry the audio waveform as radio waves. This

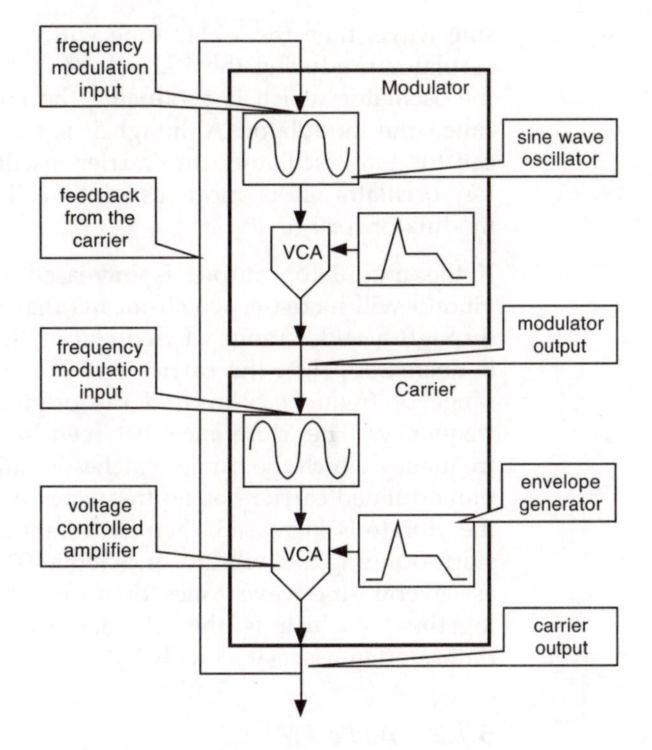

Figure 5.1.1 The terminology of FM is different from analogue subtractive synthesizers, although many of the component parts are the same. In this example, the basic FM is produced by two identical 'modules' which can be thought of as consisting of a sine wave DCO, VCA and envelope generator.

use of the word 'carrier' persists even in audio FM, where radio transmission has nothing to do with the sound that is produced.

FM synthesis is not really like any of the major synthesis techniques described so far – although it was briefly described in the context of a modulation method in Section 4 of Chapter 2. It is not a subtractive or additive method, and it does not fit easily into the 'source and modifier' model either. FM has its roots in mathematics, and is concerned with producing waveforms with complicated spectra from much simpler waveforms by a process which can be likened to multiplying. The simplest waveforms are sine waves, and FM is easiest to understand if sine waves are used for the initial explanations. In fact, unlike analogue synthesizers, where the waveshapes are often the main focus of the user controls, FM is much more concerned with harmonics, partials and the spectrum of the sound.

5.1.1 Vibrato

So, if an oscillator is set to produce a 1 kHz sine wave, and another oscillator is used to change the frequency with a 20 Hz sine wave, then the 1 kHz tone will have a vibrato effect. The oscillator producing the 1 kHz tone is called the carrier, whilst the oscillator which is producing the modulation waveform is called the modulator. Although it is technically only correct to call the two oscillators the 'carrier' oscillator and the 'modulator' oscillator, it is more usual to call them the carrier and modulator for brevity.

If the modulator output is increased then the depth of the vibrato will increase, which means that the carrier is sweeping through a wider range of frequencies. If the modulator output is decreased, then the carrier will be swept through a smaller range of frequencies around the original 1 kHz unmodulated frequency. The difference between the highest and lowest frequency which the carrier reaches is called the deviation: so an unmodulated carrier has no frequency deviation. If the speed of the vibrato is increased, then above about 30 Hz, the result will stop sounding like vibrato, and above 60 Hz it will be perceived as several sine-wave tones of different frequencies all mixed together – which is the 'characteristic 'bell-like', clangorous timbre often associated with FM.

5.1.2 Audio FM

Audio FM replaces the low-frequency modulator with another audio frequency. Suppose that the modulator level is initially zero. The output of the carrier oscillator will thus be a sine wave. As the level of the modulator is increased, then the sine wave will gradually change shape as extra partials appear. Initially two sidebands (partial frequencies either side of the carrier) appear, but as the depth of modulation increases, so does the modulation index, and thus more partials will appear. The timbre becomes brighter as more partials appear, although unlike opening up a low-pass filter, partials appear at higher and lower frequencies. The lower frequencies can alter the perception of what the pitch of the sound is – if a 1 kHz sine wave acquires an additional sine wave (it is a partial, not a harmonic, because it need not be related to the fundamental by an integer frequency ratio) at 500 Hz, then it can sound like a 500 Hz tone with a partial at 1 kHz.

As described in Chapter 2, the output consists of the carrier frequency and sidebands made up from the sum and difference frequencies of the carrier and multiples of the modulator frequency. The number of sidebands depends on the modulation index, and a rough approximation is that there are two more than the modulation index. The modulating frequency is not directly present in the output. The amplitudes of the sideband frequencies are determined by a set of functions called Bessel functions (Chowning and Bristow, 1986). So, as the level of the modulator signal increases, the output gradually changes into a much more complex timbre. The transition is a smooth addition of frequencies much as you would expect with a low-pass filter gradually opening up, but with the added complication of extra frequencies appearing at lower frequencies too.

5.1.3 Bessel functions

Mention of Bessel functions (Chowning and Bristow, 1986) normally means that mathematics takes over and the next few pages should be filled with formulae. FM is often presented as being inaccessible because of its complexity, so here I will attempt to try and describe how FM works in as simple and non-mathematical a way as possible.

We will start by taking the filter analogy a little further. Imagine an additive synthesizer which has individual envelopes for each of a number of frequencies. If we want to simulate a low-pass filter opening up, then we need some envelopes which allow first the low frequencies to appear, then middle frequencies, and

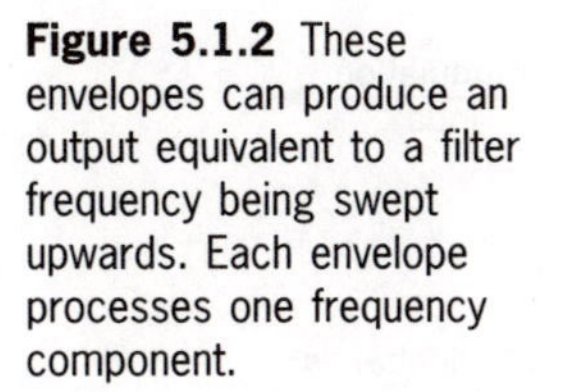

Figure 5.1.2 These envelopes can produce an output equivalent to a filter frequency being swept upwards. Each envelope processes one frequency component.

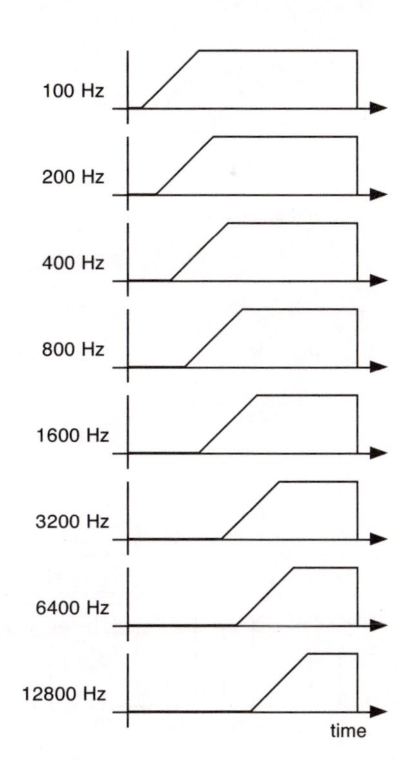

finally the higher frequencies (Figure 5.1.2). The envelopes would look like a series of delayed attack and sustain segments, where the delay in the start of the attack was related to the frequency which was being controlled – the higher the frequency the longer the delay time. Triggering the envelopes would cause the lower frequencies to appear, then the middle, and finally the high frequencies. A similar set of envelopes could be used to produce frequencies which were lower than the fundamental.

The shapes of these envelopes control the harmonic/partial structure of the sound produced by the synthesizer. By changing the shapes of the envelopes, we can change the way that the frequencies will be added as time passes. Actually we don't need to use envelopes: we could use any controller which can map one input to lots of outputs whose behaviour we can control. It just happens that an envelope is one way of using time as the controller. If we used a control voltage and lots of voltage

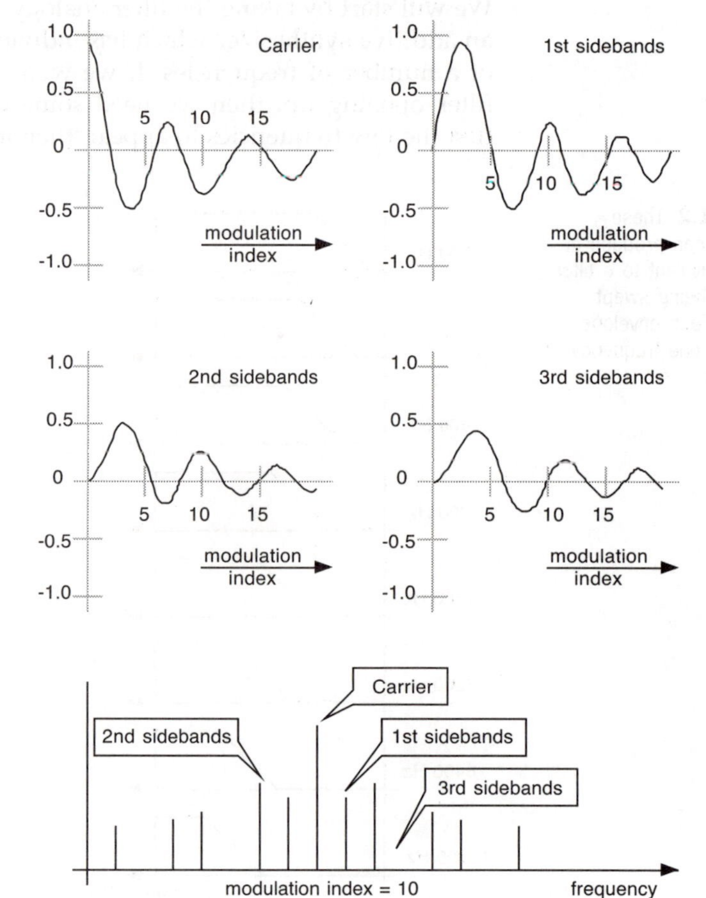

Figure 5.1.3 Bessel functions describe how the amplitude of the sidebands in FM vary with the modulation index. In this example, an FM output is produced using a modulation index of 10. Only the first four Bessel functions are shown.

modifiers, it would be possible to control the frequencies from the additive synthesizer in just the same way, and the same envelope shapes could be used to describe what would happen – the only difference would be that the envelopes are now curves which show how the frequencies change with the input voltage instead of with time.

Bessel functions are the name given to the curves which relate how the frequencies are controlled in FM. Although they are smooth curves instead of the angular envelopes, the principle is exactly the same. In much the same way as the filter envelopes have time delays built in to them, so the Bessel functions vary in a similar way – the further away from the carrier frequency, the higher the value of the control needs to be to have an effect. Instead of time or a control voltage, the control is the modulation index. So as the modulation index increases (when the modulator level increases, or the modulator frequency decreases) then the number of partials increases (Figure 5.1.3). And that is really all there is to Bessel functions: they merely describe how the level of the partials changes with the modulation index. There are one or two complications in reality, and you should look at the references if you need more details. Chowning and Bristow, (1986) is particularly recommended.

The only other thing which needs considering for FM is how the frequencies are controlled. In FM the spacing between the partials is related to the carrier and modulator frequencies, but there are really only three basic relationships:

- **Integer**: Integer relationships between the carrier frequency and the modulator frequency produce timbres which have harmonic structures which are similar to those of the square, sawtooth and pulse waveforms – harmonics at multiples of the fundamental. The only complication is that the fundamental is not always at the carrier frequency because of the extra frequencies which can appear below the carrier.
- **Slightly detuned from integer**: Slightly detuned carrier and modulator frequencies produce the same sort of 'multiples of the carrier' harmonic structures, but with all the harmonics detuned from each other too. This can produce complex beating effects, although the amount of detuning needs to be carefully controlled to avoid too rapid beating effects.
- **Non-integer**: Non-integer relationships between the carrier and the modulator frequencies produce the bell-like, clangorous timbres for which FM is famous. If either the carrier or modulator is fixed in frequency (i.e. it does not track the keyboard pitch), then the relationship will change with the pitch of the note being produced.

In all of these cases, the basic timbre produced is set by the relationship between the carrier and modulator frequencies, whilst the number of harmonics and partials which are produced is controlled by the modulation index, using the values in the Bessel functions. There are some additional complexities caused when the modulation index is so large that the spreading out of the partials causes some frequencies to go below the zero frequency point and get 'reflected' back – which can cause some additional cancellation effects.

That's really all there is to FM: you choose the timbre, and then control it – usually dynamically. FM may be different from subtractive or additive synthesis, but the controls and the way that they work are relatively straightforward: once you understand what is happening. Almost all of the other functions, like LFOs, portamento, envelope shapes, and effects, should be very similar to the same functions in other synthesizers.

FM is normally produced using oscillators which are made available as general purpose building blocks called operators. These consist of a sine wave oscillator, and envelope generator and a VCA – all in digital form, of course. The oscillator is a variable speed playback of a sine wavetable, whilst the VCA is a multiplier connected to the envelope generator.

On a larger scale, FM may use more than one pair of operators (carrier and modulator), several modulators onto one carrier, or even a stack of operators all modulating each other. It is also possible to take the output of a carrier operator and feed it back to the input of a modulator, which can be used to produce noise-like timbres, although it is still the modulation index and the frequency relationships which determine much of the timbre. Learning how to make the most of FM involves analysing the FM sounds produced by other people, and programming sounds yourself – but the model described here should provide the basic concept of how FM works.

FM is good for producing sounds with complicated time evolutions and detailed harmonic/partial structures, but it can be difficult to programme: the explanation above has been simplified, and there can be quite a lot of parameters to cope with. It is also possible to produce FM with non-sine wave oscillators, although all that happens is that each sine wave which is present in the waveform, acts as its own FM system, and so you get lots of FM happening in parallel, which can lead to very noise-like timbres because of the large numbers of frequencies which are produced. FM is especially suited for 'metallic' sounds like guitars, electric pianos and harpsichords.

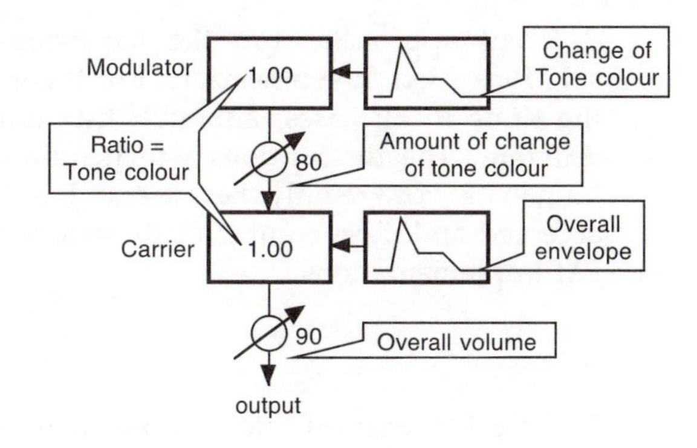

Figure 5.1.4 This overview of a simple FM synthesis system shows how the individual component parts contribute to the final sound output.

FM really only requires the following parameters to define a timbre:

- carrier frequency
- modulator frequency
- modulator level.

But most FM sounds change the modulator level dynamically by using the modulator envelope, and the carrier also has an envelope, but even then the number of parameters required to specify a given sound is less than 20. In comparison to subtractive or additive synthesis, this is a much smaller number of parameters to deal with. For this reason, FM has been investigated as the synthesis part of an analysis-synthesis resynthesis system, but there are problems in extracting the FM parameters from sounds. In particular, it is not easy to take a specified waveshape or spectrum and calculate the required FM parameters, especially if there are any partials or non-harmonic frequency components. See Section 6 of this chapter for more information on resynthesis.

Having a small number of important parameters also enables FM to be a very powerful synthesis method when using real-time control. By changing the carrier or modulator frequency, or the modulator depth with specialized MIDI commands (or front panel controls), FM can be used to produce sounds which can change rapidly and radically. On an analogue subtractive synthesizer the only comparable parameter change is the filter cut-off – and this is much more restricted in the timbral changes which it can produce.

5.1.4 Realization

The actual details of the way that FM is realized differ on different platforms. Computer-based software will probably use a different approach from the hardware-oriented custom DSP solutions used in synthesizers. But the basic elements are much the same in all cases, although the terminology may be very different. The descriptions which follow are mostly based on Yamaha's FM, mainly because it has been the most widely accepted and most commercially successful of any of the digital FM implementations.

Oscillators

Initially FM was produced using only sine waves. The mathematics behind FM are easiest to understand if sine waves are used, and early FM work was at an academic level, where both the understanding of the sound production process and the aesthetics of the resulting sounds are important. The first commercial FM synthesizers also used just sine waves, probably for reasons of price: the Yamaha DX7 was introduced at a time when digital consumer electronics was virtually unknown – the CD player was not introduced for another year. The high ratio between the features and price of the DX7 was partially due to the use of digital technology, but also to a careful minimization of functionality: after all, Yamaha were testing the market with a very different type of synthesizer. The lack of front panel control knobs shows that they were prepared to take radical design decisions in both the synthesis and user interface areas.

Implementing a digital FM sine wave generator has been covered in Section 3.3 on DCOs. With only one waveform, the size of ROM memory required can be quite small, especially if the symmetry of the sine wave is used to reduce the storage requirement – you can produce a complete sine wave cycle from just one quarter of a cycle of sine wave waveform points. The Yamaha design multiplexes the oscillator: it is used to provide the waveforms for all of the oscillators, with storage of the successive outputs used to give the equivalent of four or six separate oscillators. Further multiplexing is used to provide the DX7's 16-note polyphony, which was about twice the normally expected polyphony of polyphonic synthesizers in 1983.

Later, more advanced FM synthesizers used more complex waveforms. But this was often only achieved by deriving additional waveforms from the same sine wave ROM memory, by changing the way that the quarter cycle is re-assembled to form a waveshape. Quite minor changes to the waveshape can have significant effects on the spectrum, and using waveforms which contain additional frequencies can produce FM sounds which are very rich in harmonic and partial content – even to the extreme of becoming noise-like. The last generation of commercial FM synthesizers from Yamaha (the SY77 and SY99) also added the ability to use samples as part of the FM synthesis, but this did not prove to be very popular with users, and subsequent models in the series concentrated on sample replay technology rather than FM.

Most FM oscillators can be used in two modes. The usual mode is to allow the oscillator to track the keyboard pitch, although this need not be the normal keyboard scaling. The second mode is usually called 'fixed frequency', and here the oscillator frequency does not change. Fixed oscillators can be used in several ways. At low frequencies they can be used as carriers or modulators to produce vibrato-type cyclic timbral change effects, whilst at higher frequencies they can be used to partially emulate a very resonant system. Fixed frequencies of a few hundred hertz are often used to produce vocal sounds, since the resulting sound has many of the qualities of the formants which determine human vocal sounds. A fixed oscillator within an FM algorithm produces an output spectrum with frequency components which are related to the fixed oscillator frequency, or harmonics of it – and this can sound somewhat like a resonant tube.

In each of these cases – sine wave, sine-derived waveforms and samples – the technology used to produce the FM waveform is very similar to the advanced DCO designs described in Section 3.3. The output of a period of time (not necessarily a single cycle) for each oscillator is stored, and then used as the basis for the modulation of the next oscillator – ending with the carrier oscillator. The high precision of the sine wave, the frequency resolution, and the linearity of the frequency modulation, all enable FM synthesis to be achieved in a precise, repeatable and controllable way – a big contrast to producing FM using analogue technology.

Envelopes and VCA

Although DCOs were found in hybrid synthesizers, FM required digital control over the amplitude of the oscillator outputs, and for this they used the multi-segment rate/level type of envelope. Rate/level envelopes provide comprehensive control over the shape of an envelope by using function generator controls to set the characteristics of each segment. As the name implies, two parameters are used to control each segment: a rate and a level. The rate specifies how long the segment lasts, whilst the level sets the final level which that segment reaches. The initial level

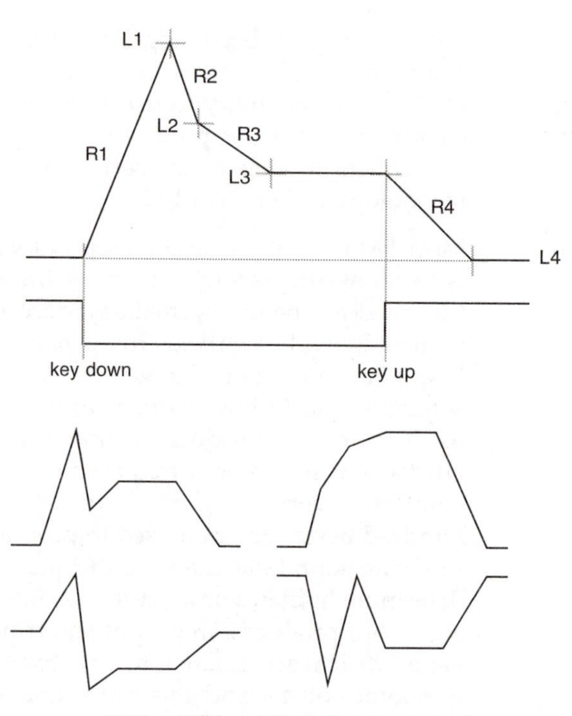

Figure 5.1.5 The envelope generators used by Yamaha in their FM synthesizers provide great flexibility because of their structure. The four levels can be anywhere in the permitted range, which allows a wide variety of envelope shapes – including inverted envelopes and pseudo-exponential attack segments.

is normally the same as the final level, although some later instruments do not have this restriction.

Yamaha had identified that conventional ADSR-type envelopes were not suitable for envelopes which had complex attack stages – especially where the start of the sound did not rise at a constant rate. The multi-segment envelopes which they used in the DX7 had three segments to cover the 'key on' part of the envelope, plus one segment for the 'key off' or 'release' portion of the envelope. Because the final level of the third segment is held whilst the key is held down, it effectively produces a separate 'sustain' segment, which is the fixed level which the attack segments end at. This produces a categorization of the segments according to their function within the envelope. The first three segments are used to control the 'attack' portion of the envelope, the level of the last attack segment sets the sustain level, whilst the final segment controls the release behaviour.

The envelope used in DX-series Yamaha 6-operator FM synthesizers is a five-segment rate/level. There are three 'attack' segments, one 'sustain' segment, and one 'release' segment. EG forced damping is found in the Mark II DX7, and forces the

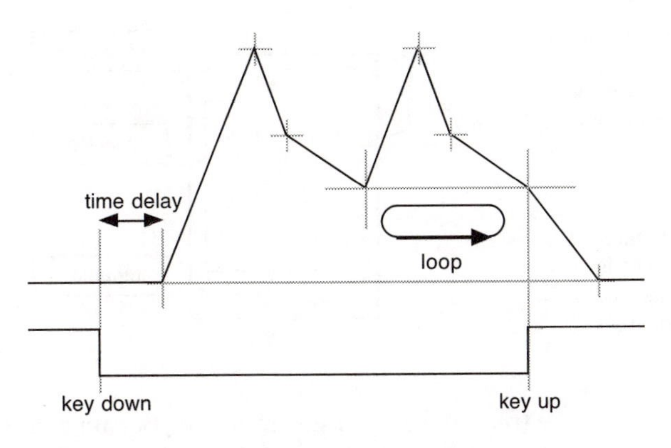

Figure 5.1.6 Loopable envelopes allow previous segments to be continuously looped whilst the envelope is in the sustain segment. In this example, the first three segments are looped. The use of delayed attack segments enables the production of echo-like and arpeggio effects.

envelope to restart when a note is reassigned because of note stealing at the limits of the polyphony.

SY series Yamaha 6-operator FM synthesizers use eight-segment envelopes which add a separate initial level, delay time, two release segments and the ability to loop the envelope whilst the key is held down. Also, the final level is not necessarily the same as the first level. The two release segments enable additional control over the end of the note, whilst the delay time is used for special effects like arpeggiated operators or allocating operators to specific parts of the final sound – using one set of operators to generate the initial portion of a sound, whilst delayed envelopes produce the remainder of the sound. This effectively increases the apparent number of envelope segments at the expense of using operators for only part of the duration of the sound. The looping enables the sustain segment to be less static: simpler envelopes just reach a sustain level and stay there as long as the key is held down.

Low cost FM implementations with four operators have simplified ADBDRR envelopes which had only six controls: five rates and two levels (break points for the decay and release segments).

Not all multi-segment envelope generators use the word 'rate'. Time and slope are sometimes used as synonyms. There is also no standardization on how the parameters relate to the duration of the segment: some manufacturers use small numbers to mean short times, whilst the converse is used by others.

The VCA in FM is almost always treated as part of the envelope generator.

Operators

The combination of an oscillator, envelope and VCA is such a fundamental building block of FM synthesis that it is often

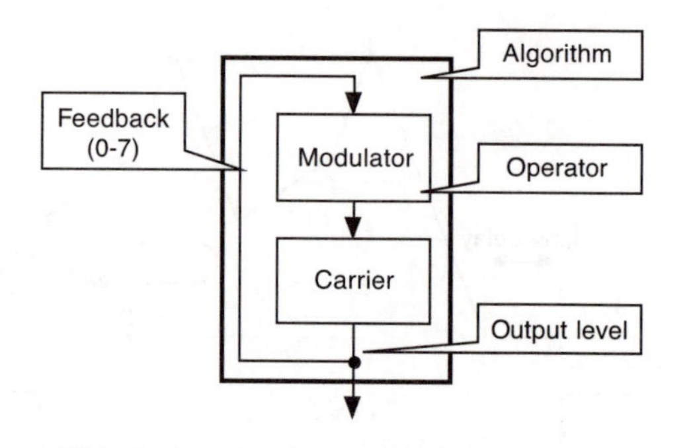

Figure 5.1.7 A Yamaha FM operator consists of a sine wave DCO, VCA and envelope generator. Operators can be connected together in arrangements called algorithms, where the description of the operator as a carrier or a modulator is determined only by its position in the algorithm.

treated as a single module. Because the first major commercial success of utilizing FM in a digital synthesizer was from Yamaha in the early 1980s, the terminology that they used has become widely adopted. Yamaha used the word 'operator' for the block formed by an oscillator, envelope and VCA.

The initial FM prototype instruments were the Yamaha GS1 with eight operators, and the GS2 with four operators, whilst the initial DX synthesizers had six operators. (The DX9 had two operators deliberately disabled to give it reduced functionality.) Lower cost FM implementations followed using four operators with restricted frequency control and limited internal calculation precision and the chips for these were made available to other manufacturers – these became a 'de facto' standard for the basic implementation of a PC sound-card. Prototype FM synthesizers like the V80 were produced by Yamaha with eight operators, although these never progressed beyond the development laboratory. Some of the Yamaha HX-series organs were released with eight operators, but these did not have the depth of user-programmability of the synthesizer products. Some FM implementations use multiple 'pairs' of operators, which do not provide the same flexibility as being able to arbitrarily connect more than six operators together.

Algorithms

Yamaha use the word 'algorithm' for the arrangement and interconnection of operators. Although there are many ways of arranging the topology of four or six operators, there are only a few important types:

- additive
- pairs
- stacks
- multiple carriers
- multiple modulators

- feedback
- combinations

Additive
Although not actually FM, six parallel operators can be used as a simple additive synthesizer – producing six frequencies. Unlike many additive synthesizers, the six frequencies need not be harmonically related, and so slightly detuned oscillators can be used to provide chorused 'additive' sounds. Each operator provides a single frequency component, or partial, with the envelope generator controlling just that frequency.

Pairs
The simplest FM algorithm (apart from a single operator, which can only produce sine waves, of course) is a pair of operators: one carrier which is modulated by one modulator. The carrier envelope generator and level control give control over the overall volume of the sound which is produced, whilst the level control and the envelope of the modulator control the modulation index of the FM. The timbral controls are thus the two operator frequencies, plus the modulator envelope and level controls.

Stacks
By taking a second modulator, and connecting this so that it modulates the modulator in an FM pair, then a stack of three operators can be produced. Additional modulators can be added, although a stack of four operators is normally the most that is required. Since the pair formed by the two modulators produces an FM sound, the carrier which is modulated by this sound (and not by the sine wave which would be produced by a single modulator) is much more complex because each frequency in the modulating signal creates FM with the carrier operator. Stacks are often used for pad sounds, where lots of slightly detuned harmonics and partials are used for producing rich, chorused sounds.

Multiple carriers
By connecting one modulator to more than one carrier, then the same modulator can be used to control two carriers. By having different frequencies for the two carriers (or different envelopes), then the output is two related but separate FM sounds. If the modulators have slightly detuned frequencies, then two similar but detuned sets of harmonics and partials are produced.

Multiple modulators
If several modulators are connected to one carrier, then each modulator can be used to produce part of the final sound, which

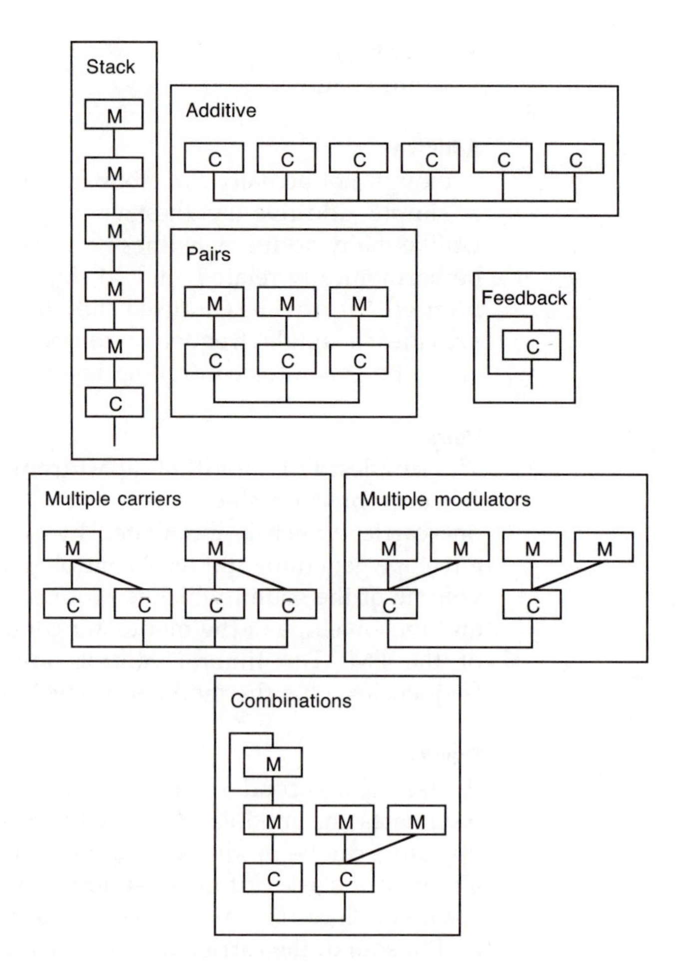

Figure 5.1.8 FM algorithms summary. In these diagrams, C indicates a carrier operator, whilst M indicates a modulator operator. There are six basic arrangements of operators, plus a seventh consisting of combinations of parts of these.

can simplify the development of sounds. Having only one carrier operator means that controlling the output envelope is easier, but it also restricts the timbral possibilities because there is less flexibility in choosing the ratio between the carrier and modulator frequencies (Figure 5.1.8).

Feedback

By connecting the output of an operator back to its frequency control input, the resulting feedback signal affects the output signal of the operator. In the simplest case, a single operator with a feedback loop will produce additional frequencies with a large amount of feedback, and this can sound similar to a sawtooth or pulse type of waveform. Feedback around several operators can be used to produce very complex sounds. If too much feedback is applied, then noise-like sounds can be produced.

Feedback has always been one of the more interesting and less well-understood aspects of FM. The basic idea is that you take the output of the operator and connect it to the frequency control input (in some algorithms this is the same operator, in others you get a loop of two or three). On an SY-series FM synthesizer you can patch several operators together and apply feedback between them. The 'feedback level' is a control over the level, and so controls the modulation index. Zero is no feedback, whilst 7 is an index of about 13 on a DX7 or SY. A modulation index of 13 is going to produce quite a lot of frequency deviation, and so the original sine wave is deviated well away from its basic frequency, but at a rate which is tied to itself. This produces lots of extra harmonics and perhaps even partials (the spectrum for a single operator on a DX7 with a feedback value of 7 has 23 harmonics) and a very contorted waveform. In fact, with a feedback value of greater than 5, the underlying precision of the FM synthesis implementation used by Yamaha begins to become significant and the output begins to get noise-like: although the operator output level also affects the feedback, since the two level controls are in series! Below 5, the sound produced is merely richer in harmonics and partials.

Although the sounds produced by the feedback are described as 'noise-like', this does not usually mean that they are like the 'white' or 'pink' noise found in analogue synthesizers. With two operators, things rapidly get out of control once you start connecting a harmonic-rich waveform from a feedbacked operator as the modulator of another operator. Aliasing and the finite precision of the FM synthesis 'engine' combine to produce a plethora of noise-like sounds, with not-so-flat spectra and lots of harmonics and partials – especially if you use non-integer ratios for the carrier and mod frequencies. Careful use of feedback level and operator output levels can keep things non-noise-like, and still in the realm of complex but interesting timbres. Because of the effects of aliasing, and the way that FM folds harmonics or partials when the modulation index is large, the resulting spectra may be rich in frequency content, but they are rarely flat: the noise is not white, nor is it really coloured – various shades of grey, perhaps!

In the SY synths, these problems with producing 'white' noise were solved by providing a noise generator. This produces white noise, and this side-steps any need to use feedback to try and get a flat noise spectrum. Feedback noise sounds tend to be slightly too structured or grainy to fool the ear, whereas a simple maximal-length pseudo-random sequence is probably used by Yamaha's noise generator to provide white noise on tap. Using

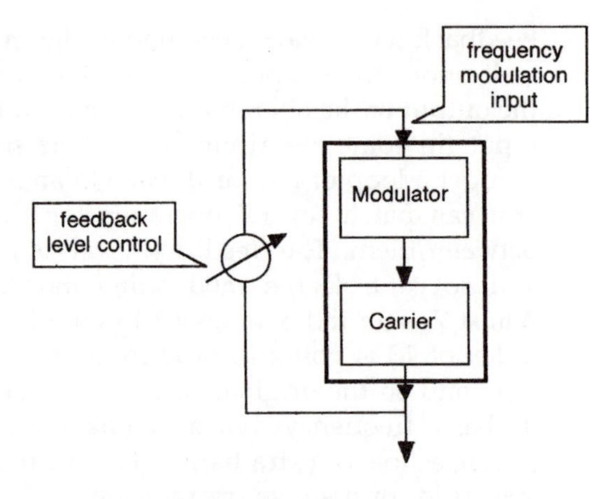

Figure 5.1.9 FM feedback. By feeding back the output of an FM algorithm to the frequency modulation input, it is possible to generate noise-like outputs.

feedback creatively with all the flexibility offered by the SY synths is worthy of further explanation, and I have only really scratched the surface here. There is some basic information on feedback in John Chowning's and Dave Bristow book: *FM Theory & Applications for Musicians*, pages 133–136 (1986).

Because the output of a 'fedback' operator is a spectrum relatively full of harmonics and partials, changing the carrier or modulator frequencies merely changes some of the harmonic and partial amplitudes and the aliasing components. The only audible effect is often a change in the timbre or 'colour' of the noise. Only with low mod indexes and low feedback values will the carrier and mod frequencies make any significant difference.

Combinations

Most FM sounds are made up from combinations of the simple algorithms. Two parallel stacks of three operators are often used because they enable two separate sounds to be combined, whilst each stack of three operators is a versatile FM sound source. Multiple modulators can produce complex sounds where each modulator contributes a distinct element to the sound, and they can each be controlled separately. The development process for FM sounds tends to be iterative, with operators being turned on and off to determine their effect on the sound each time their parameters are changed. This technique is especially important where groups of operators are used to provide different parts of the sound – unlike many methods of synthesis, FM allows the programmer to investigate the effect of minor changes both in isolation and in context.

DX-series FM synthesizers provide fixed algorithms where the topology can be selected from a number of presets. The presets provided typically include all of the possible arrangements of operators, and many of the additional possibilities provided by adding one feedback loop. SY-series FM synthesizers provide user-control of the interconnections and multiple feedback connections, as well as preset topologies. Choosing a specific algorithm is largely a matter of experience. But in many cases, starting from a simple pair of parallel stacks is a good idea, because extra modulators (or carriers) can then be added as required. Familiarity with the timbral possibilities of a simple pair or stack of operators can be very useful in helping to produce FM sounds. Examining pre-programmed sounds can also help to reveal some useful techniques.

5.1.5 History

John M. Chowning's paper The Synthesis of Complex Audio Spectra by Means of Frequency Modulation, in the *Journal of the Audio Engineering Society* in September 1973, was the first serious description of the practical use of digital technology to implement audio FM as a way of synthesizing timbres. This is very much a 'landmark' in digital synthesis – unlike additive synthesis, where the large number of required parameters made a digital realization unwieldy, FM showed that digital synthesis could be powerful and yet require only a relatively small number of controls.

5.1.6 Implementations

There are three strands of FM development from Yamaha: four, six and eight operator. Four-operator FM tends to be used in the lower-cost and computer-oriented areas, whilst six- and eight-operator FM is used in 'professional' instruments. Yamaha's use of FM spans the period from 1982 to 1992.

Table 5.1.1
FM implementations

	1982	1983	1984	1985	1986	1987	1988	1989	1990	1991	1992
	Sine waves										
					Sine variants						
									Samples		
4-operator	GS2	DX9	CX5M	DX21	DX100 FB01	TX81z 707	DX11	V50 TQ5	SY22 TG33	EVS1	SY35
6-operator		DX7	DX1 TX7	DX5 TX816		DX7II DX7S			SY77	SY99	
8-operator	GS1					HX...		V80			

Korg's DS-8 and 707 synthesizers use FM technology as a result of a temporary pooling of research facilities in the mid-1980s. Many personal computer soundcards use a Yamaha chip set to produce FM sounds.

Although Yamaha had acquired the patent rights to the commercial exploitation of FM in the early 1980s, there were several variants on FM which differ enough to be usable without actually infringing the patent. For example, phase modulation changes the phase of the carrier instead of the frequency, which produces very similar sounds (and spectra) to FM.

Casio's CZ series of synthesizers used waveshaping, but their later VZ series of synthesizers used an FM-like phase modulation method, calling it phase distortion (PD). Eight operators were available, with eight-stage envelopes.

Peavey has also used the term 'phase distortion' to describe synthesizers which appear to be S&S instruments, and not a variant on FM.

FM produced on computers using software initially offered enhanced sophistication at the price of non-realtime operation, although DSP technology now makes FM more accessible and immediate.

5.2 Waveshaping

Waveshaping is a way of introducing controlled amounts of distortion onto a waveform. This differs from the 'fuzz box' type of distortion which is used by guitarists, because it is used on the 'monophonic' outputs of the oscillators, and so it merely changes the shape of the waveform without adding in all the intermodulation distortion which happens when more than one note is passed through a waveshaper or fuzz box.

Waveshapers are non-linear amplifiers. This means that they provide control over the way that the amplifier processes incoming signals to produce an output. For an amplifier with a fixed gain of two, you expect to get an output which is twice the input. If a graph is plotted of input against output for a linear amplifier, it would be a straight line through the origin (zero) of the graph. This line is called the 'transfer function' of the amplifier, it shows the way that the input is 'transferred' to the output. In fact, the straightness of this line can be used as a measure of the quality of an amplifier, since a perfect amplifier would have a perfectly straight line transfer function. If the amplifier did not have a gain of two for high levels of input

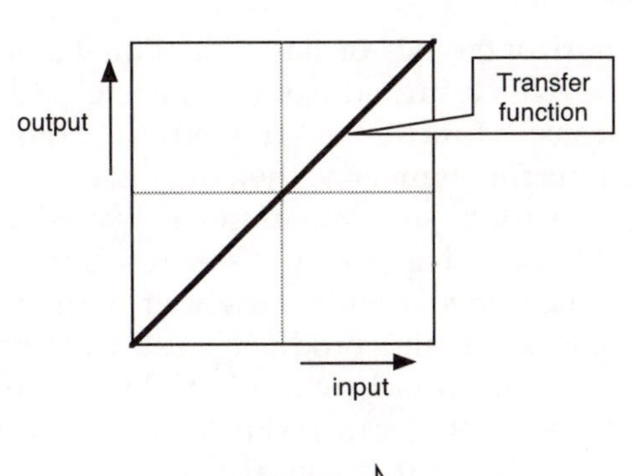

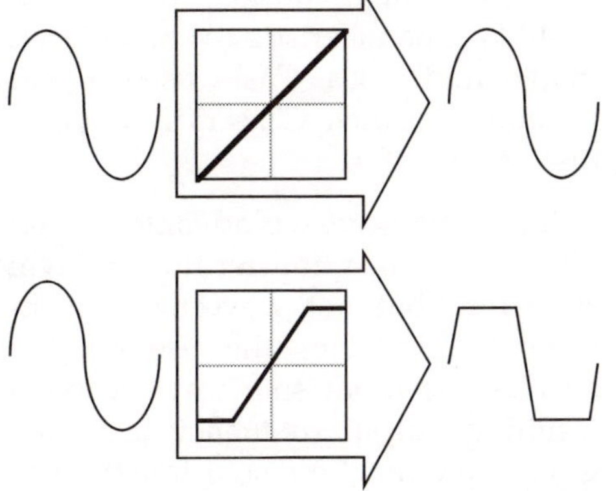

Figure 5.2.1 A transfer function is a graph which relates an input to an output. In this example, a straight line transfer function allows a sine wave to pass unconverted, whilst a transfer function which has a steeper slope and two flat zones converts a sine wave into a trapezoidal waveform.

signal, then the transfer function graph would be curved at high input levels, which means that an audio waveform which is passed through the amplifier will change shape. Changing the shape of the transfer function changes the shape of the waveform.

It is the convention that transfer function graphs always have the input plotted horizontally, and the output vertically. The scaling is also arranged so that the input and output ranges are from –1 to 1, with the zero point of both axes being in the centre of the graph. The input sine wave moves completely across the horizontal axis once per cycle: from a value of 0 to +1, then back through the zero position to –1, and then back to zero again. The output waveform is dependent on the transfer function, although the maximum and minimum outputs are normally +1 and –1 respectively.

Although this sounds like an easy way to produce extra waveshapes, it actually does rather more than that. Distorting the shape of a waveform changes the harmonic content of the waveform – in fact, in most cases, it adds harmonics rather than subtracts them. If the transfer function is symmetrical about the horizontal axis or has a rotational symmetry, then the harmonics which are added will be the odd harmonics, whilst if the transfer function is symmetrical about the vertical axis or shows a mirror symmetry, then only even harmonics will be produced. So with a sine wave and a waveshaper, it is possible to use different shapes of transfer functions curve to produce outputs which have a wide variety of harmonic contents. Now using a sine wave and producing extra harmonics from it sounds like FM, and in fact, with the right type of transfer function curve, waveshaping can produce sounds which are very FM-like in character. But it can also produce sounds which do not have 'FM-like' characteristics. When FM was at its peak of popularity in the mid-1980s, Casio used a waveshaping-based synthesis technique in their CZ-series of synthesizers, but called it phase distortion.

Using a sine wave has advantages and disadvantages. It is possible to calculate a transfer function which will produce any given spectrum (but not waveshape) from a sine wave input. The technique involves the use of Chebyshev polynomials and enables the input sine wave to be changed in frequency. The resulting output frequency is multiplied by the order of the Chebyshev function: so a fourth order function would produce an output sine wave which is four times the input frequency. By adding together several Chebyshev functions it is possible to produce a composite transfer function which will then produce any required spectrum. The calculations of the relevant Chebyshev polynomials are simplified if the input waveform is a sine wave. Because the sine wave shape has two different times when the same value occurs, then some waveshapes cannot be produced, but this restriction is often not a problem, since the harmonic content is normally more important for a specific timbre.

The input waveform to a waveshaper need not be a sine wave. If a sawtooth wave is used, then the waveshaper is little more than a look-up table for output values, and so resembles a wavetable oscillator. The positive and negative half-cycles of the sawtooth wave just map onto images of the output waveform, and so the two half-cycles can be different. Effectively, the two halves can be thought of as two separate transfer functions, although they normally share a common point at the origin. But for a sine wave, the symmetrical nature of the waveshape means that there is more redundancy, which in turn means that there is scope for more independence of transfer functions. A sine wave can only be converted into other waveforms of a particular class of shapes by using a single non-linear transfer function – basically only those where the first quarter of the cycle is the same as the next quarter cycle, but where the second cycle is time-reversed. As the sine wave input moves up and back down the transfer function horizontal axis, the symmetry is inevitable. The same applies to the third and fourth quarters of the cycle.

But by providing different transfer functions for each of these quarter cycles, then the waveshaper can be used to convert a sine wave into waveforms which do not have this first and second quarter cycle mirroring. This means that a single transfer function graph can have two separate halves for each half cycle of the sine wave, and the symmetry produces waveforms which have a large content of even harmonics. In contrast, if there are two separate transfer functions, with each quarter cycle having its own graph, then any waveshape can be produced. This type of quarter cycle waveshaping is used in the second generation of Yamaha FM synthesizers to produce additional waveforms from a wavetable ROM containing just a single high-precision sine wave.

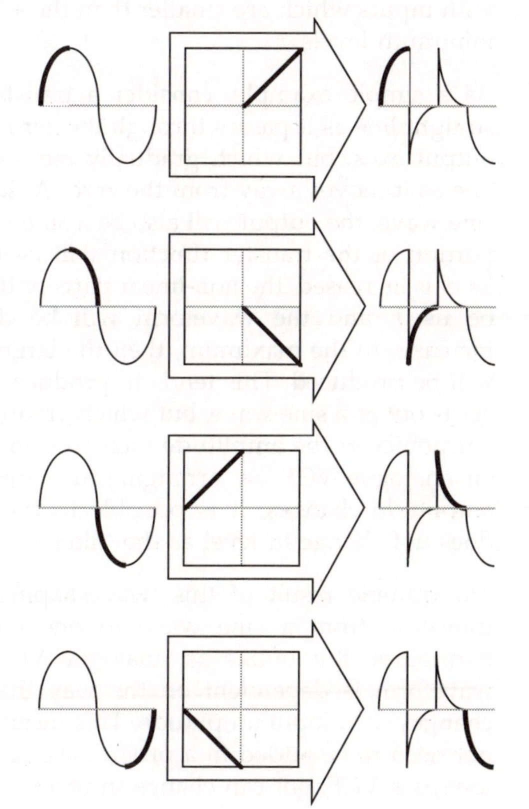

Figure 5.2.2 By using separate transfer functions for each half or quarter cycle of a waveform, it is possible to produce almost any required output waveform from an input sine wave. In this example, each quarter cycle of the input sine wave has its own transfer function. The output waveshape is the concatenation of the four output quarter cycles.

Although waveshaping can be used as a general purpose tool for changing the shape of a waveform, it can be arranged so that the audible behaviour of the waveshaper is similar to the VCF found in an analogue synthesizer. Using digital technology to emulate familiar 'analogue' characteristics is a continuing theme of most digital synthesis methods. In the case of an analogue low-pass filter, harmonics are successively added as the filter cut-off frequency is increased – so the output waveform is initially a sine wave at the fundamental frequency. As harmonics are added the shape of the output waveform will change until with the filter fully 'open', then all frequencies will pass through and the output waveform should have the same shape (and frequency spectrum) as the input.

For a basic waveshaper implementation, this 'filter emulation' behaviour would seem to imply that the transfer function is changing dynamically – which would suggest that a great deal of complex processing is being carried out. However, by designing the transfer functions carefully, and by ensuring that the transfer function curve passes through the origin (zero) of the graph, it is possible to produce waveshapers which can be used with inputs which are smaller than the +1 and –1 maximum and minimum levels.

As a simple example, consider a transfer function which is a straight line as it passes through the zero points of the input and output axes, but which gradually curves away from a straight line as it moves away from the zero. At low amplitudes of input sine wave, the output will also be a sine wave, because the linear portion of the transfer function will be used. But as the input level is increased, the non-linear parts of the transfer function will be used, and the waveform will be distorted. As the level increases to the maximum, then the largest waveform distortion will be produced. This tends to produce an output signal which starts out as a sine wave, but which gradually acquires additional harmonics as the amplitude increases, in much the same way as an analogue VCF. By arranging an amplifier to correct for the amplitude changes, it is possible to produce an output which does not change in level as the 'filtering' action takes place.

The audible result of this 'waveshaping' process is a smooth transition from a sine wave to one containing a number of harmonics. But unlike an analogue VCF, the evolution of the waveform is dependent on the way that the transfer function changes with input amplitude. This means that the harmonics do not need to be added in a progressive sequence comparable to a low-pass VCF, but can change in other ways which can be more

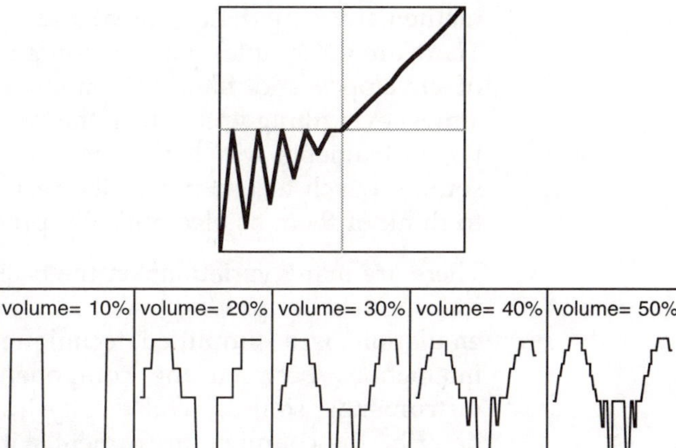

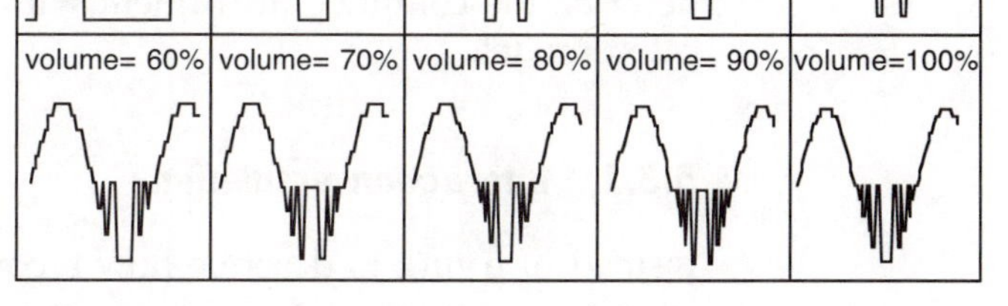

Figure 5.2.3 Dynamic waveshaping alters the input level and then scales the output to compensate. In this example a sine wave is passed through an asymmetric transfer function which is linear for positive inputs but a complex function for negative inputs. The outputs for different levels are shown – it can be seen that the output waveform changes as the input level is increased in much the same way as opening a VCF does on an analogue subtractive synthesizer. (The limitations of the computer simulation which was used to produce these diagrams is the cause of the poor quality of the waveforms shown in this diagram. In a real system, the outputs would not be stepped and distorted in this way, and would have a sine wave upper half and a resonant 'ringing' type lower half.)

interesting to the ear. These complex changes of harmonic content are also found in FM, although the evolution of FM waveforms is fixed by the Bessel functions. For a waveshaper-based synthesizer the transfer function is not fixed and so can produce more sophisticated and varied harmonic changes – at the price of an increased need for mathematical understanding on the part of the designer of the transfer function. Unlike FM, the additional frequencies which are produced by waveshaping are always harmonically related to the input frequency since the waveshaping is based around the shape of one cycle of the waveform.

Some manufacturers have used waveshaping in a much more limited sense. For example, the Korg 01 series synthesizers, although S&S instruments, do implement waveshaping, but this is a very limited form of single non-linear transfer-function waveshaping. It is used to process the outputs of the oscillators and is really limited to just adding in a few extra harmonics to the raw samples. Casio-style dynamic waveshaping is a much more powerful technique – if Korg had moved the waveshaper to after the VCF or VCA, or made the transfer curve controllable, then the possibilities for timbral change would have been much greater.

5.3 Physical modelling

Whereas other digital methods of sound synthesis tend to try and emulate the terminology of functions of analogue synthesis, mathematical modelling breaks away from these conventions. There are no samples, no function generators and much less use of envelopes and filtering – and yet despite throwing away almost everything with which the synthesizer user may be familiar, instruments which use modelling techniques can produce sounds which feel so much like real instruments that it is hard to think of them as electronically produced.

There are many variations on the basic idea of using mathematical models to produce sounds. In this section. 'Interaction emulation' is a simplified technique that concentrates on the interactions between the component parts that produce an instrument's sound, whilst 'physical modelling' attempts to describe the complete instrument with a complex and sophisticated model.

5.3.1 Interaction emulation

Instead of trying to describe how a complete instrument works in terms of equations, interaction emulation looks for a way that the important elements can be encapsulated in a form which provides control, but is easy to use. It turns out that there is a way – and it comes from research into speech. When you speak, your vocal chords are vibrated by the air which rushes past them, and this raw sound is then modified by the complex set of tubes and spaces formed by your throat, nose, mouth, teeth lips and tongue. A physical model of this would need to consider the velocity of the air, the pressure, the tension in the vocal chords, the space between them, their elasticity, etc. – and trying to work out the exact mechanisms for how they vibrate could be difficult and time-consuming. The more pragmatic approach of interaction emulation asks: what does the raw sound produced by the vocal chords sound like, what sort of filter do the throat, mouth and nose form, and how do these two parts interact with each other?

Interaction emulation assumes that musical instruments can be split into three parts:

- Drivers, which produce the raw sound. Examples are the hammer hitting a piano string, or the pick plucking a guitar string, or the reed vibrating in an oboe.
- Resonators, which colour the sound from the driver. Most musical instruments exhibit some sort of resonance: often the whole of the instrument vibrates along with the sound to some extent, and the way that it vibrates affects which frequencies are emphasized, and which are suppressed.

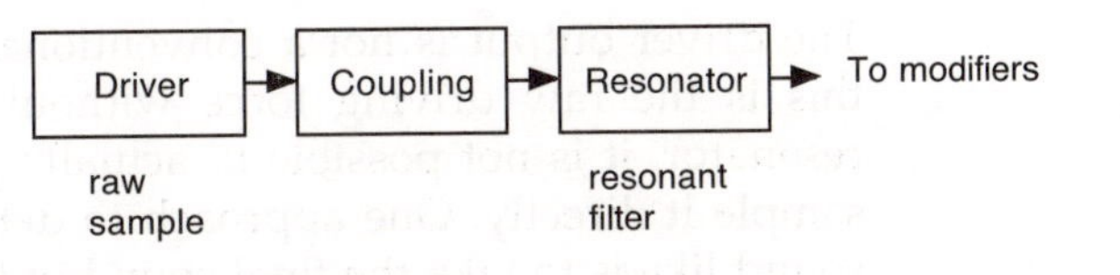

Figure 5.3.1 The driver produces a raw sample sound which has had the effect of any resonance removed artificially. This is then coupled to a resonator section through a coupler section, which allows control by the performer.

- The coupling between the driver and the resonators which determines how they two interact with each other.

In a real instrument, the drivers and resonators are very closely connected. They interact with each other: the hammer hitting a piano string causes the string to vibrate, but the vibration of the string is affected by the fact that the hammer is touching the string, has probably stretched the string slightly when it moved the string, and has added in a low-frequency thump. The act of setting the string vibrating depends on the hammer – you can't have the sound without it, but the hammer affects the sound. The two are inextricably interconnected. In interaction emulation, the two are separated, but the same interactions can be produced by controlling the way that the driver and resonator are connected together.

The basis of this technique is to separate the driver and resonator, and then couple them together so that they can interact. Instead of trying to model the driver, the technique assumes that the raw driver is more or less fixed, whilst the coupling to the resonator is the important aspect. This means that a driver 'sample' can be used to provide the stimulus for a resonator model via a coupling device – there is no need to try and create a model for the driver at all. Modelling resonators is much easier, since they are just filters, and filter theory is well understood. This means that it is easy to produce a number of driver 'samples', and resonator specifications, and couple them together. This approach means that a large number of possibilities are opened up without any need for careful research into musical instruments.

The coupling part of interaction emulation deals with the interconnection and interaction between the driver and the resonator. This is probably the major part of the technique to use the same approach as 'physical modelling'. A bowed string is a good analogy for the process. The player of a stringed instrument can control parameters like the position of the bow on the string, and how hard the bow is pressed onto the string. The resonator can be changed as well – it may be a fixed resonance, or one that changes with the playing pressure, for example. The combination of a simple model for the coupling, plus the fixed driver 'sample' and the variable resonance, produces a versatile synthesis 'engine'.

The driver output is not a conventional audio sample. Because this is the raw driving force without any modification by a resonator, it is not possible to actually place a microphone and sample it directly. One approach to determining what it would sound like is to take the final sound of the instrument, and then remove the effect of the resonances. If you listen to a raw driver signal, then it will sound very bright with an emphasized initial transient, almost like high-pass filtering. But since most resonators act as band-pass or low-pass filters, coupling this driver signal to a resonator transforms it into a sound which suddenly takes on a more normal sound. In fact, it sounds much like the sample that is normally associated with a sample – which is what a sample is, of course: the result of a driver coupled to a resonator. The difference is that by separating out the driver and the resonator, by then changing the parameters which control the resonator, you can change the timbre. This is not possible with a conventional sample at all. It is easy to design resonators which behave like strings, tubes, cones, flared tubes, drums, and even customized ones. Most will have a combination of band-pass or low-pass response, combined with one or more narrow peaks or notches.

Although this may sound like the S&S 'pre-packaged' sample concept, in fact, the combination of driver 'sample', coupling and resonator produces sounds which can change their harmonic content much more than any S&S sample which can be merely filtered. Remember that this is not a physical modelling instrument, although it is similar in some respects: especially the coupling section. What you lose is the transition between notes, so whereas an instrument based on physical modelling will move from one note to another in much the same way as a real instrument, one using interaction emulation will merely play two notes, one after the other. This is most noticeable for brass sounds, where a physical modelled instrument like the Yamaha VL1 will exhibit the characteristically 'overblown' brassy natural series of notes when the pitch is changed with the pitch-bend control, whilst an interaction emulation instrument will merely bend the note.

5.3.2 Physical modelling

The 'physical modelling' technique uses digital signal processing (DSP) chips to create a mathematical model of how some real musical instruments work. Instead of the conventional 'source and modifier' approach used by many S&S instruments, where a basic sample sound is modified by a filter and envelopes to produce a finished sound, a physical modelling instrument uses its internal model of an instrument to create the sound whole – in one operation. Because the model covers the entire instrument, it behaves like the actual thing, and so it also produces realistic transitions between notes, not just the notes themselves. It can produce sounds which emulate the behaviour of the real thing – often with astonishing realism. But the depth of detail which is required is formidable: you need to know a huge amount about the physics of musical instrument, acoustics, and mathematics – and then you need to convert this into software and electronics.

The techniques and algorithms for modelling musical instruments have only recently begun to reach the level of sophistication where they can be done in real-time without the aid of rooms full of super-computers, and currently the number of types of instrument which can be adequately described is quite small. It is not yet clear if physical modelling represents the same sort of technological leap that the GS1/DX7 did ten years ago, when analogue synthesizers were replaced by digital FM-based ones almost at a stroke.

Mathematical models

Using mathematics to make models of real world objects is common in engineering, but it is more unusual to find it used in musical applications. The underlying concept is the same for any model: you look at the inputs, outputs, their interconnections and dependencies, and then determine the equations which connect them all together. Imagine a tap and a bucket with a hole in it. Suppose that the tap can provide anything up to 10 litres of water per minute, the bucket holds 20 litres, and that the hole leaks at the rate of 1 litre per minute (Figure 5.3.2). Ignoring the leak, the fastest time taken to fill the bucket by the tap (when full on) is the time it takes for the tap to provide 20 litres of water, which would be two minutes (that's 20 litres at 10 litres per minute = 2 minutes).

When the effect of the hole is taken into account, the figures change correspondingly. In the first minute, one litre of water will escape out of the hole, and so only nine out of the ten litres supplied by the tap will be in the bucket at the end of the first minute. During the second minute, another litre of water leaks away and so there will only be 18 litres in the bucket, so it will obviously take slightly longer than the original estimate of two minutes because the tap will still need to provide just over two more litres of water.

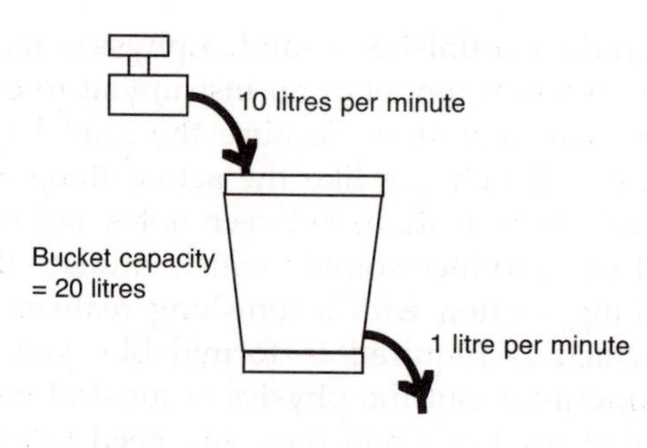

Figure 5.3.2 This 'bucket' diagram shows the power of a mathematical model in predicting the behaviour of a real world system. Physical modelling uses much more complex models of musical instruments to produce sounds.

By using this simple 'tap and bucket with hole' model, it is possible to make several other deductions based on how the system works. For example, if the tap supplies less than one litre per minute, then the bucket will never fill up because the hole leaks at one litre per minute. When there are 20 litres in the bucket, then it will begin to overflow – and if you subtract the one litre per minute leak, then the overflow rate is the tap supply rate (from just over one to ten litres per minute) minus the leak rate – so for the tap fully on, the bucket will overflow after just over two minutes have passed, and the overflow rate will be nine litres per minute.

As you can see, with just simple calculations we can make some quite complex predictions about the way that the real world works. The models of how musical instruments work which are used in physical modelling are obviously more complex than this example, but they are based on the same principles – you measure what happens, produce a description of what is happening, and then you use this information to work out what will happen.

Model types

Physical modelling synthesis falls into two distinct areas: continuous and impulsive:

- **Continuous models** deal with blown or bowed instruments, where there is a continuous transfer of energy into the instrument from the air flow, or the bow. The sound which is produced thus carries on as long as the energy is transferred. Typical examples include a trumpet and a violin.
- **Impulsive models** are for plucked or struck instruments, where a sudden 'impulse' of energy is transferred to the instrument, which then produces a sound as it responds to this input. The sound decays away naturally since energy is lost as friction, sound and movement once the initial input

is taken away. Typical examples include a piano and a snare drum.

For continuous models, the two major parts of most blown/bowed musical instruments are the bit that you blow or move and the part that vibrates. In a reed instrument, air is blown into a mouthpiece, whilst for a trumpet the lips move and control the flow of air. For a stringed instrument, the bow scrapes across the string. In all of these, the player is forcing the instrument to make a sound – and so these are called drivers, just as in interaction emulation. In contrast, the air inside a saxophone or trumpet vibrates inside a tube and so makes a sound; or the string vibrates and moves the air around to make a sound, and these make up the resonator part of the model. Whereas in a real instrument there are normally fixed combinations of drivers and their corresponding resonators, with physical modelling a reed type of driver feeding into a string type resonator is entirely possible, even though a real-world equivalent would be difficult to construct.

A variation on the 'Karplus–Strong' (Karplus and Strong, 1983) plucked string algorithm might well be the background to the impulsive models. This algorithm uses a damped resonator and a step input of energy to simulate what happens when a string or bar is plucked or struck. The resonator produces a note at its resonant frequency, with additional harmonics caused by its other resonances, and the decay of the sound occurs because the resonator has no source of power except for the initial input of energy. So as the energy leaks away, the sound decays. The way that the resonator loses energy, and the way that it produces the sound output are critical to the harmonic content and the way that it changes with time. The damping depends on the way that the string or bar is mounted or supported, whilst the mass of the string or bar, the tension in the string and the dimensions of the bar, can all affect how the sound changes with time.

The Karplus–Strong algorithm is simulated by using a time delay to model the movement of waves along the string or bar. The reflections at the end of the bar or string are set so that some energy is removed from the wave, and so the reflected wave is reduced in amplitude. The initial step input can be just a sudden change in level, but it can also be a brief pulse of noise. More complex models may also take into account more details of the initial input of energy, which need not be a sudden step input of energy, but may have an 'envelope' and other characteristics which affect the way that the energy is transferred to the resonator.

Practicalities

The complexity of the mathematical models which have been used so far in physical modelling synthesis have been such that the manufacturers of commercial units have usually chosen to present a number of fixed preset instrumental sounds. The user cannot program these sounds, other than change their response to performance controllers and change some modifiers. Although this is very different to most previous synthesizers, it is exactly how real instruments are treated – you do not take a drill to a saxophone and try making holes in the metalwork! Instead, you use the mouthpiece to control the sound through a combination of air pressure, lip pressure, throat resonance, vocal chords and your tongue.

The models which have been used in the initial physical modelling instruments are complex enough to provide exactly the sort of subtle and expressive control over timbre and pitch that you would expect from a real instrument. And there appears to be quite a lot of scope for modelling a wide range of instruments, although at least one academic paper has commented that there are only good models for a limited number of real instruments. Digital waveguides crop up several times in the research literature of physical modelling, and seem to be a very computationally efficient way of simulating a pipe or string by using DSPs, and probably form part of the continuous models.

Physical modelling can require a large amount of data to specify a specific timbre. For example, the Yamaha VL1 duophonic 'virtual acoustic' synthesizer uses 387 Kbytes to store 128 patches, which is roughly 3000 bytes per patch. For comparison, a DX7 FM patch uses only 155 bytes, and only 128 bytes in the compressed form! Even so, the size of the VL1 file is still tiny compared to the size of a sample in an S&S synthesizer, where about 88 Kbytes of storage are required for each second's-worth of sample.

Controlling the instruments provided by a physical modelling synthesizer can be difficult because of the large number of parameters which may need to be manipulated. Keyboard control is useful for pitch and velocity control, but it is not as natural an interface as a wind controller. Keyboards have the disadvantage of a naturally polyphonic keyboard, whilst a blown instrument is normally monophonic. Unfortunately, despite the advantages of a wind-instrument-like controller, the keyboard has still appeared on the first generation of commercial physical modelling instruments. Blowing can be easily simulated by using a breath controller, but lip or bow pressure,

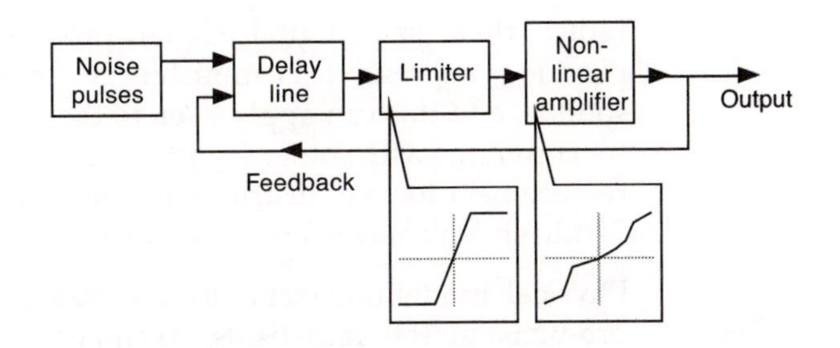

Figure 5.3.3 A delay line can be used as the basis for experimentation into physical modelling using analogue audio equipment.

muting or string damping are less obvious, and foot controllers, velocity and after-touch can be used, although it requires practice for a keyboard player to become familiar with the use of additional controllers.

Experimentation

It is not necessary to have sophisticated digital workstations to experiment with physical modelling synthesis. Using conventional recording studio equipment, it is possible to try out the underlying principles for real. All that is needed is an audio delay line with a few milliseconds of delay (almost any effects processor with an echo or delay setting will do), a limiter or compressor/limiter, a noise generator (or synthesizer with noise generator), and a non-linear amplifier. The non-linear amplifier is an analogue equivalent of the waveshaper described earlier in this chapter – almost any op-amp can be used to provide this function. (Clayton, 1975).

The basic idea is to connect the output of the delay line to the limiter, the output of the limiter to the non-linear amplifier, and then the output of the amplifier back to the input of the delay line. The noise generator should be mixed into the input of the delay line as well. The output of the delay line also serves as the output of the system (*Sound on Sound*, February 1996). By adjusting the feedback and injecting pulses of noise into the system, it should be possible to get percussive sounds which decay away as per Karplus–Strong synthesis, whilst with higher levels of feedback, sustained continuous tones should be produced, whose timbre can be changed by adjusting the non-linear amplifier settings. By sampling the results into a sampler, some of the more interesting or useful timbres can be stored for future use.

5.3.3 Summary

Physical modelling represents the first of many possible methods of digital synthesis based on sophisticated software rather than just digital signal processing hardware. It can produce expressive, astonishingly 'real' feeling instrument sounds, and this can apply even to the impossible synthetic ones. In common with other synthesized sounds, these are not a replacement for real instruments, more a whole new set of them. Synthesis will never be the same again.

Physical modelling technology began to appear in a range of products in the mid-1990s. Technics produced an interaction emulation synthesizer, whilst Yamaha and Korg produced several physical modelling products, and MediaVision produced a PC card using physical modelling techniques. If events are following their usual trend, then these instruments may be the first members of a new family. As an example, the costly and rare Yamaha GS1 proceeded the affordable and best-selling DX7, but it gave a glimpse of what would happen a year or so later.

It remains to be seen whether interaction emulation can move away from prepared driver 'samples' and move nearer to a physical model approach. Conversely, physical modelling needs to move towards programmability and polyphony. When the first physical modelling instruments appeared, they were expensive and monophonic or duophonic, whereas the first interaction emulation instruments which appeared were polyphonic for about the same price.

5.4 Granular synthesis

Granular synthesis is often regarded as an unusual technique. Unlike many of the other methods of synthesis described so far, it has not been used in commercial synthesizers, although it has been used by some composers working in the academic and research fields. It does not fit into the source and modifier model, but instead approaches the production of sound from a bottom-up point of view, which is very different to most other methods of sound synthesis.

Granular synthesis builds up sounds from short segments of sounds called 'grains'. In much the same way that many pictures in colour magazines are made up from lots of dots, granular synthesis uses the tiny sound fragments to produce sounds. The grains are of very short duration: 20 or 30 milliseconds, which is close to the 10–50 millisecond timing 'resolution' of the human hearing system – audio events which occur closer together than this tend to be heard as one event instead of two (Figure 5.4.1). The controls are relatively straightforward: the number of grains in a given time period, their frequency content

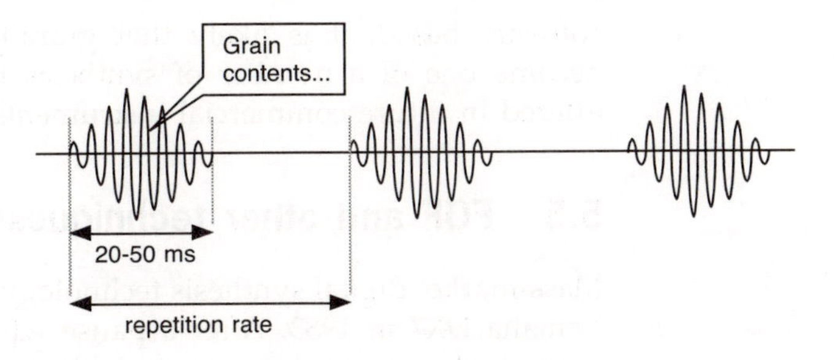

Figure 5.4.1 Granular synthesis uses small 'grains': short segments of audio which are arranged in groups. The contents can be waveforms, noise or samples. The major controls include the number of grains, their lengths and their repetition rate.

and their amplitude are the major parameters. The difficulty lies in controlling these parameters: rather like the large number of parameters in additive synthesis, manipulating a large number of grains requires envelopes, function generators and other controllers, and can become a very large overhead.

Grains are normally enveloped so that they start and finish at zero amplitude, so that sudden discontinuities are avoided; any sharp change in the resulting waveshape would create lots of additional unwanted harmonics and the result would sound like a series of clicks. Grains may contain single frequencies with specific waveforms, or band-pass filtered noise, and each grain can be different. In some ways, granular synthesis can be considered as the limiting case of wavetable synthesis, where the table of waveforms is swept very rapidly to give a constantly changing waveshape – but few wavetable synthesizers have the control of wavetable selection and the zero-crossing smoothly enveloped grains that are found in granular synthesis. In fact, granular synthesis is normally produced by software, and so the grains can be produced using a number of techniques from additive sine waves to filtered noise or even processed samples of real sounds. Some experimenters have worked on coupling granular synthesis with mathematical systems like chaos theory, John Conway's 'life', and fractals.

Granular synthesis seems to be somewhat analogous to the way that film projectors work. By presenting a series of slightly different still images at a rate which is just about the limit of the eye's response to changes, the impression is one of a smooth continuous movement. In granular synthesis, the rapid succession of tiny fragments of spectra combine into an apparently continuously changing spectrum. This constant change of grains is reflected in the timbres which are produced by granular synthesis: words like 'glistening' or 'shimmering' are often used to describe the complex and busy sounds which can result, although the technique is also capable of producing more subtle, detailed sounds too. As digital synthesizers become increasingly software based, it is likely that granular synthesis may well become one of a number of synthesis techniques that will be offered in future commercial instruments.

5.5 FOF and other techniques

Mass-market digital synthesis technology first appeared with the Yamaha DX7 in 1983. After a pause whilst the other manufacturers looked around for other viable methods of digital synthesis, the additive and S&S instruments began to appear. Over the next ten years, S&S gradually took over until by the early 1990s, it was virtually the only digital method of synthesis. After such a slow and steady development over ten years, the mid-1990s marked a sudden change when a number of sophisticated instruments were released which could utilize combinations of additive, subtractive and FM synthesis, and these were soon joined by instruments based on physical modelling techniques. It seems likely that commercial S&S instruments will be joined by a large number of techniques which have so far only been used in academic research.

This section looks at some of the synthesis techniques which may well be incorporated into the digital synthesizers of the next century, and perhaps even before then. They all have a common theme, which is derived from a combination of research into musical sounds, acoustics and human speech and singing. They are the result of a fusion of the worlds of telecommunications, computing and music.

5.5.1 *Formants*

All of these methods are focused around the sounds which are produced by strong resonances – wherever you get a fixed set of 'formant' frequencies. The human voice is one example of this sort of system; the mouth, nose and throat can be thought of as a complicated tube-like arrangement where particular frequencies are emphasized whilst others are suppressed, so the resulting frequency response is a series of peaks. The vocal chords produce a spiky pulse-like waveform which has lots of harmonics in it, and this is then processed by the vocal tract (the mouth, nose and throat) which acts as a filtering mechanism. The result of the filtering is to produce an output which contains predominantly those frequencies, from the original pulse sound, which match the resonant peaks of the filter. Since you can only make minor changes to the physical shape of the tubing formed by the mouth, nose and throat (e.g. changing the size and shape of your

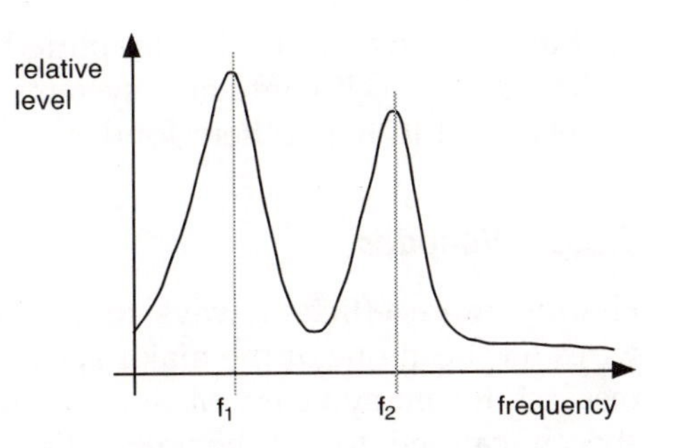

Figure 5.5.1 Formants are peaks in the frequency spectrum of a sound. This example shows two large peaks.

mouth cavity with your tongue), then the peaks are mostly fixed, and so what comes out is a set of harmonics which have peaks which are fixed by the formant frequencies, regardless of the pitch of the note being sung! The only things that do change are the fundamental and the underlying harmonics.

This can be regarded as another type of 'source and modifier' model, where the source is the vocal chords, and the modifier is the filter formed by the mouth, noise and throat. The vocal chords can be emulated by using a short burst of sound whose frequency is fixed, and then triggering this at the rate of the fundamental frequency that you want to produce. The pulse repeats, producing the harmonics associated with the fixed resonances of the formants that it represents, whilst the pitch that you hear is the repetition rate. The modifier part can be emulated by combining several band-pass and notch filters – although since changes of the shape of the 'tube' can happen, these filters need to be dynamically changeable in real-time. In fact, the human ear is very sensitive to exactly these changes in formant structure.

Instruments exhibit the same sort of formant structures: the analogy between the human vocal apparatus and some of the woodwind and brass instruments is probably the strongest. The abstraction of a source of sound connected to a 'resonant set of formants' acting as a modifier can be applied to almost any instrument. For string instruments the formants are determined by the string characteristics, its mountings and the structure of the body of the instrument. For some instruments other external factors can be very important: an electric guitar is designed to provide a rigid support for the vibrating string, and the heavy wooden body is not a very strong resonant system. But the combination of the guitar string, amplifier, speaker, speaker cabinet and feedback between the acoustic output and the guitar pickups forms a very complex resonant system which is often exploited to great effect in live performance. In contrast, synthesizers and most other amplified musical instruments tend to be used as self-contained systems, and the amplification is merely used to make them louder.

5.5.2 *Vocoder*

Finding more efficient ways to transmit human speech along wires has been one of the major activities of telecommunications research for many years. Most of the raw information content of speech can be found between 300 Hz and 3400 Hz, and so telephone systems are designed with a bandwidth of about 3 kHz. Frequencies outside of this range add to the clarity and personality of the voice – which is why it is difficult to distinguish between an 's' and an 'f' on the telephone, or why people may sound very different in real-life to hearing them over the telephone.

Research at the Bell Telephone Laboratories in New Jersey, USA, in the early 1930s, looked at how different parts of this 3 kHz bandwidth were used by speech signals. By using band-pass filters, the speech could be split into several separate 'bands' of frequencies, and the contribution of each band to the speech could then be determined. By using an envelope follower, the envelope of the contents of each frequency band could be determined. Once split into these bands, the audio signal could be mixed back together again in different proportions, and even have new envelopes applied to each band. As basic research into the properties of speech, the results were interesting (you need all of the 3 kHz bandwidth: removing bands alters the timbre of the speech too radically to be useful for telephony), but they had no practical application at the time. It was not until digital processing techniques became available in the 1960s and 70s that vocoders were to find re-use in telecommunications.

But the vocoder proved to be a powerful tool for processing audio signals. By splitting an audio signal into separate bands, analysing the contents and then allowing separate processing of these bands, it allows sophisticated control over the timbre of the sound. More importantly, by separating the analysis and processing functions of the vocoder, it is also able to extract the spectral characteristics of one sound and apply them to another. The fidelity with which this can happen depends on both the number of bands and the characteristics of the envelope followers. As the bandwidth of the bands decreases, more filters are required to cover the audio spectrum. For 'octave' bands, each covering a doubling of frequency, only eight filters are required: six band-pass, one low-pass and one high-pass. This produces

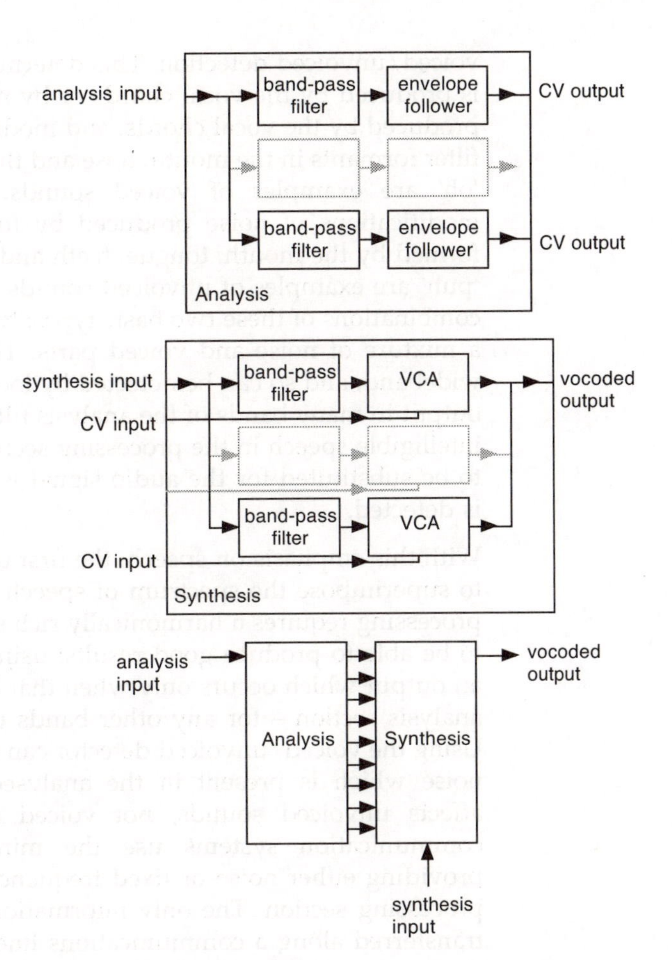

Figure 5.5.2 A vocoder is made up of two sections: analysis and synthesis. The analysis section converts the incoming audio signal into frequency bands and produces a control voltage proportional to the envelope of the contents of that frequency band. The synthesizer section has identical band-pass filtering, but this time it acts on a different audio signal. Each band is controlled by a VCA driven from the analysis section. The characteristics of the analysed signal are thus superimposed on the synthesized signal. Although this diagram shows analogue blocks, implementing a vocoder is now easier in digital circuitry or on a DSP chip.

only a coarse indication of the spectral content of the audio signal which is being analysed, and correspondingly coarse changes to the signal which is being processed. For 'third-octave' bands, 30 or 31 filters are required, and the resulting finer resolution significantly improves the processing quality. The envelope followers determine how quickly the spectrum can be imposed on the processed signal – if the time constant of the envelope follower is too long then the bands will not accurately follow the changes in the signal which is being analysed, whilst if the time constant is too short then the controlling of the amplitude of the bands can become noticeable.

Vocoders began to be used to process musical sounds in the 1950s. The basic vocoder structure had some features which were specific to processing speech – most importantly the voiced/unvoiced detection. This determines if the speech sound is produced by the vocal chords or by noise. Voiced sounds are produced by the vocal chords, and modified by the the resonant filter formants in the mouth, nose and throat: 'ah', 'ee', 'mm' and 'oh' are examples of voiced sounds. Unvoiced sounds are modifications of noise produced by forcing air through gaps formed by the mouth, tongue, teeth and lips: 'sh' and 'f', 't' and 'puh' are examples of unvoiced sounds. Many vocal sounds are combinations of these two basic types: 'vee', 'kah' and 'bee' have a mixture of noise and voiced parts. The noise tends to be be wideband, and so can be detected by looking for a simultaneous output in many bands of the analysis filters. In order to produce intelligible speech in the processing section, a noise signal needs to be substituted for the audio signal when an unvoiced sound is detected.

With this emphasis on speech, the first uses of the vocoder were to superimpose the spectrum of speech onto other sounds. The processing requires a harmonically rich source of sound in order to be able to produce good results: using a sine wave will give an output which occurs only when that band is activated by the analysis section – for any other bands there will be no output. Using the voiced/unvoiced detector can be used to substitute for noise which is present in the analysed signal, but this only affects unvoiced sounds, not voiced sounds. Some military communication systems use the minimalistic technique of providing either noise or fixed frequencies in the bands for the processing section. The only information that then needs to be transferred along a communications line are the parameters for the bands and the voiced/unvoiced detection. This results in a very robotic sound which has high intelligibility but almost no personality. Using a vocoder to superimpose the spectral changes of speech onto musical instruments has a similar effect: the output has a robotic quality and sounds synthetic. This has been used for producing special effects like talking pianos, brass instruments and even windstorms.

Implementing large numbers of filters in analogue circuitry is expensive, and so analogue vocoders tend to have restricted numbers of filters, whereas digital vocoders can have much finer resolution. Digital vocoders can also extract additional information about the audio signals in the bands, and the 'phase vocoder' is one example – it can work with narrow, high-resolution bands and can output both amplitude and phase information, which improves the processing quality and enhances the creative possibilities for altering musical signals.

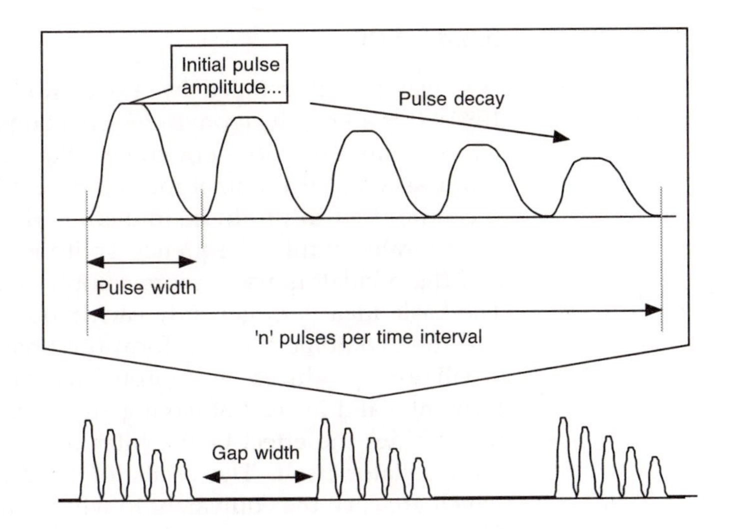

Figure 5.5.3 VOSIM produces pulse trains with controllable pulse width, repetition rate, amplitude decay and gap width. It is similar to FOF in many ways.

5.5.3 VOSIM

VOSIM is an abbreviation for voice simulation, and uses a simple oscillator to produce a wide range of voice-like and instrumental timbres – although the original intention was to use it for speech synthesis. The original hardware was developed in the 1970s at the University of Utrecht, and has since been adapted for software-based digital generation. The oscillator produces asymmetrical waveforms which are made up of repetitions of a series of raised sine-squared waveforms – called a 'pulse train'. The series of waveforms reduces in amplitude with time, and so only a small number of parameters are required: the width of the pulses; the decay rate of the amplitude; the number of pulses; and the repetition rate of the pulse trains. Because the spectrum that is produced is dependent only on the parameters which control the pulse trains, and not on the repetition rate, the harmonic content is independent of the pitch. This is exactly the opposite of a sample playback system, and is useful for simulating the fixed formant frequencies which are found in vocal and instrumental sounds.

The simple controls, versatility and small number of parameters used in VOSIM are ideally suited to the real-time control requirements of a speech synthesis system. In many ways VOSIM has a similar 'minimal parameter' interface to FM, although whereas FM has been commercially successful in musical applications and has only seen limited use as a speech synthesis method, VOSIM is more suited to speech synthesis and has not been used for mass-market musical applications.

5.5.4 FOF

FOF was first developed by Xavier Rodet in Paris in the early 1980s. It is a French acronym for Fonctions d'Onde Formantique, which translates to something like 'formant-wave-function synthesis', and it is sometimes referred to as formant synthesis. It can be used to produce simulation of vocal type sounds, and incorporates similar frequency splitting elements to vocoding, and the oscillators use a more complicated variation of VOSIM. The basic idea is to generate each required formant separately, and then combine them to form the final output. Each formant 'oscillator' produces an output which deals with just one formant – and instead of having an oscillator and resonant filter, it combines the effect of the filter on the oscillator output into the oscillator itself. The oscillator produces a series of pulses which are each the equivalent to what would be output from the filter if a single rapid step signal was passed through it – called the impulse response of the filter. The pulse contents are thus derived from the impulse response of the filter, and if a series of these pulses is then output, the resulting sound is the same as if the filter was still processing the original step signal. More importantly, the rate of outputting these pulses can alter the frequency of the sound which is produced, but the filtering will remain the same, since it is the shape and contents of the pulse that determine the apparent 'filtering', not the repetition rate of the pulses.

The output from a typical FOF oscillator is a succession of smoothly enveloped (as in granular synthesis) audio bursts which happen at a repetition rate which is the same as the pitch of the required sound. Each burst of audio has a peak in its

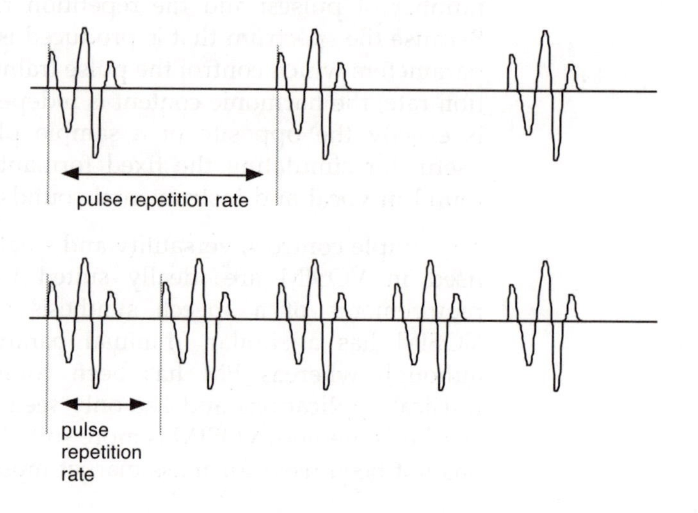

Figure 5.5.4 FOF produces pulses whose shape is determined by the impulse function of the sound which is required. The repetition rate determines the frequency of the sound, whilst the pulse contents determine the formants of the sound.

spectrum that is the same as the required formant frequency. If the repetition rate is above 25 Hz then these bursts produce the effect of a single formant with spectral characteristics determined by the audio burst itself. For lower repetition rates then it provides a variant of granular synthesis. Digital implementations of FOF normally provide both FOF and granular modes, and this allows continuous transformations to be made between vocal imitations and granular textures. Each FOF oscillator produces a single formant, and the output of four or more of these can be combined to produce sounds which have a vocal-type quality.

FOF can be produced using conventional synthesizers by taking a sound which has a fast attack and decay time, with no sustain or release, and then triggering it repeatedly so that it produces a rapid series of short bursts of audio. If the synthesizer produces these short audio bursts at 100 Hz, then the fundamental frequency of the output will be at 100 Hz, but the apparent filtering of the signal will be determined by the contents of the sound itself – and so changing the repetition rate will change only the pitch – the formants (filtering) will remain the same because the sound which is being repeated is also staying the same.

In MIDI terms, this usually means choosing a single note and making a simple and very short enveloped sound which has the right harmonic content and then sending note on and off messages very rapidly for just that one note, where the repeat rate sets the fundamental frequency, and thus the pitch, of the resulting sound. This is easiest to do by creating lots of messages and then changing the tempo of playback! Unfortunately, MIDI is too slow to create high frequency note repetitions. This limits the maximum frequency which can be generated using this method to monophonic sounds at just under 800 Hz under ideal conditions.

Producing suitable sounds for FOF involves throwing away some of the instinctive approaches that many sound programmers have. In fact, it is not necessary to use sounds which approximate to the impulse response of a filter. All you need is a quick burst of harmonics. For simulating real instruments and voices, then you need to have something which sounds like a single click processed by whatever it is you want to sound like, whilst for synthetic tones almost anything will do.

5.5.5 Dynamic filtering

There are a large number of techniques that utilize the same model of the throat, mouth, nose and vocal chords as the other methods in this section, but which approach the design from the opposite viewpoint. Most were originally developed for use in telecommunications speech coding applications, but they can also often be used to synthesize formant filter-based sounds. One of the best known is LPC, which is an acronym for Linear Predictive Coding. LPC techniques can also be used in resynthesis to help design suitable filters. Other techniques include CELP, PARCOR and the filters used by Emu in their Morpheus and UltraProteus products.

To generalize the dynamic filtering method, a digital filter is used to approximate the formants, and this filter is used to process a source waveform into the desired output. This is very different to extracting the formants and synthesizing them individually, since a single multi-formant filter can produce the equivalent of several separate FOF oscillators simultaneously. The filter shape is controlled by a number of parameters, and can usually be changed in real-time to emulate the changes which can occur in a real-world resonant system like the mouth, nose and throat.

5.5.6 Software

For stand-alone instruments, digital synthesis is a combination of digital hardware and software, although strictly there is usually an analogue output stage and low-pass filter connected to the output of the DAC. But it is also possible to use digital synthesis to produce sounds using a general purpose computer. In this case the software is normally independent of any hardware constraints – the use of specialized DSP chips to carry out the digital signal processing is often only required to improve the calculation speed. The output of such software is in the form of 'sound files', which are encoded audio files. Sound files come in a large number of formats, many specific to a particular computer platform, although there are also generic file formats. Some of the common formats are shown in Table 5.5.1.

These sound files can be used as the basis for further processing, transferred to samplers for replay, or replayed using a computer sound card or built-in audio facilities.

This software-only synthesis comes in several forms. Commercial software tends to be either simple sample editing programs, or else sophisticated audio processing software. The Freeware and Shareware software is much more varied, ranging from complete digital synthesis systems to sample processing programs – although there is less emphasis on the detailed audio editing that is found in the commercial software.

Table 5.5.1 File formats for sound files

Suffix	*Type*	*Format*	*Amiga*	*Atari*	*Mac*	*PC*	*Unix*
.aif	Audio	AIFF	•		•	•	
.aifc	Audio	AIFF	•		•	•	
.aiff	Audio	AIFF	•		•	•	
.au	Audio	μ-law	•		•	•	•
.au.gsm	Audio	GSM μ-law					•
.avi	Data	Intel Video				•	
.gm	Data	MIDI	•	•	•	•	•
.gmf	Data	MIDI	•	•	•	•	•
.mid	Data	MIDI	•	•	•	•	•
.mov	Movie	QuickTime			•	•	•
.mp2	Audio	MPEG Audio	•		•	•	•
.qt	Movie	QuickTime			•	•	•
.ra	Audio	Real Audio			•	•	
.sds	Audio	MIDI Sample Dump Standard	•	•	•	•	•
.smf	Data	MIDI	•	•	•	•	•
.snd	Audio	SND: System Resource			•		
.voc	Data	SoundBlaster			•	•	
.wav	Audio	WAV				•	

5.6 Analysis-synthesis

Analysis-synthesis techniques are the basis for the resynthesizer, which takes a sample of a sound, extracts a set of descriptive parameters, and then uses these parameters to recreate the sound using a suitable synthesis technique. There are two major problems in achieving this:

- converting the sample into meaningful parameters
- choosing a suitable synthesis method.

The conversion between a sample of a sound and a set of parameters which describe that sound is not straightforward. There is also the issue of mapping those parameters to the chosen synthesis method.

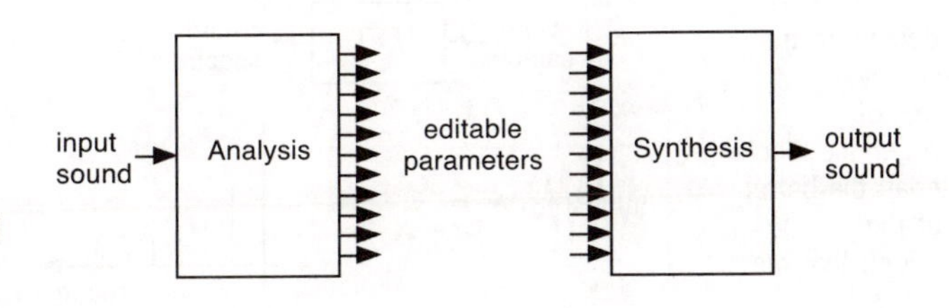

Figure 5.6.1 Resynthesis takes an existing sound sample, and analyses it to produce a set of parameters. These parameters can then be edited and used to control a synthesizer which produces an edited version of the original sample.

5.6.1 *Analysis*

The first stage is to analyse the sample. Parameters which might be required to describe the sound adequately to allow subsequent synthesis include:

- pitch information.
- pitch modulation: LFO and/or envelope?
- harmonic structure.
- formant structure.
- envelope of complete sound
- envelopes of individual harmonics
- relative phase information for individual harmonics
- dynamic changes to any parameter in response to performance controls.

There are a number of techniques which can be employed to produce this information. Fast Fourier transforms (FFTs) are a way of transforming sample data into frequency data, and they are widely used for spectrum analysis. FFTs require considerable computation in order to convert from the time domain (a waveform) into the frequency domain (a spectrum). The detail which can be obtained from an FFT is inversely proportional to the length of the sample which is analysed. So short samples have only coarse frequency resolution, whilst long samples can have fine resolution: if a sample of 20 milliseconds is converted, then the resolution will be 50 Hz. If the harmonic content of the sample is changing quickly, then a compromise will need to be made between the length of sample which is analysed, and the required frequency resolution. Successive FFTs can move the sample 'window' in time, overlapping the previous sample, and so build up detailed spectrum information, even though the majority of the sample data is the same. An alternative approach is to use interpolation between the spectral 'snapshots'.

Linear predictive methods can be used for formant analysis, since they output the parameters which describe a filter which emulates those formants.

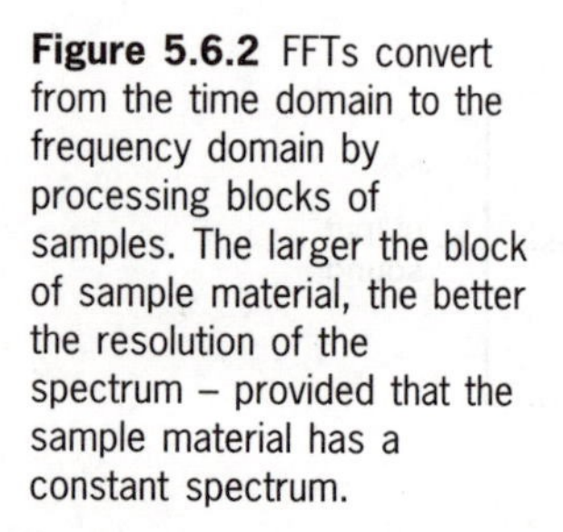

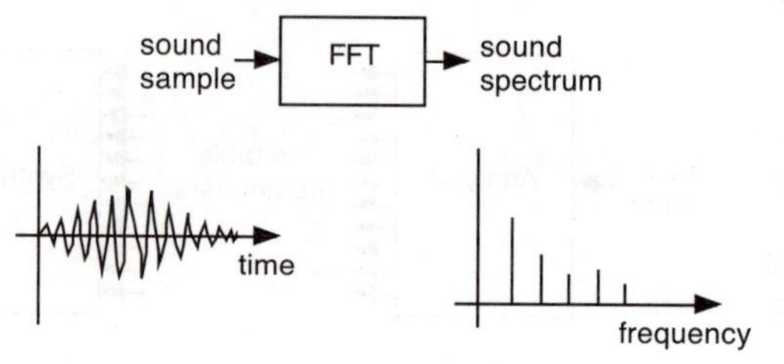

Figure 5.6.2 FFTs convert from the time domain to the frequency domain by processing blocks of samples. The larger the block of sample material, the better the resolution of the spectrum – provided that the sample material has a constant spectrum.

Pitch extraction employs a number of techniques in order to determine the pitch of a sampled sound. Because the perceived pitch of a sound is concerned more with the periodicity rather than the frequency of the fundamental, pitch extraction can be difficult. Methods include:

- **Zero crossing**: The simplest is to count the number of zero-crossings, but this is prone to errors because of harmonics causing additional zero-crossings. Filtering the sample sound to remove harmonics and then counting the zero crossings can be more successful, but a better technique is to use the peaks of the filtered sample since the harmonics have been removed and a simple sine-like waveform is all that should be left after the filtering. This method has problems when the fundamental frequency is weak, since filtering the harmonics still leaves a noisy, low-level signal.
- **Auto-correlation**: Auto-correlation is a technique which compares the waveform with a time-delayed version of itself, and looks for a match over several cycles. When a delay equal to the periodicity of the waveform is reached, then the two waveshapes will match. This assumes that the sample sound does not change rapidly and that there are no beat frequencies or large inharmonics.
- **Spectral interpretation**: Spectrum plots derived from FFTs can be used to determine the pitch. The spectrum is examined and the lowest common divisor for the harmonics shown is calculated. For example, if harmonics at 500,

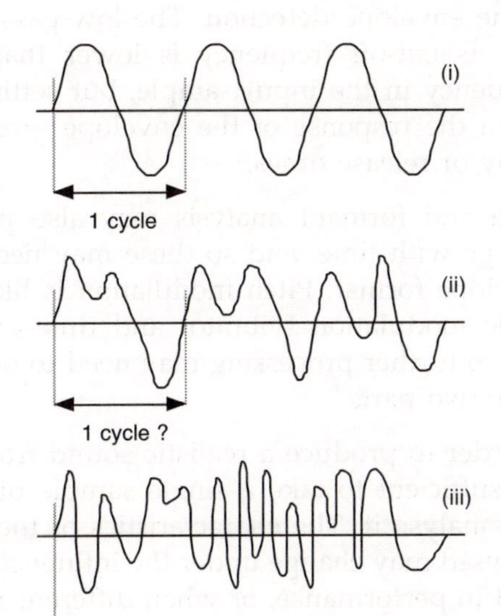

Figure 5.6.3 Pitch extraction needs to be able to cope with a range of inputs: from simple sine waves (i) which can be processed by a zero crossing method; through waveforms which change slightly from cycle to cycle (ii) where autocorrelation or cepstral analysis can produce useful pitch outputs; and finally noise (iii), where the pitch extractor should indicate that it is noise rather than a rapidly changing pitch. Although the human ear can readily achieve this, the process is less straightforward for electronics and computers.

600, 1000 and 1200 Hz were present, then the fundamental frequency would probably be 100 Hz. Again, beat frequencies and large inharmonics can produce significant errors with this technique – normally producing fundamental frequencies which are too low (a few hertz or tens of hertz).

- **Cepstral analysis**: By further processing the spectrum, it is possible to produce plots which quite clearly show peaks for the fundamental. The process involves converting the amplitude axis of the spectrum into a decibel or logarithmic representation instead of the normal linear form, and then calculating the spectrum of this new shape – using an FFT to treat the spectrum as if it is a waveform! The resulting 'cepstrum' (a reworking of the word 'spectrum') will show a peak in the upper part of the time or 'quefrency' axis that indicates the fundamental frequency of the sound. The cepstrum merely indicates the underlying spacing of the harmonics shown in the spectrum, and so spectra with only odd harmonics, or very sparse harmonics (like a sine-wave!) can cause errors.

Envelope following

Extracting the envelope from a sample sound is relatively straightforward in comparison to pitch extraction. The sample sound is low-pass filtered, and then a 'leaky' peak detector is used to produce a simple curve that approximates to the original volume envelope. The setting of the low-pass filtering and the peak detector decay time constant govern the effectiveness of the envelope detection. The low-pass filter should be set so that its cut-off frequency is lower than the lowest expected frequency in the input sample, but setting it too low can slow down the response of the envelope – resulting in slow attack, decay or release times.

Pitch and formant analysis may also produce outputs which change with time, and so these may need to be converted into envelope format. Pitch modulation is likely to be in two parts: cyclic modulation (vibrato) and time-varying (pitch bending), and so further processing may need to be employed to separate these two parts.

In order to produce a realistic sound from a resynthesizer, it is not sufficient to take a single sample of the instrument sound and analyse it. The characteristics of the sound which is being analysed may change under the influence of external parameters used in performance, or when different notes are played. There is thus a need to take into account any changes caused by performance controls and different playing pitches. One example is the change in timbre which happens when an instrument is played harder or more vigorously: hitting a piano key harder, or bowing a string with more pressure. Other examples include damping strings or muting a brass instrument. Several samples will be required in order to measure the dynamic changes to parameters which occur in response to these performance controls. Different pitches can be dealt with by making several samples of the instrument throughout its playable range. The outputs of these dynamic measurements can then be interpolated to give approximations for all notes and performance control settings.

5.6.2 Synthesis

Almost any synthesis technique could be a candidate for the synthesis 'engine' for a resynthesizer. The most important consideration is how the parameters of the technique map to the parameters that can be extracted from the sample. The mapping needs to be complete and unambiguous, but it also needs to produce a parameter set which can be manipulated by the end user of the resynthesizer.

Additive

Additive synthesis appears to offer perhaps the simplest approach to resynthesizing sounds from parameters. The only parameters which are required are detailed pitch, amplitude and perhaps phase information for each of the harmonics which are present in the sample sound. Unfortunately, this is likely to be a large number of harmonics, each with complicated multi-stage envelopes for the changes in the pitch, amplitude and phase parameters with time and performance controller settings. So, although the extraction of the parameters is relatively straightforward, presenting them to the end user in a manageable form is more difficult.

FM

Frequency modulation has a much smaller set of required parameters than additive synthesis. In this case, the problem is how to convert the extracted parameter information about pitch, amplitude and phase for each harmonic, into suitable parameters to control FM. There is no simple way to work backwards from a sound to calculate the FM parameters which produced it – a process called deconvolution. An iterative process which tests possible solutions against the given parameters might be successful, but it is likely to require considerable processing power as well as time.

Subtractive

Subtractive synthesis requires more parameters than FM, but it provides a smaller set of controls than additive synthesis. The major problems with using subtractive synthesis are the fundamental limitations of the technique: the filtering is often a simple resonant low-pass filter and there is a limited set of source waveforms. The combination of these problems means that subtractive synthesis has a very limited set of possible sounds, and this seriously restricts the possibility of being able to resynthesize a given sound.

FOF/VOSIM

Formant synthesis techniques like FOF and VOSIM have small numbers of parameters, and the conceptual model is similar to subtractive synthesis. But unlike subtractive synthesis, formant synthesis techniques are not restricted to simple filtering, but can recreate complex and changing formant structures. Although the source waveforms may be simple to control, the dynamic formant filter presents a considerable problem to a user interface designer.

In fact, FOF is part of a complete software package called CHANT, written at IRCAM in Paris by Xavier Rodet and others in the early 1980s. CHANT can be used to analyse a sampled sound and extract the harmonic peaks, and then use these formants as the basis of an FOF resynthesis of the sound.

Physical modelling

Physical modelling can be considered to be a type of analysis-synthesis technique, although the analysis process is more sophisticated since it involves a study of the physics of the instrument and its sound, and then the building up of a physical model of that instrument. The synthesis part is then relatively simple – just run the model to simulate the instrument's behaviour. At the moment, the process of analysing a real instrument is a time-consuming one, although the commercial development of physical modelling may facilitate the development of software tools for this task.

5.6.3 Resynthesis

Any resynthesis technique requires a compromise between the depth of required detail to describe the original sound, and the ability of the user to make meaningful changes to the sound. With enough parameters, it should be possible to resynthesize a specific sound very accurately – but it may not be possible for a user to make any useful changes to that sound because of the complexity of the controls and the number of parameters.

Because software can cope with large amounts of data easily quickly, whereas complex mathematical processing often involves additional time, the two techniques which seem to offer the best resynthesis engine are additive and FOF/VOSIM. In both cases, the software would need to present some sort of abstracted user interface to the synthesis engine to avoid displaying all of the parameters.

Commercial resynthesizers have not been very successful. Whilst the idea has been talked about for a long time, only a few minor manufacturers have attempted to produce a resynthesizer. None have succeeded in combining a practical user interface, rapid analysis, and a versatile synthesis engine at a reasonable cost.

5.7 Hybrid techniques

With a wealth of powerful techniques becoming available, digital synthesis has increasingly used software-based methods. Instruments are gradually relying less on a specific technology, and more on a mixture or combination of synthesis techniques. This provides a wide range of sounds, and avoids any specific limitations of a particular technique. One example is the FM synthesis implementation found in the first generation of Yamaha instruments like the DX7 – the 'weak' areas include rich string or pad sounds, as well as filter sweeps. By combining more than one synthesis method, there is also scope for producing sounds which are not possible using any of the separate methods in isolation.

Examples

- The Yamaha SY99 and SY77 mix together FM (AFM or Advanced FM) and AWM2 (Advanced Wave Modulation 2), which makes the most of FM's flexibility and S&S's realism – and adds resonant filtering. By allowing the S&S waveform to modulate the FM operators, the S&S sound can be processed as part of the FM synthesis – Yamaha call this real-time convolution modulation or RCM. FM with non-sine shaped waveforms produces lots of harmonics, and RCM is useful for adding harmonics and then removing them using the digital filtering. This is an under-exploited technique – few of the sounds produced on the SY99 and SY77 make use of RCM.

- Korg's Prophecy mixes several digital techniques to give a sophisticated monophonic 'lead-line' instrument which has a very 'analogue' feel to some of its sounds. It provides a conventional 'analogue' synthesis emulation; FM, physical modelling of brass, reed and plucked instruments; and three variations on sync/cross-modulation and ring modulation analogue emulations. To control these methods, it has a wide range of performance controllers, and can be incorporated as a plug-in for Korg's polyphonic S&S instruments like the Trinity.
- Technics WSA mixes bits of 'physical modelling' with S&S to give a simplified 'driver and resonator' interaction emulation synthesis instrument which has the advantage of being polyphonic at a time when other physical modelling instruments are monophonic or duophonic.
- Kurzweil's variable architecture synthesis technique (VAST) provides many resources, but they are more like a modular approach to an S&S synthesizer than any combination of separate synthesis techniques.

The future seems to lie with a combination of techniques, since none of the available methods offers a complete solution. As hardware becomes more powerful, the software functionality increases and also becomes more flexible. The limits are more likely to be the user interface and the processing power rather than the synthesis methods.

Future synthesizers are likely to be general purpose synthesis engines which can be configured to produce a number of different techniques, although it is unlikely that any standardized way of controlling these techniques will emerge in the near future. This means that even though the synthesis methods will converge, the user interfaces and sound storage formats will not. Hybrid instruments are thus similar to the pre-MIDI analogue instruments – where interconnecting synthesizers was not possible without sophisticated hardware. With complex software-based synthesis, the possibilities for interfacing become more remote – which is very useful for commercial synthesizer manufacturers, but not as good for users.

5.8 Example instruments

Casio CZ-101 – waveshaping (1985)

The Casio CZ-series of synthesizers are one example of a commercial use of a full waveshaping implementation to produce sounds. Although it is called 'phase distortion', it uses waveshaping, but presents it in a way which is intended to emulate the operation of an analogue synthesizer. Two DCO oscillators provide the raw pitched sound source, and two parallel sets of modifiers follow. Each DCO has a separate envelope generator for controlling its pitch, although vibrato is provided by a single LFO. The DCO output passes through the DCW waveshaper, again with an associated envelope generator, and finally through a digital VCA, or DCA. Ring modulation and noise can also be added. By using just one of the two sets of DCO and modifiers, the polyphony is doubled.

The DCW or waveshaper behaves in much the same way as the VCF found in an analogue synthesizer. Harmonics can be gradually added to a sine wave, so that it changes into one of the eight waveforms, and this can be controlled by an envelope generator as well as tracking the keyboard note. This implies that the transfer function is changing dynamically, which would suggest that a great deal of complex processing is being carried out. However, by working backwards from the waveform, it is possible to work out what is really happening. Because waveshapers tend to add harmonics, not take them away, then the only way that a sine wave can be produced is if the basic waveform at the input to the waveshaper is a sine wave. The waveshape selection is thus used to change the transfer function of the waveshaper, not the waveform produced by the DCO. The waveshapes shown represent the final output of the waveshaper

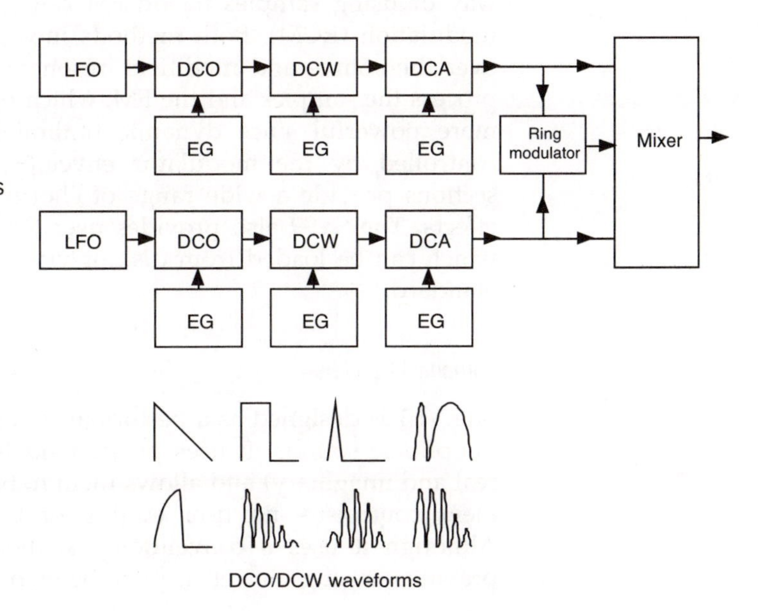

Figure 5.8.1 The Casio CZ101 uses waveshaping – but presented to the user as a DCO followed by a DCW (a digitally controlled waveshaper). The waveshapes provided include the sawtooth, square and pulse waves of conventional synthesis, plus five more unusual ones.

Figure 5.8.2 The Yamaha SY99 gives a comprehensively equipped set of FM and sample replay synthesizers, but also allows the samples to be reprocessed through the FM.

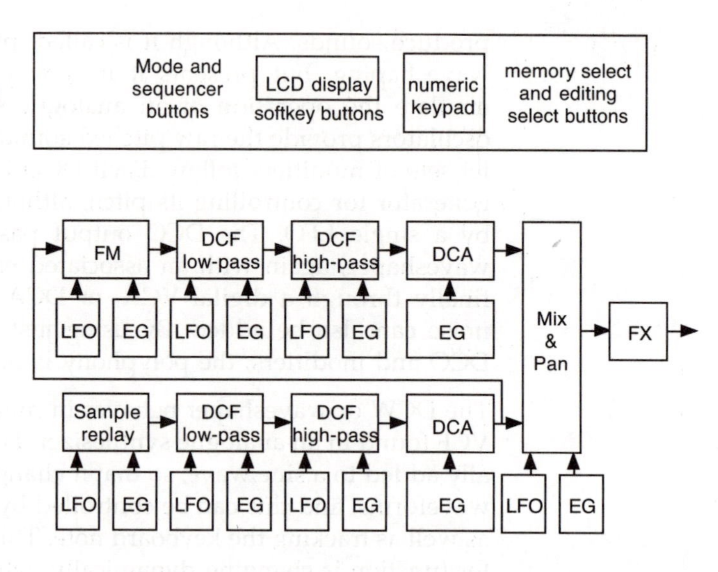

when the full range of the transfer function is being used. The waveshapes which are provided reinforce this: the sawtooth, square and pulse shapes are joined by a 'double sine', 'saw pulse' and three 'resonant' waveshapes.

Yamaha SY99 and SY77 (1991, 1990)

The 'RCM' synthesizers incorporate advanced versions of Yamaha's FM and sampling (AWM) technologies, as well as a way of using samples inside FM called real-time convolution modulation (RCM). Both methods incorporate detailed control over the source and modifiers: resonant filters can be used to process the samples and the FM, which makes the FM synthesis more powerful since dynamic timbral changes are not only controlled by the modulator envelopes. The built-in effects sections provide a wide range of chorus, reverb, EQ and echo effects. The SY99 also provides user RAM for storing samples which can be loaded from disk or via the MIDI Sample Dump Standard.

Yamaha VL1 (1994)

The VL1 is designed as a performance instrument and provides duophonic sounds. It uses preset models of instruments (both real and imaginary) and allows them to be controlled via instrument controls – no user editing of the models is allowed. Although it uses a conventional keyboard with velocity and pressure sensing, as well as pitch-bend, dual modulation wheels,

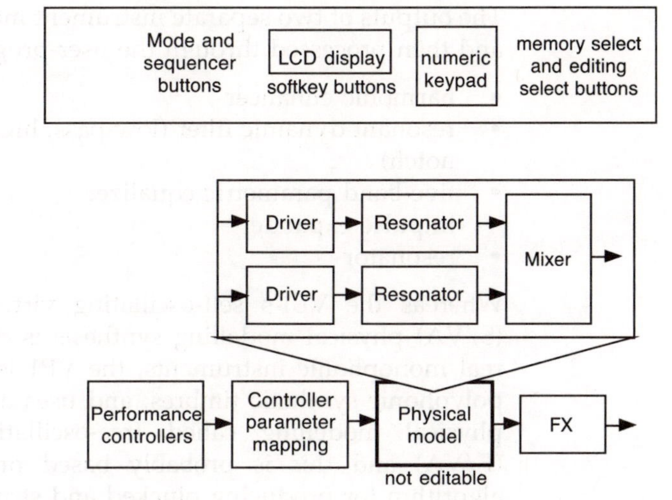

Figure 5.8.3 The Yamaha VL1 uses a fixed physical model to produce the sounds, but the user control of the model via MIDI controllers is very sophisticated.

pedals and breath controller inputs, these can be mapped to a large number of instrument controls, including:

- pressure (or bow speed)
- embouchure (tightness of lips or bow pressure on the string)
- pitch (the length of the tube or string)
- vibrato (affects pitch or embouchure via an LFO)
- tonguing (simulates half-tonguing damping of saxophone reed)
- amplitude (controls volume without changing the timbre)
- scream (drives the whole instrument into chaotic oscillation)
- breath noise (adds widely variable breath sound)
- growl (affects pressure via an LFO)
- throat formant (simulates the players lungs, throat and mouth)
- dynamic filter (controls the cutoff frequency of the modifier filter)
- harmonic enhancer (changes the harmonic structure of the sound)
- damping (simulates air friction in the tube or on the string)
- absorption (simulates high-frequency loss at the end of the tube or string)

As you can see, most of the controllers are very specific to real instruments – even though the sound which is being controlled may be one that doesn't and can't exist! There are individual scaling curves and offsets for each controlled parameter, so you can adjust the effect of a controller like breath control to do exactly what you want with great precision.

The outputs of two separate instrument models can be combined and then processed through the user-programmable modifiers:

- harmonic enhancer
- resonant dynamic filter (low-pass, high-pass, band-pass and notch)
- five-band parametric equalizer
- impulse expander
- resonator

Whereas the VL1's self-oscillating virtual acoustic Synthesis (S/VA) physical modelling synthesis is designed to synthesize real monophonic instruments, the VP1 is intended to produce polyphonic synthetic timbres, and uses a different variation of physical modelling called free-oscillation virtual acoustics (F/VA) and this is probably based on the Karplus–Strong algorithm for producing plucked and struck sounds.

Technics WSA1 (1995)

The WSA1 takes the best parts of physical modelling and combines them with the familiar parts of S&S. Rather than model a complete instrument, it takes the driver and resonator split, and provides preset driver 'samples', and connects them to a programmable resonator – and includes physical model-like interaction control of the resonator. The output of the resonator passes through a conventional DCF, DCA synthesis section.

The driver 'samples' in the WSA1 are not really equivalent to the samples you find in S&S synthesizers. They do not have the

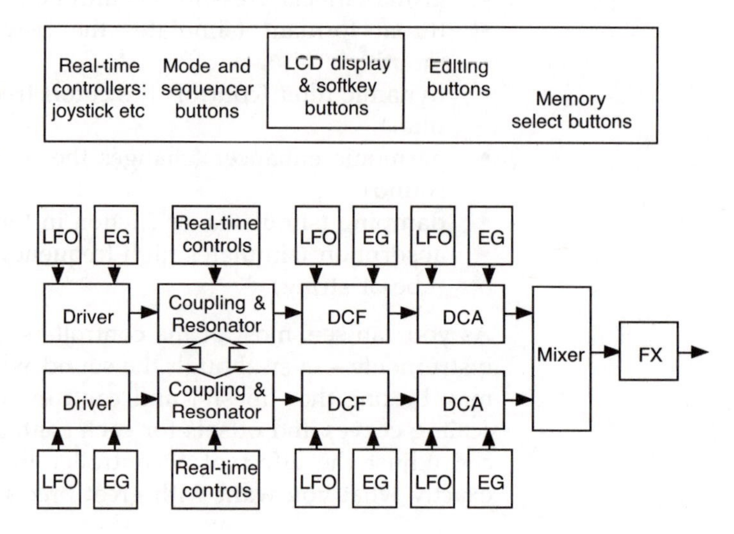

Figure 5.8.4 The Technics WSA1 uses interaction emulation to provide 'physical model' like capabilities. Two of the four synthesis sections are shown, and the resonators in each of these can be coupled to provide more complex resonators. Two front panel real-time controllers provide 'live' user control over the timbre: a conventional joystick and a tracker ball.

same emphasis on length/time and multi-sampling that conventional samples have because of the use of resonators to modify the sound of the driver 'sample' in a way that would normally require multi-sample. The driver/resonator model works extremely well – the bass and snare drums are an excellent example where although the basic driver sounds usable on its own, putting it through a resonator suddenly makes it more 'drum-like'.

5.9 Questions

1. Compare and contrast the major features of analogue and digital methods of synthesis.
2. What are the two common artifacts which can result from digital synthesis?
3. What happens to the output of the carrier oscillator as the level of the modulator oscillator is increased in an FM synthesis system?
4. What are the three basic parameters which define a static FM timbre?
5. What is the difference between waveshaping and a guitar 'distortion' pedal?
6. What is important about the relationship between the input frequency and the additional frequencies which are produced by a waveshaper?
7. Why is an audio signal always sampled at a rate of a least twice the highest frequency component?
8. Multi-sampling each note on an instrument is one approach to obtaining maximum realism from a sample replay instrument. Suggest an alternative way of using several samples to enable accurate reproduction of the dynamics of an instrument.
9. Describe one way of splitting a musical instrument into separate parts that may be useful in producing a physical model of the instrument.
10. What is the connection between FOF, VOSIM and human speech?

Time line

Date	Name	Event	Notes
1877	Thomas Edison	Inventor of the cylinder audio recorder – the 'phonograph'. Playing time was a couple of minutes!	Cylinder was brass with a tin foil surface – replaced with metal cyclinder coated with wax for commercial release
1878	David Hughes	Moving coil microphone invented	
1888	Emile Berliner	First demonstration of a disc-based recording system – the 'gramophone'	Disc was made of zinc, and the groove was recorded by removing fat from the surface, and then acid etching the zinc
1898	Valdemar Poulsen	Invented the telegraphone, which recorded telephone audio onto iron piano wire (also known as the dynamophone)	30 seconds recording time, and poor audio quality
1903	Double-sided record	The Odeon label release the first double-sided record	Two single-sided records stuck together?
1915	E. C. Wente	Produced the first 'condenser' microphone using a metal plated insulating diaphram over a metal plate	Now known as a 'capacitor' microphone
1920	Louis Blattner	The first magnetic tape recorder	Blattner was a US film producer
1920–1950	Musique Concrète	Musique concrète	Tape manipulation
1920s	Harry Nyquist	Developed the theoretical basis behind sampling theory	Nyquist frequency named after him
1920s	Microphone recordings	First major electrical recordings made using microphones	Previously, many recordings were 'acoustic' – using large horns to capture the sound of the performers
1930s	Record groove direction	Some dictation machines record from the centre out instead of edge in	This pre-empts the CD 'centre out' philosophy
1930s	Run-in grooves	Run-in grooves on records invented	Previously, you put the needle into the 'silence' at the beginning of the track
1935	AEG, Berlin	AEG in Germany used iron oxide backed plastic tapes produced by BASF to record and replay audio	Previously, wire recorders had used wire instead of tape
1937	Tape recorder	Magnetophon magnetic tape recorder developed in Germany	The first true tape recorder
1940s	Wire & ribbon recorders	Major audio recording technology uses either steel wire or ribbon	High speed, heavy and bulky – and dangerous if the wire or ribbon breaks!
1948	Pierre Schaeffer	Musique concrète	
1949	C. E. Shannon	Publishes book *The Mathematical Theory of Communications*, which is basis for subject of information theory	Shannon's sampling theorem is basis of sampling theory
1950s	Tape recorder	Magnetic tape recorders gradually replace wire and ribbon recorders	There were even domestic wire recorders in the 1950s!

Date	Name	Event	Notes
1951	Herbert Eimert	Northwest German Radio NWDR in Cologne starts experimenting with sound using studio test gear	Used oscillators and tape recorders to make electronic sounds
1958	Charlie Watkins	Charlie Watkins produces the Copicat tape echo device	
1958	RCA	RCA announces the first 'cassette' tape – a reel of tape in an enclosure	Not a success
1960s	Mellotron	The Mellotron, which used tape to reproduce real sounds	Tape based sample playback machine
1963	Philips	Philips in Holland announces the 'Compact Cassette' – two reels plus tape in a single case	A success well beyond the original expectations!
1978	Philips	Philips announce the Compact Disc (CD)	This was the announcement – getting the technology right took a little longer
1979	Fairlight	Fairlight CMI announced. Sophisticated sampler and synthesizer	The start of the dominance of computers in popular music
1980	E-mu	Emulator – first dedicated sampler	
1980	E-mu	E-mu Emulator released. This was the first widely available digital sampler	8-bit, unusual in appearance and with limited facilities
1981	Roger Linn	The Linn LM-1: world's first programmable digital drum machine	Replays samples held in EPROMs
1985	Akai	The S612 was the first affordable rack-mount sampler, and the first in Akai's range	12-bit, Quick-Disk storage and only 6 note polyphonic
1985	Ensoniq	Introduced the 'Mirage', an affordable 8-bit sample recording and replay instrument	
1988	Akai	S1000. One of the first professional quality stereo 16-bit samplers	S1100 was expanded and enhanced version
1988	Yamaha	TX16W sampler released. Rack-mount, stereo, 12-bit model with sophisticated filters	Lots of editing facilities. Now supported by 3rd party software
1992	Akai	S01 released. Low-cost 16-bit, monophonic sampler	A move away from the top-of-the-range pro Akai samplers
1992	Kurzweil	The K2000 is launched. A complex S&S instrument, which mixes sampling technology with powerful synthesis capability	
1993	Akai	CD3000. 16-bit stereo sampler with sampling from the built in CD/CD-ROM drive	A reflection of the growth in popularity of sample CDs
1993	Roland	S760 sampler released. A low-cost 16-bit stereo sampler with optional external TV monitor display	
1995	E-mu	ESI32 sampler released. 32 note polyphonic, 16-bit stereo sampling at a low price	A contrast to E-mu's top-end samplers

6 Using synthesis

The chapters so far have concentrated on the theory behind the methods of synthesis. This chapter deals with the use of synthesis to make music and other sounds, whilst Chapter 7 looks at how synthesizers are controlled.

6.1 Arranging

At the simplest level, the user interacts with the synthesizer to produce sounds. This often means directly controlling the synthesizer directly by using a keyboard or other controller. The synthesist is thus carrying out a role which is analogous to that of a player in an orchestra.

But when the control is indirect, as in remotely controlling a synthesizer by using MIDI or other computer-based means, then the role is much closer to the conductor of an orchestra – but with the option of simultaneously being able to act as an individual performer as well. This dual role actually provides the synthesist with something much closer to the detailed control and freedom which is available to an arranger, rather than the decreasing degrees of freedom which are available to the conductor and the performer.

Synthesis thus provides the ability to control three layers of a performance:

- choice and control over the use of timbres (arranger)

- precise control of the timing and dynamics of individual sounds in context, as they are produced (conductor)
- very low-level detailed control over the production of timbre (player).

For an arranger, working with a conductor and an orchestra requires a transfer of instruction via the score, demonstration or the spoken word. But for a synthesist, changes to the performance or the conducting require only a change of role. (This can make the task of a synthesist working as part of a larger orchestra a difficult set of compromises regarding control over the synthesizer.) Many multi-timbral synthesizers provide this 'three layered' level of control within one instrument, since they can produce several different independent timbres simultaneously. The skills of arranging are thus applicable from single synthesizers up to large 'synthesizer orchestras', or even a combination of real and synthetic instrumentation. In this section, the emphasis will often be focused onto individual sounds, but the same principles equally apply to larger structures.

The sales literature for many synthesizers concentrates on the production of sounds, and not the wider aspects of arranging. This chapter is not intended to be a guide to arranging, since that is a complete topic of study, but instead to show some of the tools and techniques which are available to the synthesist. There are three major divisions:

- **Arranging**: This covers topics like stacking, layering and hocketing.
- **Timbres**: This covers topics like multi-timbrality and polyphony, general MIDI and the use of effects.
- **Control**: This covers topics like the use of performance controllers and editing.

Many of the terms which are frequently used in connection with synthesizers are only loosely defined – or even redefined for this specialist usage. Stacking and layering are often confused and given different or overlapping definitions, and they are often used interchangeably in manufacturer literature. In this book, stacking will refer to a single composite sound produced from

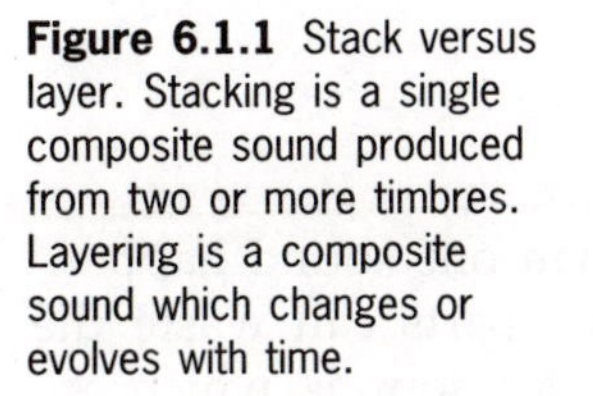

Figure 6.1.1 Stack versus layer. Stacking is a single composite sound produced from two or more timbres. Layering is a composite sound which changes or evolves with time.

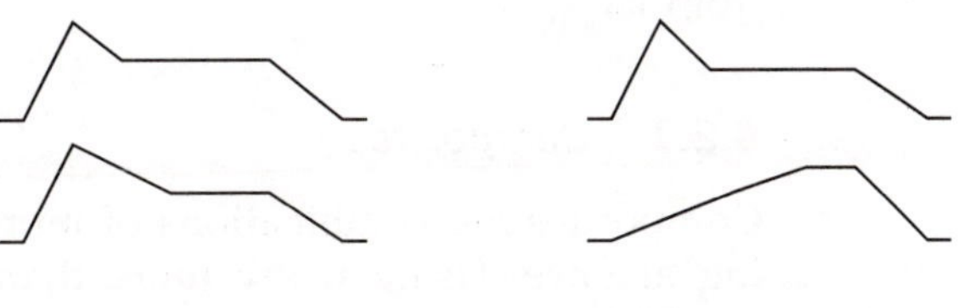

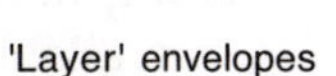

two or more timbres, whilst layering implies a composite sound which changes or evolves with time, and thus exhibits dynamic changes of separate 'layers' of sounds (Figure 6.1.1).

There are two different methods which can be used to produce stacking or layering: several separate synthesizers can be used, or alternatively, a single multi-timbral synthesizer can be used. These two ways of achieving stacking or layering are rather more than just differences in scale or equipment.

- Physically wiring synthesizers together with MIDI cables requires a MIDI patchbay and specific synthesizers – which can make it difficult to store or move from one location to another unless the same equipment is available in both places. This is thus a large-scale structure, typically within the MIDI network – and stored in many different locations, each of the individual sounds within the synthesizers, plus the MIDI setup held in the patchbay or a sequencer setup.
- Multi-timbrality is implemented as part of the operating system functions of a synthesizer and so is achieved using software. Stacks or layers produced by utilizing multi-timbral features can thus be stored as part of a performance memory inside the synthesizer, which makes transportation simple – everything required for the sound is contained within one instrument.

In this chapter, this distinction between the sources of sounds will be largely ignored so that the details of the techniques are not obscured. The phrase 'synthesizer sound source' or the word 'part' will be used whenever a sound can be produced by a separate synthesizer or by a section of a multi-timbral synthesizer. But the practicalities of employing the methods described above should still be considered when they are being used in practice.

6.2 Stacking

Stacks can be divided into two types: composites and doubles. Composites are concerned with multiple sounds or timbres, whilst doubles take one sound and use multiple pitches derived from it.

6.2.1 Composites

Composites are combinations of more than one sound happening at once. Using many more than two parts can waste the available polyphony, and can also produce sounds which are over-complicated. The best sounds are often simple but distinctive – which is harder to achieve than you might think. There are several methods of producing composite sounds.

Additive

Simply combining two or more sounds at random rarely produces useful results. By choosing simple sounds it is possible to use the same abstraction as additive synthesis to produce composite sounds which are made up from much simpler sounds. This technique is particularly useful when the synthesis techniques used are limited in their timbral possibilities: for example, subtractive synthesis often only provide low-pass filtering – but by adding together two sounds produced by subtractive synthesis it is possible to produce more complex timbres which do not have the limitation of a single low-pass filter.

Hybrid

Hybrid sounds are produced using contrasting or complementary synthesis techniques. Although this often implies the use of physically different synthesizers, some multi-timbral instruments do allow different synthesis techniques to be employed for each part. The range of 'contrasting and complementary' techniques is large, but some possibilities include the following:

- **Analogue with digital**, where the 'natural' sound, variations in timbre and slight tuning often attributed to analogue synthesis can be used to complement the more precise and controlled 'digital' sound.
- **Imitative with synthetic**, where the resulting sound has some of the characteristics of the real instrument, but with enough artificiality to ensure that it is not mistaken for a purely imitative sound.
- **Familiar with alien**, where the final sound has some elements that are familiar to the listener, but which also includes additional elements which are unfamiliar. This can be useful for avoiding the over-use of a lush string sound as a generic pad or backing timbre. Examples include mixing violin samples with slightly pitch enveloped sawtooth waveforms, or adding munchkinized vocal sounds to conventional choral sounds.
- **Additive with FM**, where FM sounds are used to counter the inharmonic weakness of the additive synthesis technique.

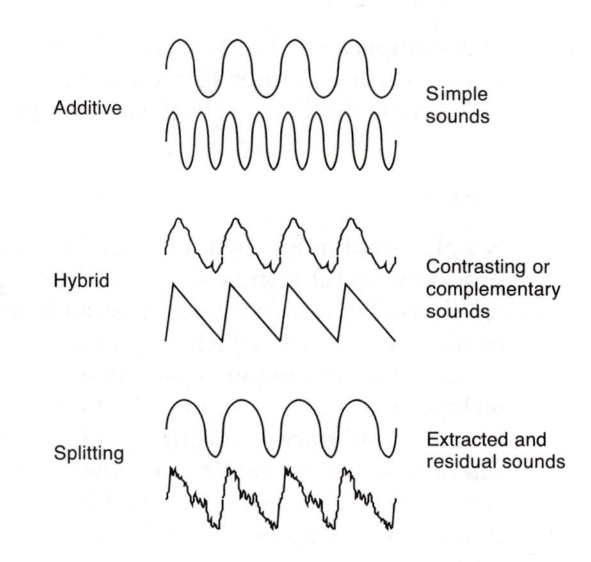

Figure 6.2.1 Composite stacks.

- **Sample with imitative**, where the basic sample is enhanced with additional imitative sounds to make it sound more like 'the real thing'. Curiously, many people prefer sounds which are made hyper-real in this way rather than exact copies.

Splitting

The process of splitting a sound into component sounds, then producing those sounds using separate sound sources, and finally combining them together to produce the final sound, is related to the analysis-synthesis methods of digital synthesis. But in terms of stacking, the usual method is to choose sounds which approximate to the required components and then iteratively combine them – with any analysis being done intuitively by the synthesist. Techniques such as residual synthesis, where a sound with a similar spectrum is 'removed' from a source sample, and the residual source sample is then used as the basis for further iterative extraction of sounds, do exist, and may be used in future synthesizers.

6.2.2 Doubling

Doubling is a hi-tech musical term which is used to describe the re-use of the same musical information. In contrast, the word 'double' is used in classical music to mean a variation. Doubling implies either a transposition or tuning change, with both the original and the doubled parts then playing together. Doubling using transpositions thus produces fixed parallel intervals.

Detuning

Detuning the two parts produces a 'richer' sound because of the chorusing providing additional 'movement' and interest in the timbre. There are two approaches to detuning parts: detune both away from 'in-tune' by opposite amounts, or detune just one. These two methods have different results when used in combination with other sounds, and the optimum is best chosen by experiment.

Octaving

Transposing a part up or down one or more octaves can be used to 'thicken' up a sound, although as the interval increases the parts will tend to be heard as two separate sounds. Octaving can also change the harmonization of the music.

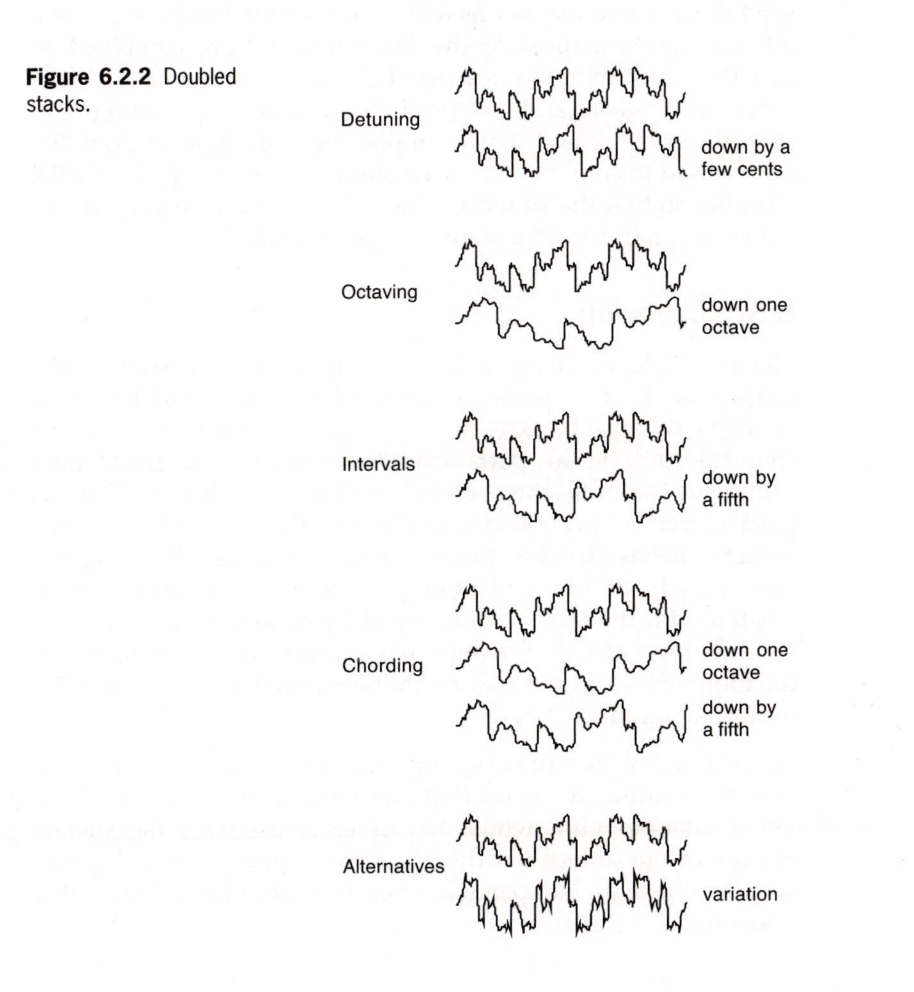

Figure 6.2.2 Doubled stacks.

Intervals

Transposing a part up or down by an interval other than an octave produces parallel pairs of notes. Fifths are commonly used, although this can change the harmonization of the music.

Chording

Transposing two parts away from a third part produces parallel chords. Although useful as a special effect, the constant chording can completely upset the harmonization of the original music.

Alternatives

Although not strictly doubling, it is possible to stack two very similar sounds from different synthesizer sound sources. This is used in the same manner as audio 'double-tracking', where two different performances of the same material are combined so that the result incorporates the slight imperfections and differences from each and so produces a more interesting and sonically 'rich' sound. For example, this variation on doubling can be used to combine the string sounds from two general MIDI modules so that the sound is 'thicker' and not as readily identifiable as coming from a standard GM source.

6.3 Layering

The principle of layering has already been discussed in the context of the two parts of many S&S sounds, but the same principle can also be extended to using more than one complete sound – individual parts can be layered, and then these composite sounds themselves combined in layers. This is particularly useful when equipment which is full of preset sounds needs to be made more personal. By layering customized sounds with the presets, new textures can be created with the minimum of effort by making hybrid sounds. For this type of use, conventional sounds are not as useful as the more unusual ones that might be rejected as being unusable in normal circumstances.

Layering involves a time separation of the sounds (Figure 6.3.1). This conventionally means that one sound is used to provide an initial attack sound, whilst the other is used for the sustain portion of the sound. But this is just one approach to using two sounds which are independent of each other in time. Some other possibilities include:

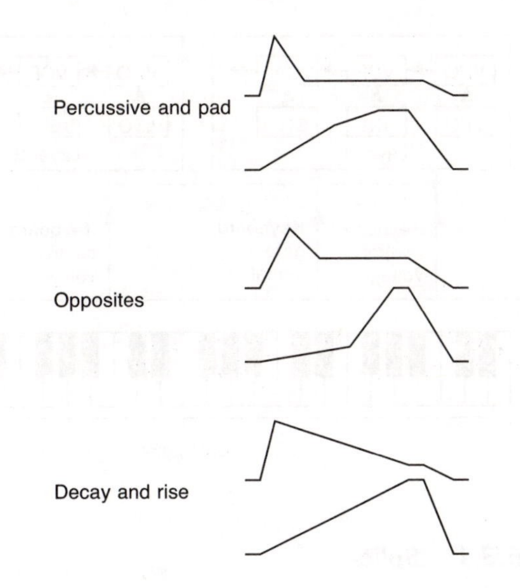

Figure 6.3.1 Layering is concerned with changes in envelopes over time.

- **Percussive and pad**: The combination of a fast attack/rapid decay sound with a slow attack, slow release pad sound can be very useful for providing accompaniment timbres from a single performance keyboard. By altering the playing from staccato to legato, or by using the sustain pedal, the level of the pad sound can be controlled.
- **Opposites**: By providing the two sounds with opposite envelopes, the result is a complex sound which dynamically changes between the two timbres. This is particularly effective with slowly evolving sounds, or sounds which have some common elements and some contrasting elements.
- **Echo/reverb and dry**: By layering two sounds which have different acoustic 'spaces', it is possible to dynamically change the apparent 'position' of a sound. Dry sounds tend to be perceived as being close to the listener, whilst echoed or reverberant sounds are interpreted as being further away. Two sounds which have different timbres and different echo timings can be very useful for creating polyrhythmic textures.
- **Pan position**: Layering sounds which have contrasting pan positions can be used to produce composite sounds which change their stereo position dynamically.
- **Slow decay and slow rise**: By using a sound which decays slowly and another sound which rises slowly whilst the first is decaying, the composite sound can have a 'sustain' sound which is not static, but which changes as the relative balance between the two layers changes.

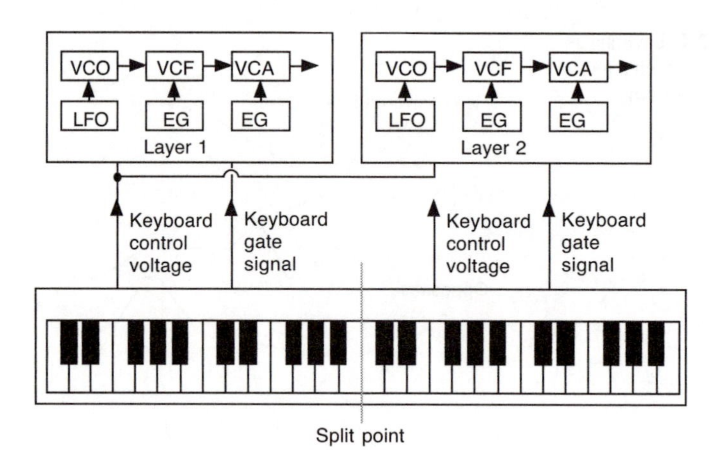

Figure 6.3.2 By splitting a keyboard and sending the keyboard CV to two 'layers' of a sound, it is possible to use the separate gates to control the second layer independently.

6.3.1 Splits

Although layering can be achieved by assigning both sounds to the same key, many synthesizers provide the ability to 'split' the keyboard, which allows different sounds to be used in different areas of the keyboard range. A typical application for jazz use might place a piano sound on the upper half of the keyboard, with a string bass sound on the lower half, perhaps with an additional brush sound. Less typically, the parts of a split can be used to produce different layers of a sound instead of different notes – this is easy to set up on an analogue synthesizer, although it can also be produced using MIDI equipment with specialized software. This allows control over which layers of a composite sound are produced on a note by note basis, which is more akin to the sort of control available to an orchestra and not a keyboard-based synthesist.

6.4 Hocketing

Hocketing is the name given to the technique of sending successive notes from the same musical part to different instruments, instead of to the same instrument (Figure 6.4.1). With the use of different (complementary and/or contrasting) timbres, this can produce very complex sounding arpeggios and accompaniments, and combined with doubling and layering, can give the impression of a very detailed arrangement, when in fact all that has happened is that a few notes have been moved from one track to another, and a few tracks have been copied and pasted.

Hocketing can be achieved in several ways:

- by note

Figure 6.4.1 Hocketing involves changing the instrument that plays a note based on criteria like the MIDI note number, or the MIDI velocity, or the position of the note in the bar or in time. In this example, the four notes are hocketed to two instruments using the note number.

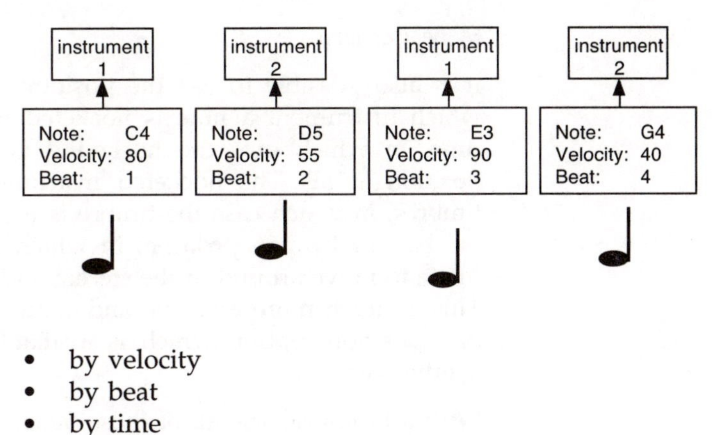

- by velocity
- by beat
- by time

By note

Hocketing by note is the obvious one; you put the first note of the arpeggio or melody to the first instrument, the next note to a second instrument, and so on, with perhaps the third or the fourth note being allocated to the first instrument again. With similar timbres for each of the hocketed instruments, the effect is quite subtle, and by using slight detuning of the instruments, can produce very 'non-synthetic' ensemble effects. Using different pan positions for the hocketed instruments can be used to provide movement in a sound which would otherwise be static. With contrasting timbres, the effect of the hocketing is more like splitting the part into separate parts.

By velocity

Using velocity to determine which instrument plays the sound involves selecting only those notes which have a velocity value above a specific value (half-way, 64, is a good starting point) and then allocating those notes to a different instrument. This is like the velocity splitting available on some samplers and S&S instruments, although the split point can be edited as part of the score or sequencer information instead of as part of the sound. Hocketing by velocity can be particularly effective on sounds which move between an accompaniment and melody role. By using alternative sound sources for the velocity-hocketed notes, it is possible to have different snare sounds selected from different drum machines or samplers merely by editing the velocity. Reducing the velocity sensitivity of the sounds allows the velocity to be used as a mixing control rather than a dynamics control.

By beat or time

It is also possible to use the position in the bar to determine which instrument a note is hocketed to, or even the absolute time, in which case the hocketing is not related to the bar position at all. The hocketed instruments can have different timbres, in which case the timbre is reflected by the location in the bar, or by pan position, in which case the sounds can be made to move around in the stereo field in time with the music. This is much more effective and controllable than the 'random pan position' option which is available on some commercial synthesizers.

With a little practice, all of these 'hocketing' edits are relatively quick and easy to make on a computer sequencer, and yet they can greatly improve the detail and quality of the finished music.

6.5 Multi-timbrality and polyphony

Multi-timbrality is the name given to the ability of a single synthesizer to produce several different timbres at once. This has the effect of turning a single physical synthesizer or expander module into several 'virtual' ones – although some of the functions remain common: normally the effects, the overall control and the management of the sounds.

Multi-timbrality is an extension of the concept of 'stacking'. If an instrument can produce two different timbres at once when a key is played, then it is logical to extend this capability so that the two different timbres can be played independently. The history of multi-timbrality is closely connected to the development of polyphonic synthesizers and the differences between analogue and digital synthesis.

Many sounds and instruments are naturally monophonic: one sound at once. Most people can only sing one note at once, and many acoustic instruments will only produce one note: flute, tuba and triangle are some examples of naturally monophonic instruments. The availability, price and complexity of early modular analogue synthesizers meant that they tended to be used as monophonic instruments too. Recording onto tape allowed the single notes to be combined into complete 'polyphonic' performances. Even when the synthesis resources were available, trying to keep two or more notes in tune and with the same timbre could be harder than using tape.

The first true polyphonic synthesizers were based around simplified analogue monophonic synthesizer VCO/VCF/VCA circuits to provide the notes. These cards are often called 'voice' cards, since each card provides a single 'voice' – rather like independent singers in a choir. These synthesizers could be operated in two modes: either with common control of timbre and independent control of pitch, or with independent control of timbre and pitch. The mode with common control of timbre allowed true polyphonic operation, where several notes with the same or similar timbre could be played simultaneously. The mode which provided the independent control of timbre allowed stacking and layering of sounds, which reduced the number of different notes which could be produced simultaneously. Because of the problems of tuning and controlling voice cards, there was a practical limit of about eight-note polyphony. Even with eight voice cards, the minor variations in tuning and timbre can produce a distinctive open 'feel' to the timbres produced – rather like the ensemble playing of a violin section in an orchestra. Eight-note polyphony is very restricting for producing multi-timbral music, and so analogue polyphonic synthesizers were used more for their polyphony than their multi-timbrality.

Polyphonic digital synthesizers like the Yamaha DX7 started out as mono-timbral in 1983, although digital technology provided larger polyphony than many of the analogue mono- or multi-timbral polyphonic synthesizers: 16 notes instead of the 4, 5, 6 or 8 note polyphony found in the analogue instruments of the time. But the precise tuning and identical timbres gave a very different 'feel' to these early digital instruments, and it was not until the early 1990s that the technology for emulating the imperfections of analogue polyphonic synthesizers began to be implemented in digital synthesizers. But the availibility of larger polyphony meant that utilizing the multi-timbrality was easier and more effective. The introduction of 8-part multi-timbrality in 16-note polyphonic synthesizers was quickly followed by 16-part multi-timbrality, which has formed the basis of most synthesizer specifications ever since.

6.5.1 Definitions

- **Polyphony** is the total number of different pitches or sounds that an instrument can make at any one time. For a single sound, the polyphony is the number of different pitched notes that the instrument can play simultaneously.
- **Multi-timbrality** is concerned with the number of different sounds or timbres that can all happen at once, and played by at least one note each.

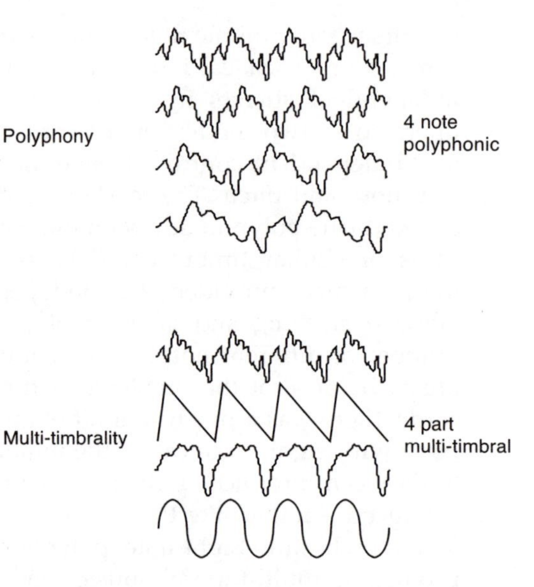

Figure 6.5.1 Polyphony is the total number of different pitches or sounds that an instrument can play simultaneously. Multi-timbrality is the number of different sounds or timbres that can happen simultaneously.

The word '**part**' is frequently used to indicate a separate timbre, and although it can be used in the same sense as a musical 'part', the two terms are not necessarily synonymous.

The word '**voice**' is also sometimes used to mean a single sound generating section: a monophonic part.

So a 16-note polyphonic instrument can play a sound using up to 16 different pitches at any one time, whilst a 16-'part' multi-timbral instrument can play up to 16 different sounds or timbres at any one time. The two different numbers are independent, although the specifications for synthesizers frequently quote the same number for both the polyphony and the multi-timbrality. This has continued even when the multi-timbral parts have exceeded the 16 which can be carried by a single MIDI cable. Instruments with 32-note polyphony are frequently quoted as having 'up to' 32-part multi-timbrality, even though using all 32 parts requires the use of two separate MIDI in sockets and a multi-port MIDI interface.

As a general rule, the polyphony should always equal or exceed the multi-timbrality, so a 16-note polyphonic, 16-part multi-timbral instrument is feasible, whilst a two-note polyphonic, 16-part multi-timbral instrument is not possible: it can only ever make the sounds for two parts. Actually utilizing the 16 independent monophonic parts from a 16-note, 16-part multi-timbral instrument can be harder than it might initially appear because of the need for true monophonic sequences of notes for each part: any overlap can increase the polyphony, which automatically reduces the available multi-timbrality.

The important element of the definition of multi-timbrality is the simultaneity. With a monophonic synthesizer it is possible to make a large number of different timbres, but it is always mono-timbral: one timbre at once. If a synthesizer has two separate sound generating sections, then it has two-part multi-timbrality – and so it is two-part multi-timbral. But by adding in a third sound generating section, then each separate part that can happen simultaneously with other parts only adds one to the multi-timbrality. The number of available timbres does not affect the multi-timbrality – there may be a large number of ways of combining two timbres, but the multi-timbrality is fixed by the polyphony and the number of simultaneous parts, not the timbres.

6.5.2 Notes per part

Using multi-timbrality can require a surprisingly large polyphony: even a 64-note polyphonic instrument can only play four simultaneous pitches on 16 multi-timbral parts. Trying to produce 16 multi-timbral parts using only 16-note polyphony can result in only one note per part if everything plays at once – and many synthesizers have limitations as they approach their polyphony or part limits. Producing music using only one note per part also has problems; although orchestral arrangements use monophonic melody lines played on 'one note at once' instruments like oboes, flutes and trombones, they use several players to get the required polyphony (and volume, of course). And an orchestra also has fixed limits to polyphony: if there are four violin players, then you can only have four separate violin parts.

The transitions between the notes are the source of the difficulty when making music using only one note per part on a polyphonic keyboard. Staccato playing, with gaps between the notes, gives controlled use of polyphony. But legato playing, tied notes, or overlapping notes can all cause short overlaps. Each overlap uses up two notes of the available polyphony – and because of the way that music tends to concentrate events around bar or beat intervals, so the overlaps 'cluster' across parts. So it is likely that an overlap between two notes within one part will happen at the same time as overlaps in other parts. The required polyphony can thus be much higher than the apparent polyphony of the music: anything up to double in normal circumstances, and even higher in more exceptional cases. A piece of music written for a quartet of four monophonic instruments can have a peak polyphony requirement of eight notes.

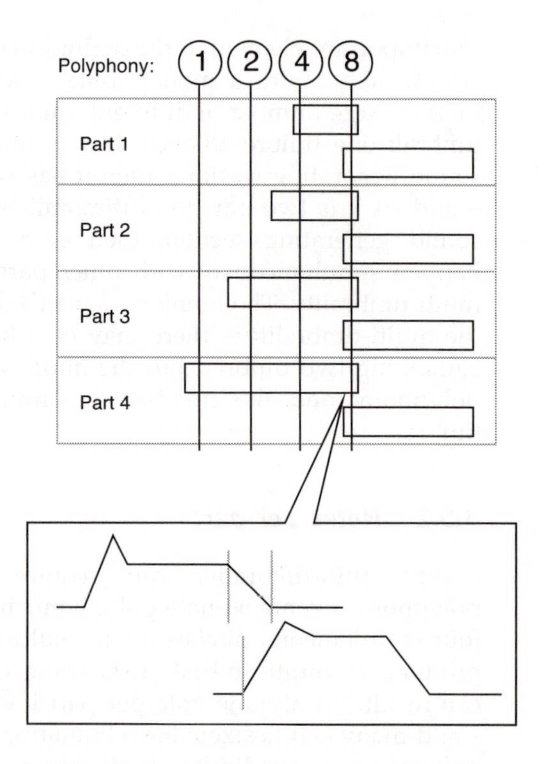

Figure 6.5.2 The peak polyphony depends on the overlaps between notes. Even if there appears to be a gap between the 'gate' signals, the release of one sound can carry on past the start of the attack of the next note.

When the polyphony limit is reached, notes are inevitably lost. If there aren't enough resources to play all the required notes at once, then some that are already playing will have to stop to enable the new notes to be heard. This produces the effect known as 'note stealing', where 'old' notes make way for 'new' ones. The behaviour of synthesizers when note stealing is taking place varies: it depends on the method used to assign the notes to the available sound-making resources inside the instrument – in an analogue synthesizer these would be the voice cards, and so these resources are often called 'voices'. A 'voice' is thus an imaginary part of a synthesizer which is capable of playing only one note, rather like a virtual polyphonic synthesizer voice card.

6.5.3 Note allocation

The way that the required notes are assigned to the available 'voices' in a synthesizer is called note allocation or voice assignment. There are a number of techniques for doing this, and they all extend the basic idea of re-using the voice resources on demand. This underlying process is called cyclic assignment, and it assigns the incoming notes to voices in order: the first note is assigned to the first voice, the second note to the second voice,

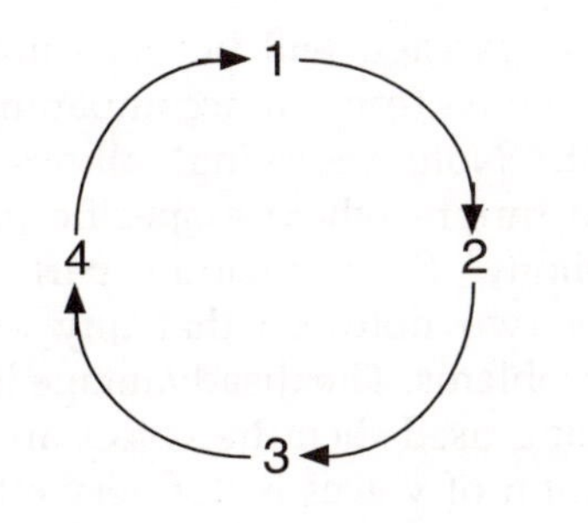

Figure 6.5.3 Cyclic assignment assigns incoming notes to available voices in sequence. In this example, there are four voices available, and the first four notes are assigned to these. The fifth note will replace the first note on voice 1.

etc. The notes which are played could come from either the instrument's keyboard, from an internal sequencer, or via MIDI from an external keyboard, instrument or sequencer.

An eight-note polyphonic synthesizer would contain eight of these 'voice' resources, and so after all eight voices had been assigned, then the next available voice would be the first one again. The ninth note would thus cause the first voice to stop playing the first note, and this voice would be used to play the ninth note. Exactly the same 'stealing' would occur with the seventeenth note – the first voice would again be used.

Cyclic assignment does not take into account the status of the voices. If the first note is sustained, whilst the next seven are short staccato notes, then seven of the eight voices will be 'available' for the ninth note since they are not actually producing a sound. A pure cyclic note assignment strategy would ignore this and assign the ninth note to the first voice, thus stopping the sustained note. This approach thus wastes the voices, since the assignment is arbitrary and does not use the available resources efficiently.

There are many ways to improve the assignment strategy by making it responsive to the incoming required notes and the availability of voice resources. Some of these 'dynamic voice allocation' (DVA) approaches include:

- Note reserving
- Part priority
- Voice status
- Envelope status
- Volume status
- Repeated note detection
- Sustain pedal detection.

Note reserving

Sometimes a particular part needs to always have a specific polyphony. For example, a note stolen from a sustained melody line, or a solo passage will be very noticeable to the listener whilst a note stolen from an accompaniment pad sound will be less apparent. 'Note reserving' allows a fixed allocation of polyphony to timbre where a specific part will always have a given polyphony. So a melody part might be allocated a polyphony of two notes so that any overlaps will not cause assignment problems. The disadvantage is that when the melody part is not being used then the voices are still not available, and so the utilization of voices is not very efficient.

By using fixed polyphony, it is possible to make eight-note polyphony sound like 12 notes. Two contrasting timbres are allocated to two parts, and assigned to the same pitch control channel (or MIDI channel). Note reserving is then used to allocate the voice resources asymmetrically: six notes and two notes for example. The result is like a six-note polyphonic version of the two-timbre stack, because the listener will tend to hear the last two notes played, which are always played with the two timbres – and not the remaining four notes which are played on only one timbre. This is very effective when the two timbres are detuned relative to each other, since the result sounds like 12 detuned notes when only eight are actually being used.

Part priority

Assigning priorities to parts allows a melody or a drum part to be relatively immune from note stealing, without the need for the permanent allocation of note reserving. By assigning a priority number to parts, it is possible to force notes to be stolen from the 'less important' voices or parts which have lower priorities. The highest priority parts are only stolen from when all other resources are in use. Time priority also needs to be considered – a higher priority needs to be given to the most recent notes. Despite the complexity of allocating priorities, this method gives a more efficient use of the available polyphony than fixed reservation.

Voice status

By monitoring the status of the voices, it is possible to alter the cyclic assignment so that incoming notes are only allocated to voices which are not playing sounds. This means that some mechanism is needed to keep track of the 'playing/not playing' status of each voice. Note stealing then only needs to occur when there are no 'not playing' voices available.

Envelope status

Some portions of the envelope of a sound are less important than others. Stealing a note from a sound which is in the release segment is probably less audible than stealing from a sound which is in the attack segment. By monitoring the envelope status of each voice, it is possible to arrange to only steal sounds which are in the appropriate segments: normally the release or sustain segments.

Volume status

Volume, either controlled by the note velocity, overall volume or envelope position, can also be used to determine if a voice is a suitable candidate for stealing. Lower volumes or audio levels will be less audible if they are stolen by a voice assigner.

Repeated note detection

If the same note is allocated using a cyclic assignment, then the allocation will run through each of the available voices in turn, even though the same pitch is being played each time. By detecting the repeated note, and re-assigning it to the same voice, perhaps with a retriggering of the envelope, the remaining voices will not be disturbed.

Sustain pedal detection

The sustain pedal on a synthesizer or MIDI device causes the notes which are being played when it is activated, or are played whilst it is active, to be held at the sustain segment of their envelope. If the number of incoming notes exceeds the available polyphony, then notes will be stolen, and because of the time-based nature of the sustain pedal event, the note stealing decision should probably be on a cyclic 'oldest-note' basis.

In order to allow the reservation, priority, status and other parameters to be tracked, the synthesizer needs to maintain running lists of all the parameters which might affect the note allocation. This changes the 'note assigner' circuitry, from the simple counter that is required for a cyclic assignment technique, to one which has to maintain detailed records for several separate voices. Such an 'intelligent' assigner can add a considerable software overhead, which may mean slower response times from the synthesizer to incoming notes. There is thus a compromise between the complexity of the note allocation scheme and the response of the synthesizer.

6.5.4 Using polyphony

Multi-timbrality and polyphony tend to encourage the use of the layering, splitting and hocketing techniques described in this chapter. Polyphony can be either over-used or under-used: too much doubling or detuning can over-thicken a sound and instead make it dense and obscure the harmonic structure; too little available polyphony can be improved by reserving or prioritizing parts, and perhaps even thinning out the chords used for accompaniment or reducing the number of simultaneous drum events.

There are a wealth of low-cost synthesizer expander modules which are available, both new and second-hand, which can be used to increase the available synthesis polyphony. As has been mentioned several times, a hybrid or mix of several instrumental sounds using different synthesis techniques can be very powerful.

6.6 General MIDI

General MIDI (White, 1993) is an extension to the MIDI specification, and is a shared set of specifications and guidelines agreed by the MIDI Manufacturer's Association and the Japan MIDI Standards Committee.

General MIDI uses the idea that most music can be reproduced using a small set of commonly used instrumental timbres. In effect, the concept provides a minimalistic 'electronic' orchestra. Many of the sounds chosen are orchestral or 'popular' in origin, although some additional synthetic sounds are also included. General MIDI is abbreviated to GM, and there is a distinctive General MIDI logo which can be found on compliant instruments.

There are 128 sounds, plus a drum kit mapping, in the basic GM sound set. The sounds are divided into 16 categories or 'groupings', each containing eight sounds:

- keyboards
- chromatic percussion
- organ
- guitar
- bass
- strings
- ensemble
- brass
- reeds

- pipes
- synth lead
- synth pad
- synth effects
- ethnic
- percussion
- sound effects

The numbering and names of the groupings and sounds are given in the GM specification. The drum kit (percussion) receives MIDI messages on channel 10, whilst the remaining channels can be used for any purpose: the allocation of channels to parts is left to the user. The GM specification requires that the polyphony of the GM playback device should be a minimum of either '24 fully dynamically allocated' notes, or 16 'dynamically allocated' notes for instrumental sounds and eight notes for percussion. Some GM modules cannot always meet this requirement. GM playback devices should also be 16-part multi-timbral.

The many General MIDI sound sets which are produced by manufacturers are all intended to sound as similar as possible, so that a MIDI file created using one General MIDI device will sound much the same when played back using another GM expander. This extends to how they combine together, how they respond to MIDI performance controllers like pitch-bend, velocity and volume. The result is a largely uniform range of instruments which can reproduce the same sounds in a predictable way.

The actual sounds that are included in the GM sound set are not intended as anything more than a means to an end: making a form of 'painting by numbers' music, where the MIDI file contains the information for both the musical events and the musical sounds. With minor differences, a given MIDI file will produce very similar results on any GM playback device. For use as a synthesis medium, GM offers very little. The sounds are chosen for their ubiquity of application, and many GM playback devices do not offer any editing of the sounds – sample replay is frequently used for producing the GM sounds. GM playback devices can be used in combination with synthesizers to produce composite sounds.

GM playback devices can range from professional synthesizers, through small expander modules, to personal computer sound cards and even software-only versions. Several manufacturers have introduced their own additional extensions to the basic GM specification, and these allow extra functionality to be provided by specifying additional sounds and controls.

6.7 On-board effects

In the 1970s built-in effects processors were the exception rather than the rule. Reverb, chorus and echo were added to the output of the synthesizer in the studio by using 'outboard' effects units. Built-in effects processors first began to appear in polyphonic hybrid synthesizers in the 1980s as a low-cost method of disguising the use of single DCOs instead of dual VCOs. String and piano sounds were significantly improved by the use of small amounts of chorus effect – and in fact, this spelled the end of the 'string machine' as a separate instrument. Whilst simple effects processors were commonly added to hybrid and S&S polyphonic synthesizers throughout the 1980s, it was not until the end of that decade that digital FM synthesizers were fitted with built-in effects processors. The 1990s have seen the effects unit change from being a simple 'after-thought' to an integral part of the synthesizer.

6.7.1 Effects history

Although effects processors have been used to process most electronic (and amplified) musical instruments for many years, the addition of effects processors to synthesizers has been a gradual evolution. The effects types which follow are organized into a rough chronological order of introduction.

- Reverb
- Chorus
- Echo
- Phasers and flangers
- Ring modulation
- Exciters, compressors and auto-wah
- Pitch-shifters
- Distortion

Reverb

Reverberation is the effect produced by almost any acoustic environment: a short delay (the 'pre-delay') is followed by a series of echoes from the boundaries (the 'early reflections') and then echoes of those echoes (reverberation) which decay in volume (the 'reverb time') as energy is transferred from the audio to the environment. Small containers have short delays whilst large rooms have long delays. The timbral quality of the reflections and reverberation are determined by the boundaries of the environment: smooth, shiny concrete walls will give long reverberation times and only slight high-frequency attenuation, whilst sound absorbent material like curtains will give short reverberation times and significant high-frequency attenuation.

Sounds without reverberation sound empty and synthetic; the human ear expects to hear sounds played in real spaces, and removing the effect of the space can sound very 'wrong'. Synthesizers without reverberation emphasize this perception, so adding reverberation is almost essential unless the synthesizer will be heard in a naturally very reverberant acoustic environment. Many commercial synthesizers suffer from having too much reverberation on their preset sounds – based on the assumptions that 'more is better' and that the synthesizer will be auditioned using head-phones. (It is easy to show that large amounts of reverberation can detract rather than enhance a sound: listening to a piano being played in a ballet or dance rehearsal room with lots of mirrors to reflect the sound is an excellent tutorial exercise.)

In the 1970s, reverb usually meant either a spring-line reverberation unit or a tape-based fedback flutter echo unit. The spring-line was prone to mechanical interference (they were microphonic, and knocking the casing was audible in the audio output), whilst the tape machines were prone to mechanical failure (tape loop splice failures or wearing out of components because of long periods of continuous use). In the 1980s, analogue bucket-brigade delay lines were replaced by digital effects, and the quality of reverberation improved markedly. The 1990s saw sophisticated DSP-based reverberation algorithms, with features like early reflections, pre-delays, and room size simulations. Reverberation has now become so ubiquitous that it is provided even on low-cost minimalistic GM modules.

On a synthesizer, reverberation can also be approximated by using an envelope release curve which starts out decaying rapidly, but then lengthens the release time as the level or volume of the audio signal decreases.

Chorus

The chorus effect is a cyclic detuning of the sound, mixed with the original sound. This causes the same type of phase cancellations that are associated with several performers playing the same notes on acoustic instruments: the violin section in an orchestra is a good example.

Chorus is normally achieved by delaying the audio signal in time, and then changing the time delay dynamically. This has the effect of changing the pitch as the delay time is altered, and so produces the detuned chorusing effect. Chorus can also be produced by deliberately detuning two VCOs, oscillators or sounds.

Echo

Echo is a repetition of the original audio signal after a specific time delay, and at a lower volume. It simulates the effect of having a remote large object from which audio sounds will be reflected. Rapid echoes are often called 'flutter echoes', whilst the timbral quality of echoes is largely determined by the object from which the audio sounds are reflected.

Echo is produced by time-delaying the audio signal, and feeding back the output to the input via an attenuator to produce additional repeat echoes. Echo can also be simulated by retriggering envelopes using an LFO.

Phasers and flangers

Phasers and flangers are variations on the chorus effect, with a mixing of an undelayed with a delayed audio signal, but with feedback from the output to the input. Phasers use a phase shift circuit, whilst flangers use a time delay circuit. In both cases, cancellations occur when the delayed and undelayed audio signals are out of phase, and so a series of narrow cancellation 'notches' is formed in the audio spectrum. As the phase shift or time delay is changed by an LFO, the notches move up and down in frequency. Phasers produce notches which are harmonically related because they are related to the phase of the audio signal, whilst flangers produce notches which have a constant frequency difference because they are related to the time delay.

Ring modulation

Ring modulation takes two sounds and produces sum and difference frequencies based on the frequency content of the input sounds (see Section 2.4.3). In effects processors, ring modulation normally uses an LFO or audio oscillator for one source sound, and the signal to be processed as the other. Ring modulation can make major changes to the timbre of a sound, normally making it metallic or robotic in sound. It is particularly useful for processing drum and percussion sounds, whilst for special effects it can provide sounds which are often suitable only for science fiction genres.

Figure 6.7.1 Effects summary. Reverberation and echo can be produced using a tapped delay line. Chorus, flanging, phasing and pitch shifting can be produced using a delay line whose delay time is modulated using an LFO. Exciters and distortion use non-linear amplifiers to produce harmonic distortion. Compressors and auto-wahs use an envelope follower to filter or attenuate the signal.

Input
Delay line
time delay taps
Output
Reverberation and echo

Input
Delay line
Output
LFO
Chorus, flange, phase and pitch shift

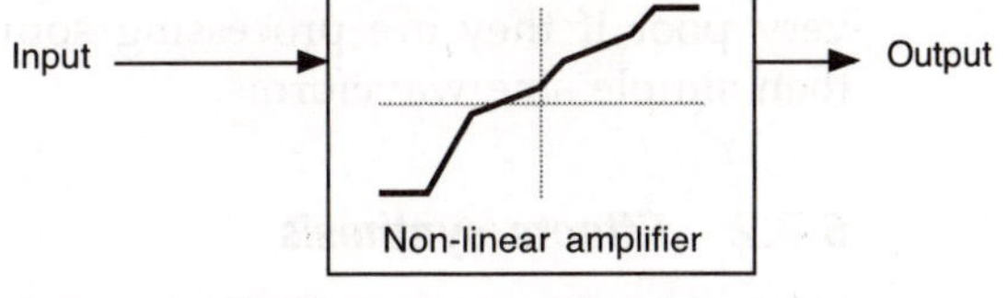

Exciters and distortion

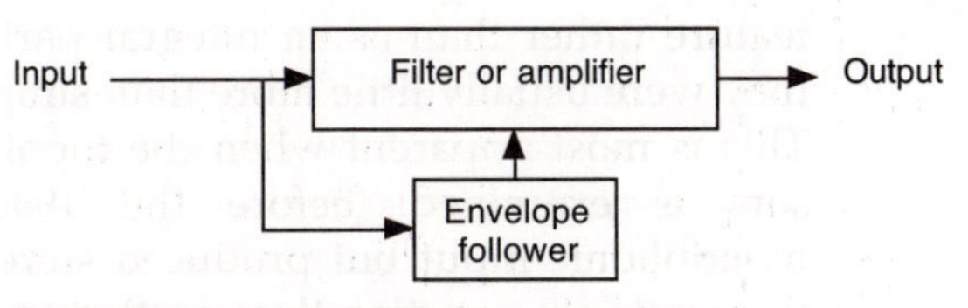

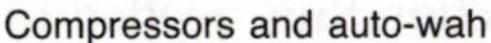
Compressors and auto-wah

Exciters, compressors and auto-wah

By the 1990s, the effects processors which were being incorporated into leading edge commercial synthesizers were similar to the top models of stand-alone effects processors from the same manufacturers. The range of effects programs expanded to include additional effects which were not based around time delays such as: exciters (small amounts of frequency-dependent harmonic distortion), compressors (dynamic range reduction) and auto-wah (envelope following/triggered filters). By the middle of the 1990s the first effects processors designed specifically for use with 'multi-timbral' synthesizers began to appear.

Pitch-shifters

Pitch-shifting is a variation on the detuning that takes place in a chorus unit, and is sometimes called harmonization. Instead of cyclically changing a time delay, the audio signal is stored at one rate and read out at another which produces a fixed pitch-shift. Slight pitch-shifts produce subtle chorus effects, whilst larger pitch-shifts produce transpositions (often at the loss of audio quality). Feedback from the output to the input can produce sounds which transpose up or down repeatedly from the initial input.

Distortion

Guitar-type 'fuzz' ranging from subtle harmonic distortion through to gross distortion can improve electric guitar and solo synthesizer sounds. Small amounts of distortion can also help to enhance any subsequent effects processing of sounds with limited harmonic content: filters, phasers and flangers can sound very poor if they are processing sounds which are little more than simple sine waveforms.

6.7.2 Effects synthesis

Until the early 1990s, the majority of effects processors which were built into synthesizers were still designed as an additional feature rather than as an integral part of the synthesizer – and they were usually little more than simple reverb or chorus units. This is most apparent when the topology of the effects processors is examined: before the 1990s, most had a single monophonic input but produced stereo outputs. The output of the synthesizer section thus needed to be combined into a single

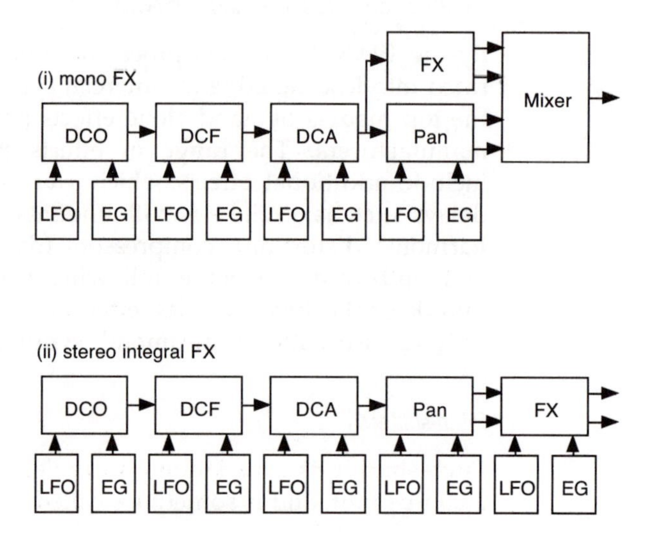

Figure 6.7.2 (i) Mono input effects processors are added in to the output of some synthesizers, but they cannot use the stereo output, and so are mixed in with the output of the pan processor. (ii) Stereo input effects processors can be integrated into the synthesizer, including using LFO and envelope generators.

channel before being processed by the effects – despite most synthesizers producing stereo outputs directly. True stereo effects, where the input and output are stereo, and the effects are part of the synthesis, only began to appear in the 1990s.

Using effects only for post-processing of sounds produced by a synthesizer is probably a consequence of the use of simple reverb or chorus units in early polyphonic hybrid synthesizers. As the effects have increased in complexity and become an essential part of the timbre-forming modifier section of the synthesizer, the need for linking the effects to the synthesizer controllers has correspondingly increased. By having some of the important effects parameters controllable in real-time from the synthesizer, the effects processing can be made an integral part of the synthesizer. Chorusing or flanging whose depth changes with filter cut-off, or panning which also changes the reverb time, are just two of the many possibilities which are often not possible with stand-alone 'out-board' effects processors which are not part of a synthesizer. This idea is not new: many modular analogue synthesizers have dynamically controlled effects, but it has only recently become available on commercial digital synthesizers.

When the effects become part of the 'sound', then the 'traditional' isolation between the effects and the rest of the synthesizer does not work. Instead of choosing a sound, and then choosing an appropriate effects setting, the sound and the effect need to be linked together. This is feasible whilst the synthesizer is only producing one sound, but poses a problem when it is being used multi-timbrally. Whereas most synthesizers can be split into 'voices' which each produce one note of a specific timbre by multiplexing the same circuitry, an effects processor requires a separate complete dedicated audio processing section for each timbre that it is processing. Most stand-alone effects processors are designed to process just one stereo signal, rather than anything up to 16 stereo signals simultaneously.

This means that an effects processor intended for use as part of a multi-timbral synthesizer is not just a minor revision of an existing stand-alone – it requires additional real-time control inputs and the duplication of a large proportion of the audio processing. A partial solution developed during the late 1980s and early 1990s was to provide several simple effects processors and to provide independent inputs and outputs via effects send and return controls in much the same way as on an audio mixing console. Since many effects processors were already being designed to produce several simultaneous effects, this avoided

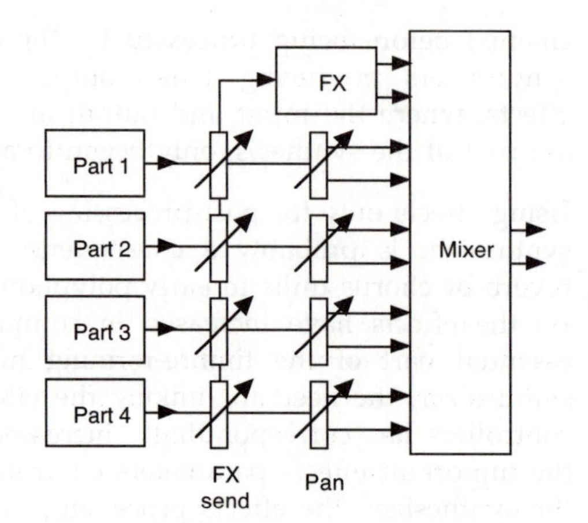

Figure 6.7.3 Some multi-timbral synthesizers use global fx busses to enable any of the parts to have an effect applied to them. This means that only one effect is available globally. More sophisticated synthesizers may provide two or more separate effects processors, but it is rare for there to be the same number of effects as parts.

any need for a major reworking of existing stand-alone designs. For multi-timbral synthesizers, this meant that two, three or perhaps four separate effects would be available, but that these were shared amongst all the parts. The topology of the effects processors was often predetermined by their original conception as a series of effects, with the result that a chorus or echo effect would always be connected to a reverb, and only limited combinations of chorus and reverb would be available – and then only globally.

This limitation was first addressed in the mid-1990s by providing global reverb and chorus effects processors, and additional separate processors which could be applied to individual parts. This 'global and individual' approach changed the nature of the effects processing inside a synthesizer away from the stand-alone 'outboard' studio use and started the development of a new type of effects processor. This process was completed when the first 'multi-timbral' effects processors were introduced into high-performance workstations in the second half of the 1990s. Although still incorporating some global reverb and chorus effects, separate effects processing for each part meant that the effect could be treated as part of the modifier section of the synthesizer.

6.7.3 Using effects

The user interface metaphor which is almost always used for the output of multi-timbral synthesizers, especially in the context of effects processing, is the mixing console or mixer. This is an example of the way that synthesizers have gradually incorporated external equipment: first the effects were built into the synthesizer, and then the mixer that would have mixed together several synthesizers becomes part of a multi-timbral synthesizer.

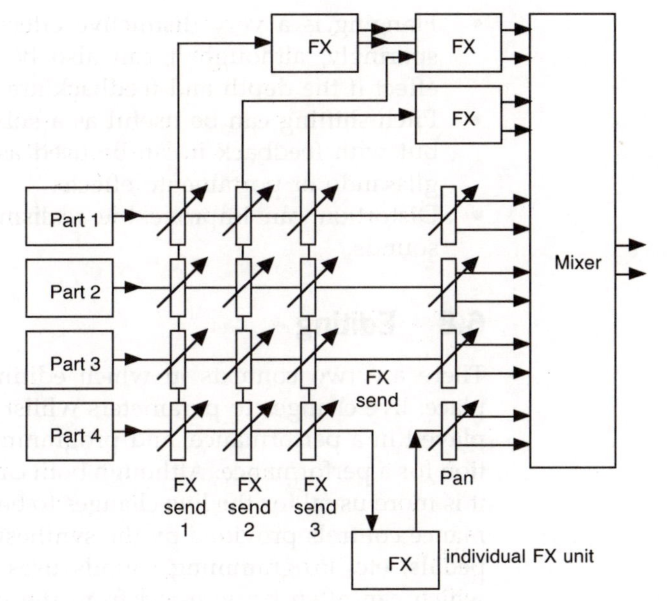

Figure 6.7.4 Some synthesizers allow access to the individual sections of multi-effects, so that a chorus followed by a reverb can then be used as a chorus and reverb as well as just a reverb. Individual effects processors that can be inserted into just one part are also found in some synthesizers.

The mixer takes the outputs of the voices and mixes these parts down into a stereo signal. Individual pan and level controls, and sometimes equalization, are included – but except on synthesizers with 'multi-timbral' effects processors, the effects provision is often implemented as several effects send and effects return controls. These are summed across the parts and sent to the specific effect, and then returned via the return level control – in some cases only the send control is implemented since the stereo outputs of the effects are mixed into the main stereo output and the effects output level control can be used to replace the effects return control. Individual send and return controls are used when an effects processor has one or more independent effects processors rather than a full 'multi-timbral' capability.

Making effective use of effects involves careful choice and frugality: using a deep flange on everything may sound impressive at first, but this initial impression soon fades.

- Reverberation is very good for providing a sense of space and size, as well as placing instruments closer to (less reverb, with short pre-delay and short reverb time) or further away (more reverb, with a longer pre-delay and longer reverb time) from the listener.
- Chorus can add extra movement to a static pad sound.
- Echo is useful for enhancing rhythmic elements, and can be used to provide syncopation effects by arranging the echo time so that it either falls on, or in between beats.

- Flanging is a very distinctive effect and needs to be used sparingly, although it can also be used as a chorus-type effect if the depth and feedback are kept deliberately low.
- Pitch-shifting can be useful as a subtle chorus replacement, but with feedback it can be used as a means of producing glissando or portamento effects.
- Distortion can improve the realism of electric guitar-type sounds.

6.8 Editing

There are two contexts in which editing of a sound can take place: live changes to parameters whilst the instrument is being played in a performance; and programming sounds in preparation for a performance. Although both can use the same controls, it is more usual for the live changes to be made using the performance controls provided by the synthesizer's keyboard, wheels, pedals, etc. Programming sounds uses more detailed controls which can often be accessed from the front panel or via MIDI messages from a computer.

6.8.1 Live parameter control

Making changes to a sound during live performance requires that the edits be easy to make and constrained to a specific parameter. This is usually set up by assigning a performance controller like a modulation wheel or foot pedal to a parameter which can be controlled using a MIDI controller message. The performance controller can then be used to change that single parameter with no concerns about changing any other parameters inadvertently. This is frequently used to provide live control over filter cut-off or a global envelope release time control when this is not available as a front panel control. Some synthesizers allow front panel controls to be set up in a similar way, although this is often as part of the sound editing facilities and is more prone to accidental editing of other parameters.

6.8.2 Programming sounds

Programming sounds is normally achieved either from the front panel or via MIDI messages from a computer running specialized editor software. The front panel method has the advantage of being rapid and requires no additional equipment, but the display is normally small and not well suited to the display of large numbers of inter-related parameters or detailed graphical information. Computer-based editing has the advantages of a

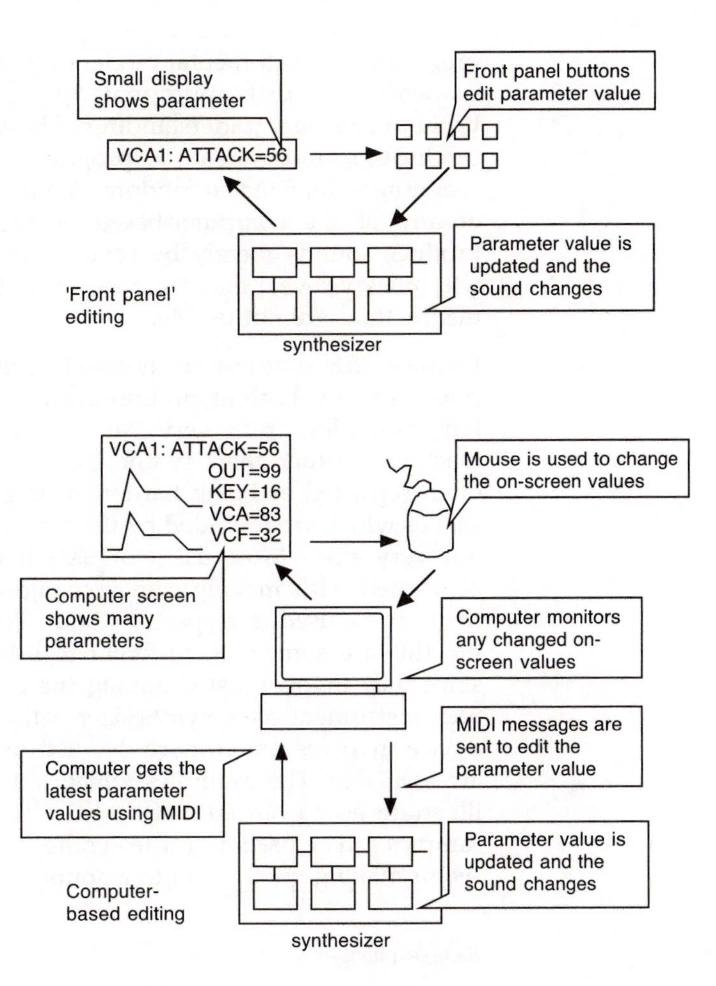

Figure 6.8.1 'Front-panel' editing involves a very direct interface between the user and the synthesizer – the front panel controls alter the parameter value which changes the sound. Computer-based editing uses the computer to display the parameter values which it has obtained from the synthesizer by using MIDI messages. Any changes to the parameters are made on the computer screen using a mouse or keyboard commands, and the changed parameters are transmitted to the synthesizer using MIDI messages.

larger screen and a more sophisticated user interface, but requires a computer and extra cabling, and the speed is limited by the screen update speed and the transfer of editing commands to the synthesizer. Both methods have their strengths and weaknesses – in general, the smaller the screen, the greater the advantage of the computer-based method for the beginner, whilst for the advanced user who requires quick edits, then the front panel may well be preferred.

6.8.3 Editing techniques

There are a number of techniques which are used by sound programmers to produce useful sounds. Most of these involve an iterative process where a parameter is changed, and the result noted, and then another parameter is altered, and its effect on the sound is listened to. One essential requirement is that the programmer has a mental model of how the method of synthesis works, so that appropriate parameters can be changed. Without any clear understanding of how the synthesis technique produces sounds, then the programmer would effectively be changing parameters at random. A critical audition of the results of any of the computer-based editing programs which can produce sounds merely by randomizing the parameter values will quickly reveal that the majority of the sounds produced by this method are not usable.

Because S&S synthesizers normally contain a large number of preset sounds, both as programmed sounds, and as the underlying samples, it is very easy to use them as replay-only machines, rather than synthesizers. For this reason, they are often regarded as being harder to program, since avoiding the clichés which are provided by the samples and modifiers is often not very easy. Most users of S&S instruments will be more concerned with making minor changes to the sounds, so that they can be used in a specific idiom. The techniques for achieving this are somewhat different to other types of synthesizer, since they involve just changing the coarse modifier. Using an S&S instrument as a synthesizer rather than a sample replay device involves much more detailed use of the facilities which are available. The examples which follow are thus designed to illustrate how even an instrument which is designed to replay samples can be used as a true synthesizer – with specific emphasis on making quick edits to a sound.

Sample changes

Choosing a different sample probably the simplest technique for making changes to an S&S sound. The 'sample selection' parameter is located and then changed. This changes the raw sound source of the S&S sound – and so should radically change the overall sound. The source samples are normally arranged in groups with similar timbres, so that several piano samples will be followed by a range of string samples, which will, in turn, be followed by another group of sounds. Choosing a different sample from the same grouping will only make subtle changes to the final timbre, whereas a sample from a different group may have a very big effect on the final sound.

Most S&S synthesizers have two sample sources which are processed separately and then combined either at the filter or at the output. This means that it is simple to leave one of the two samples alone, and just change the other. For example, many piano sounds are chorused by detuning two piano samples – but by replacing one of the piano samples with another sample, then the result is two different instrument sounds with the same envelopes, but detuned away from each other. The user is not restricted to choosing another piano sound, and in fact, there is no need to use a percussive sound at all – often string or brass samples, or even single cycle synthesizer waveforms can make a very effective contrast to a piano sample. Just leaving the detuned piano preset and changing both piano samples for other samples can produce some unusual timbres.

Often, the mixing of two contrasting timbres can give a new 'composite' instrument which has some elements of the two component timbres, but has a new character all of its own. The relative levels of the two components may need some adjusting to maximize this effect – the volume of each is changed until the sound 'gels' into one new timbre. It is a very striking effect: two separate sounds suddenly become one. This technique is widely used in orchestral music.

If a 'familiar' timbre is mixed with an unusual or 'alien' timbre, then the extra information in the 'alien' sample can give a sort of 'halo' around the existing sound – it expands and enhances, often in an unexpected way. Adding bell-like timbres to electric piano sounds can emphasize the metallic nature of the sound and give it a new 'edge'. Often, a slight change to the envelope of the 'alien' sample section can enhance the result – reducing the sustain to zero and shortening the decay time will give a short 'blip' of the 'alien' sample at the start of the sound. Increasing the velocity sensitivity of the 'alien' part, so that it only sounds when a key is played hard, can produce a more realistic feel to the sound, since it mimics that way that harmonic structures tend to become more complex as the velocity of playing increases. Preset sounds which use velocity fading or switching to make changes to a sound depending on how hard they are played are very good for exploring a combination of 'familiar' and 'alien' samples – 'slap bass' sounds are a good place to start an exploration of this type of editing.

A common technique is to make up a sound from the attack part of one sample, and the sustain part of another. This first found commercial success in instruments like Roland's D-50 synthesizer, where the technique was somewhat confusingly called LA synthesis. Changing the samples for the attack or sustain can produce large changes to the sound. Even swapping the two around – so that the attack comes from the sustain sample, and the sustain from the attack sample, can give some unexpected timbres.

It should be apparent from the above examples that editing S&S sounds is not the same process as with most other methods of sound synthesis. The limited set of available samples means that more sophisticated editing techniques are required if the resulting sound is to avoid being merely a replay of one of the existing samples. The use of envelopes to control separate parts of the sounds is a particularly useful method.

Pitch changes

Pitch can easily be overlooked when editing a timbre. Shifting the pitch of a sound up or down an octave is merely a transposition, but some very useful and interesting timbres can be found if you explore the boundaries – which means going as low in pitch, or as high in pitch as possible. Shifting the pitch of samples down low usually produces dark and mysterious timbres, whilst shifting the pitch upwards can produce aliasing effects and noise-like sounds.

How the samples are constructed can also be important. Short loops can give a pitched sound and drum sounds can provide huge resources of suitably short sounds, provided that they can be looped. Longer looped sounds produce pitches when they are pitch shifted upwards, but the complex rhythmic patterns that they produce when shifted down can form a useful background to a pad sound. If possible, fixing the pitch of this type of sample when it is transposed by a large amount, so that it does not track the keyboard, is the best idea, since then the rhythmic part stays at a constant tempo, and does not speed up as notes are played towards the right hand end of the keyboard.

Pitch envelopes are often overlooked by inexperienced sound programmers. Some brass sounds will have a quick rise to the note, but the only other commonly encountered use is a much more exaggerated pitch chirp on a lead sound, when it is often labelled as a 'funny' sound. But there are quite a few more creative opportunities with pitch envelopes. Subtle slow envelopes on vocal sounds can give a very realistic sound, especially if this is mixed with a more conventional 'choir' sample. Putting a pitch envelope onto only one of the two parts of a sound can give a very useful 'detune' type chorus effect at the start of a note, and which is then followed by the two parts of the note coming into tune for the sustain portion of the envelope. By tuning the envelope range so that it covers one or more octaves, then it is possible to create sounds which have octave doubling for only part of the envelope.

LFO modulation of pitch is often restricted to providing vibrato. But a slow LFO applied to one part of a sound will then produce a cyclic 'detuned in-tune detuned in-tune' effect, which can be very effective on string sounds, particularly if this is layered with a more 'realistic' string timbre. Using a sawtooth waveshape on the LFO is one of the elements of many string-like synthetic timbres.

Envelopes

Envelopes tend to have the reputation of being suitable only for detailed editing. Quick edits are restricted to merely changing the attack or release times. But many S&S instruments provide more than one envelope, and it is often possible to change the envelope which is being used by a part very easily. The quick edit advice is thus to use the wrong envelopes. Shortening or lengthening attacks, decays and releases, can turn sustained sounds into percussive sounds, and vice-versa. String timbres sound very different if they have piano-type envelopes.

For an S&S instrument with two separate parts to the sound, contrasting envelopes can be used for the two parts. This opens up possibilities like using the sustain part of a string sound to provide the attack, and the attack sample (looped) to provide the sustain sound. Since these sounds can be layered with presets, long, slow envelopes can be used to provide evolving sounds which are much more interesting to listen to than static pads. Using envelopes to cross-fade from one sound to another is usually used just to paste together attack and sustain samples, but this can also be used with contrasting sounds, and altering the cross-fading can provide more dynamic changes. Sounds which vanish in the middle and then re-appear in a different form may appear quirky, but they are very useful for adding a bit of interest into washes and introductions. They can also provide some interesting song or melody ideas when they are used with a sequencer, since the timing will probably not coincide with the sequencer playback, and this can create complex poly-rhythms.

Filters

The filter controls probably take the second place after the sample choice, as the main focus of quick edits. The filter frequency is probably the most powerful of all the modifier control for determining the timbre of the sound. Filtering should be used to try and remove or enhance the sample 'fingerprint'.

Removing parts of the spectrum of the sample may well disguise it, whilst emphasizing other parts of the spectrum may also help to lose some of the distinctiveness. The more complex the filter, the more control that it will give over the harmonic content of the sound, so simple low-pass filters are not ideal. Low-pass, high-pass and band-pass filters are a good starting point, whilst notch and comb filters can be very useful, and any other shapes are a bonus.

As a general rule, the resonance control acts as a selectivity control – the higher the resonance, the more changes the filter will make to the sound. This is especially noticeable with a low-pass filter which has a resonance control which gradually makes it more and more like a band-pass filter – the sound becomes more interesting and synthetic sounding as the resonance approaches the point where the filter is about to self-oscillate.

The over use of low-pass filters can tend to give a very thick and mushy sounding bottom end to sounds, whilst high-pass filters tend to 'thin out' the sound. Filters should always be used in combination with the other elements of the sound, rather than hoping that they can provide a stand-alone sound on their own. The one exception to this is the 'synth brass' filter sweep sound, where an ADSR envelope is used to sweep the filter cut-off frequency with a sawtooth cycle or brass sample. The resulting sound is one of the strongest synthesizer clichés of the mid- to late-1970s!

Summary

Although these modifications to sounds have been described in the context of S&S synthesizers, almost all of the techniques also apply to samplers. Many of the techniques also apply to other methods of synthesis, particularly subtractive, hybrid and FM synthesis.

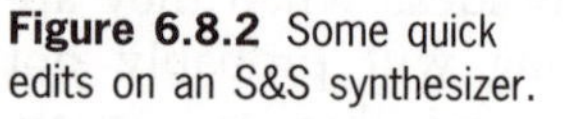

Figure 6.8.2 Some quick edits on an S&S synthesizer.

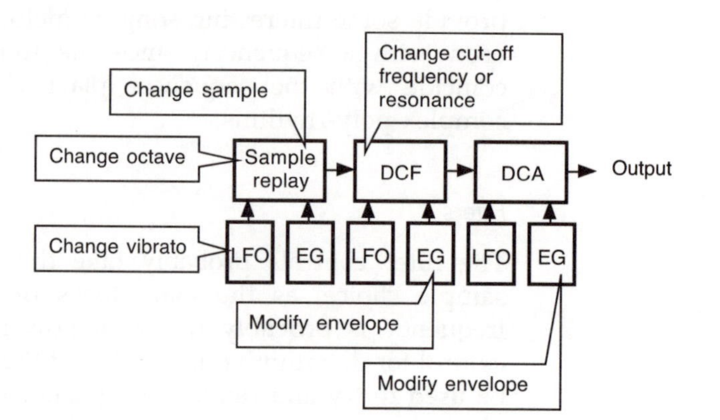

6.8.4 Managing edited sounds

Although producing, controlling, manipulating and arranging sounds is the 'foreground' activity of the synthesist, a number of associated background management functions need to take place in order for these performance-related activities to happen. Since sounds and music are the major resources for a synthesist, they need to be catalogued, stored and made available for subsequent retrieval.

Although a large number of synthesizers now offer digital storage of their operating parameters, this is not always possible for analogue instruments or for some effects processing equipment, nor for custom-built apparatus. Notes on these items may well be stored on paper in notebooks or binders, but the principles of resource management remain the same.

Sorting

Part of the process of learning how to use a synthesizer is the production of sounds. Some synthesists prefer to start their investigations from the preset sounds which are provided, whilst others ignore or remove any presets and work from the basic 'initialized' sound, which is normally a simple sine wave.

The programming of sounds can be approached in many ways. A methodical exploration of the available parameters may suit one synthesist, whilst an iterative process based on making changes to the parameters more or less at random and observing their effects may work for another person. In this learning phase, a large number of sounds are typically produced, and these can either be used as finished sounds, or they can provide the raw material for further programming. After some programming, the synthesist begins to develop a mental 'model' of the operation of the synthesizer, and this will be confirmed by further programming: followed by checking of the expected results against the reality. Once this stage is complete, the synthesist should feel confident of their ability to produce almost any required sound.

Regardless of their origin, the sounds which are produced by the programmer need to be filed in a way that allows them to be reused at a later date. Few synthesists have the ability to rapidly produce a specific sound when required – instead, they use these pre-prepared sounds or edit them in context. In order to file the sounds they need to be sorted and categorized, and a number of systems can be used to provide this function. Perhaps the most useful technique in cataloguing the sounds and music is to split the pool into categories. There are many different methods of naming sounds, ranging from ones which describe them in generalized terms like slow, smooth, bright or dark, whilst others use references to conventional instruments: brassy, piano-like, flute-like or bell-like. Some composers have used references to the moods or emotions which the timbres suggest. Most of the commercial software packages use a mixture of generalized terms and references to specific instruments.

Whilst categories are useful for grouping sounds and retrieving them at a later date, they can be unwieldy for everyday use. Simple names for sounds are a much better way of ensuring that they will be remembered easily in day-to-day use. Most synthesizers use some sort of 'bank and number' system for numbering their sounds: often based around banks containing 8 or 16 numbered sounds. The banks can either be numbered or assigned letters, so 4-12 and B5 are both typical examples. Numbers of this form are not readily associated with sounds, and so the naming of sounds is very important. There are a number of methods used for naming sounds. The simplest is to attempt to describe each sound as accurately as possible given the limitations of the number of character which can be used. Since this is normally between 8 and 16 characters, names like 'chorused electric piano with bell-like attack sound' will need to be abbreviated. One useful technique is to divide the names into two parts. The first two or three characters are used to indicate the rough grouping of sound: piano, string, brass, etc. The remaining characters can then be used to provide more detailed description, albeit in an abbreviated form. This method produces names of the form 'EP:ChrBelAtt' where the descriptive part is split into meaningful parts by using capital letters instead of spaces.

More egotistical methods of naming sounds exist. These sometimes use any additional available characters to provide a graphical name: '-=[V]=-' being a typical example. Other programmers use in-jokes or clever plays on words: one rather obvious example being 'sdrawkcab' for a sound which gives the impression of being played backwards. Foreign language words are often popular because they have a mystique due to their unfamiliarity, or they may be further plays on words: 'bête noire' (literally French for 'black beast', but English for a weakness) and 'koko wa?' (Japanese for 'where am I?') are two examples. For this type of name, the more memorable a name is, then the more effective it is.

Finding

Once filed away with a suitable set of category words, name and perhaps grouping, the sound is ready to be used. Finding the right sound involves using the reverse process of naming: deciding on the type of sound to be used, and then choosing category words and perhaps groupings so that a 'short-list' of potential candidate sounds can be produced. Auditioning of these sounds can then begin, with the user testing each sound in context with other sounds which will be used. Once the sound-set has been chosen, then the sounds will probably be stored as a single entity temporarily whilst the project is in progress.

Librarians

Software which carries out the naming, categorization, grouping, sorting, and finding functions is called librarian software. A number of programs exist for this purpose, and they are often associated with programs which can be used to edit sounds, called editors. Two types of these programs exist. Generic or 'universal' versions can be used with almost any synthesizer or related equipment since their behaviour can be programmed and new equipment can be added merely by loading in new programs. Specific versions are designed to be used with only one synthesizer – although more limited they are often lower in cost than the generic versions.

Storage

Storing sounds can be achieved in several ways. Most synthesizers have internal RAM memory storage, often with additional card-based or floppy disk-based storage. System exclusive MIDI messages can be used to send and receive the same sound data, and this method is used by librarian software. Librarians can then store the sound data on floppy, hard or optical disk storage. In general, the storage of important sound data should be in several different locations and on different media, so that the chances for losing sounds are minimized whilst the possibilities for recovering sounds are maximized.

6.9 Questions

1. What types of control over a performance does synthesis provide?
2. Outline three approaches to stacking.
3. Describe three methods of layering sounds.
4. What is hocketing?
5. What is the difference between multi-timbrality and polyphony?

6. Outline a simple note assignment strategy. How could this be improved?
7. Describe the major types of audio effects, and their uses.
8. Describe the differences between an effects unit intended for use in a monophonic synthesizer, and one designed for use in a multi-timbral polyphonic synthesizer.
9. Outline some of the ways of controlling a synthesizer during performance.
10. Which editing techniques would be common to an S&S synthesizer and a subtractive analogue synthesizer?

7 Controllers

Musical performance is a combination of the instrument and a player. The interaction between the two produces the music, and the interfacing between the player and the instrument involves the control of the instrument. There are many ways that this control can be achieved. In a conventional instrument the control interface is often predetermined by the instrument itself. For example, a guitar has six strings, a fretboard, bridge, etc. and the player can pluck, strum or tap the strings, which may be open or fretted. But for a synthesizer, there is no fixed set of controllers because a synthesizer is not constrained by any physical limitations. The control is thus determined by the flexibility of the synthesis technique and its physical implementation.

The organ-type keyboard that is used as the major controller on many synthesizers seems to have been chosen for all the wrong reasons. Whereas early synthesizers were monophonic, the keyboard is naturally polyphonic since it is all too easy to play more than one key at once. The only opportunities for expressive control using the keyboard are attack velocity when the note is initially pressed, and after-touch or key pressure once it has been pressed – which does not match the continuous and diverse expression capabilities of a synthesizer. But the keyboard was easy to wire up to produce simple control voltages and triggers for testing the first synthesizers, and this is probably why it was adopted. Since then, the keyboard has continued to be used as the major control interface to the synthesizer.

A synthesizer with just a keyboard is very limited in control terms. Most synthesizers add a number of additional controllers to augment the keyboard's limited capability. The pitch-bend wheel is used to provide changes in pitch similar to those produced by pulling the strings on a guitar – although in the case of a keyboard, the use of the pitch-bend wheel requires one hand to be removed from the keyboard – usually the left hand, since the convention is that the performance controls on a synthesizer are always on the left hand side. The modulation wheel is used to provide control over additional performance parameters like vibrato, tremolo or filter cut-off, and again requires the left hand for operation. On a violin modulation effects like vibrato are produced by moving the fretting position of the strings by the fingers. Modulation effects can be controlled by using the after-touch or key pressure on a keyboard, but this limits their use to when a key can be held down and sufficient pressure applied to activate the pressure sensing circuitry; vibrato can thus only be applied after the note has started, and cannot continue uninterrupted when the note changes, nor can it be used when a rapid run of notes is required. Foot pedals and breath controllers are just some of the additional controllers that are employed to try and provide more flexible and expressive control over sounds that are merely pitched and triggered by the keyboard.

Many of these control problems also apply to the use of piano-style keyboards to synthesizers, although this does improve the interaction between the player and the synthesizer if a piano-type of sound is being controlled. For almost all other types of sound, the piano keyboard is less suitable.

In contrast, a synthesizer which is controlled by a stringed instrument controller requires a much more sophisticated interface to be able to extract the pitch, trigger and performance information, but the resulting control is much neater and more intuitive. Pitch-bend can be applied using the same hand which is fretting the strings by pulling the strings away from their rest position, and vibrato can be combined with pitch-bend using the same fingers – in fact, each finger can be used to apply different pitch-bend or vibrato, if necessary, which is difficult for most keyboard-controlled synthesizers where the modulation is global to all notes being held down.

The complexity of the user interface for a wind instrument as the controller for a synthesizer is simpler because only a single monophonic pitch needs to be determined, but this ideally suits a monophonic synthesizer. The control over pitch is now spread between the hands and the mouth, with modulation control coming from the mouth, tongue, teeth and lips. The volume of the sound produced by the synthesizer can be controlled by breath pressure and can be continuously changed during the course of a note – something which is very difficult using a keyboard-controlled synthesizer without using a foot pedal or wheel controller (or a breath controller!).

7.1 Controller and expander

Separating the controller from the sound generating parts of a synthesizer produces two separate devices. The sound producing part of the synthesizer is the expander module – a 'keyboardless' version of the synthesizer, where the word 'synthesizer' almost always implies the inclusion of a keyboard. Because of this ubiquity of the organ-type keyboard as the 'master' keyboard controller for several expander modules, any other form of controller is normally referred to as an 'alternative controller': wind instruments, guitars and drums are three of the commoner forms.

Regardless of the type of controller, there are a number of elements which are used in all of them. Each needs to provide control signals or voltages which produce some or all of the following information:

- pitch of a note event
- start and end of a note event
- dynamics of the note (volume)
- changes in pitch
- modulation changes
- sustain
- additional expression controls.

Some controllers combine several of these into one composite controller, whilst others provide just one. A master keyboard may provide nothing other than the basic pitch and dynamics information, whilst a more sophisticated version may provide pitch bend, modulation and sustain information as well. A guitar controller will provide up to six separate sets of information (one set per string) whilst a keyboard will normally provide either monophonic or polyphonic information. Wind controllers attempt to interface blown instruments to synthesizers, and are usually monophonic. The form of a controller may change depending on the application. The pitch bend wheel on a keyboard controller may function identically to the 'tremolo arm' on a guitar controller, although the physical realization is very different.

No one controller can provide complete control over all the available performance parameters: each has specific advantages and disadvantages. A guitar controller gives detailed control over the pitch, level and modulation of up to six separate 'strings', but is limited in its expression capabilities for sustained sounds, and control over the release segment of envelopes. In contrast, a keyboard provides pitch and dynamics information, and by using after-touch or key pressure, it can control modulation and expression whilst the notes are sustained, and the sustain can be modified by using a sustain pedal. But keyboards do not provide control over pitch-bend without using an additional controller (or using after-touch, in which case the modulation control ability is lost) and the control over the level of the sound is limited to the attack dynamics of the key velocity when it is pressed, or the release dynamics when it is released.

It is possible to mix the individual controllers to form larger controllers. Guitar synthesizers have been produced which provide strings to determine pitch, but use keys to trigger the sounds. Conversely, 'string' controllers consisting of six short strings mounted on top of a small box have been produced – they can be used to produce strumming and plucking effects in conjunction with a keyboard (or alternative pitch controller) that provides the pitch information.

Controllers have been developed to utilize most of the available resources of the performer. The pitch wheel and mod wheel occupy the left hand (there is an underlying assumption that keyboard players are all right-handed), whilst the right hand is presumably playing the keyboard, plus controlling after-touch. One foot is used for the sustain foot-switch, whilst the other foot is usually employed controlling volume or 'expression'. This leaves the mouth to blow into a breath controller, with elbows and knees still available for future development (some organs apparently use knee controllers). The left hand may not be continuously occupied with pitch band or modulation control, and so a few extra controllers can be provided for the left hand: some knobs or sliders which can be set to control parameters like brightness or release time, and which may well need adjusting in the course of a performance. With all these possible controls, the synthesizer player can begin to look rather reminiscent of a one-man band.

7.1.1 Performance

Regardless of how the control is achieved, the important consideration is that the player of a synthesizer should be able to interact with the instrument in a way that allows expression to be conveyed to the listener. The more natural and intuitive that the controls are, the more that they need to avoid imposing limitations on the player. Although the keyboard imposes many limitations on the player in performance, it is still the most common interface, and careful use of the additional performance controllers can minimize the problems.

Familiarity with the instrument's controllers can help considerably. The use of a single 'master' keyboard is useful in this context because the player will be comfortable with the keyboard and its controls, and will not need to try and locate wheels or pedals in poor lighting conditions on stage. Careful preparation of the music can help to identify points where additional performance controllers like pedals, or even alternative controllers like a string or wind controller may be required. For a player of an orchestral instrument, the instrument and its controllers are one and the same, and so no consideration is required for how to accomplish a specific musical result, whereas a synthesizer and its controls can be separate and can be changed to suit the player.

Some of the possible controllers are described in the remainder of this chapter.

7.2 MIDI control

7.2.1 MIDI

Although MIDI (Rumsey, 1994) provides a large number of controllers, the basic underlying assumption is that a keyboard will be used to provide the main source of pitch and performance information – since the pitch and velocity parameters are tied together in the note on and off messages, whilst volume information, pitch-bend, modulation and other parameters all require separate continuous controllers.

MIDI is capable of detailed and precise control in some circumstances, but it has some limitations which reflect its keyboard-based origins. MIDI is poor at specifying the transitions between notes, which makes controlling guitar sounds difficult because the re-use of a string cannot be controlled properly, and attempting to control polyphonic vibrato requires the use of several channels of MIDI's mono mode, which is not widely supported by commercial synthesizers. Glissando and portamento effects are only controllable for parameters such as time via MIDI, and MIDI expects that each change in pitch associated with a note will initiate a new envelope.

MIDI provides two ways to control synthesizer (or sampler) parameters: controller messages and system exclusive messages.

7.2.2 MIDI controllers

MIDI controllers are a sort of general case of the pitch-bend or pressure MIDI messages. They are performance controls – modulation, expression, vibrato, and many more. There is provision for a large number of possible MIDI controllers: more than 64 000, but only a few are typically used. The most common is the modulation wheel (controller number 1), often found next to the pitch-bend wheel on the left side of a synthesizer keyboard.

Controller messages have three bytes. The first byte identifies the message as a controller message, and also indicates which of the 16 MIDI channels the controller is on. The controller number is the second byte. The 128 possible controller numbers are organized into three basic types:

- continuous controllers: mod wheels, pedals etc
- switches (theoretically just two values: on and off)
- mode messages to control note reception conditions.

Continuous controllers are divided into 14-bit (from 0 to 63) and 7-bit (from 64 to 95). Switches are a special case of a 7-bit continuous controller, and true on/off switches occupy a few numbers (from 64 to 69). Controllers 96 to 119 are split into two sections: the first covering 96 to 101 is the registered and non-registered parameters section – where huge numbers of parameters can be assigned to MIDI controller messages without using up valuable 14 or 7 bit controller numbers. In fact, it is possible to use just controllers 06 and 38 to control the MSByte and LSByte of all these extra parameters.

In summary, the controller numbering looks like this:

0 to 31	14-bit controllers (MS byte)
32 to 63	14-bit controllers (LS byte)
64 to 69	Switches (pedals and footswitches)
70 to 95	7-bit controllers
96 to 101	Registered and non-registered parameters
102 to 119	Undefined
120 to 127	Mode messages

Not all controllers numbers are assigned to specific devices like pedals, wheels, etc.; in fact, most aren't. Those that are seem to have been driven by historical precedent.

14-bit and 7-bit controllers

MIDI controllers come in several varieties. Some act as switches for just on and off controls – sustain pedals for example. Some act as 7-bit controllers with 128 different values – volume pedals, and even some slightly esoteric sustain pedals. But there are also 14-bit controllers where detailed control is needed. Using 14 bits allows up to 16 384 different values, which is enough for most purposes – and in fact, very few manufacturers take advantage of 14-bit controllers. For volume pedals, 128 values is often too much precision, and only 8 or 16 different volume settings are actually used in some cases. Modulation wheels often use all 128 possible values to give smooth transitions for vibrato etc. The most common use of 14-bit precision is the pitch-bend message, which is technically not a controller at all! The ultimate limit on the number of values is the switches, where there are effectively only two values that are used.

Seven-bit and 14-bit controllers can work together. The range (maximum to minimum values) of a MIDI controller is normally thought of as being 0 to 127, which makes sense for a 7-bit controller, but what about 14-bit controllers? Where are those 16 000 values? In fact, the 14-bit controllers 'fill in' the gaps between the 7-bit values, which provides much finer resolution. So the 7-bit controllers give coarse control, whilst the 14-bit controllers allow much more precise adjustment.

The 7-bit and 14-bit controllers share messages. The first 32 MIDI controllers appear to be 7-bit controllers, but they are actually just the most significant part of a 14-bit controller which uses another controller number as the fine resolution part. Using computer-speak, the 'most significant byte' and 'least significant byte' are referred to by acronyms: MSByte and LSByte. The MSByte controllers can be used on their own as 7-bit controllers, whilst the LSByte controllers just add 32 to the MSByte controller number – so volume (MSByte controller number 7) can also be finely tweaked with LSByte controller number 39, if your equipment supports this feature.

Note that an MSByte controller message on its own resets the LSByte value to zero.

Controller numbers

- **Controller 1**: Modulation wheel MSByte
  Wheels are not the only source of control for this message. Levers, joysticks, sliders and pressure sensitive plates can all provide the physical control.

- **Controller 2**: Breath controller MSByte
  Breath controllers convert breath pressure into a control signal. They are especially useful for controlling monophonic synthesizers which are playing melodies.
- **Controller 4**: Foot controller MSByte
  Usually a pedal – although some foot controller inputs will accept a control voltage instead.
- **Controller 5**: Portamento time MSByte
  This control can be used as a switch, or as a continuous control over portamento time.
- **Controller 6**: Data entry MSByte
  The Data Entry front panel control may be a slider, a rotary dial or a keypad.
- **Controller 7**: Main volume MSByte
  Often only the MSByte is used to control volume, which can cause 'zipper' noise because of the large jumps in volume caused by just 127 steps in volume.
- **Controller 8**: Balance MSByte
  This controls the volume balance between two sounds, and is taken from 'organ' terminology.
- **Controller 10**: Pan MSByte
  This controls the position of the audio in the stereo sound field.
- **Controller 11**: Expression controller MSByte
  Typically a pedal, the expression controller provides a volume boost in addition to the main volume, and is intended to be used for accenting.
- **Controller 12**: Effect control 1 MSByte
- **Controller 13**: Effect control 2 MSByte
- **Controllers 16 to 19**: General purpose controllers 1 to 4 MSBytes
  These are intended to be general purpose controllers, which means that they will be defined by individual manufacturers to suit specific parameters in their instruments.
- **Controllers 32 to 63**: LSBytes

Controllers 32 to 63 are the LSByte equivalents to the above controllers.

- **Controllers 64 to 69**: Switches (7 bit controllers)
  64 Damper pedal (sustain)

65 Portamento on/off
66 Sostenuto
67 Soft pedal
68 Legato
69 Hold 2

These are actually 7-bit controllers – although most MIDI equipment will expect them to be only two values: on and Off.

- **Controllers 70 to 74**:. Sound controllers 1 to 5
  Controllers 70 to 74 are the defined sound controllers. These have a default descriptive name, which is shown below, and are designed to provide simple 'global' quick editing facilities. They are intended to allow slight changes to a sound, probably during a performance, rather than editing actual parameter values permanently.
  The names are guides to suggested usage, but they can be redefined to suit a particular application. They are in two groups:

  Timbre controls
  70 Sound variation
  71 Timbre/harmonic content
  74 Brightness

  Envelope controls
  72 Release time
  73 Attack time

- **Controllers 75 to 79**: Sound controllers 6 to 10
  Controllers 75 to 79 are the undefined sound controllers. These do not have any predefined name, and can be assigned freely by a manufacturer.

  75 Sound controller 6
  76 Sound controller 7
  77 Sound controller 8
  78 Sound controller 9
  79 Sound controller 10

- **Controllers 80 to 83**: General purpose controllers 5 to 8

- **Controller 84 $54:** Portamento control
  This is an unusual controller in that it indicates the note number from which the currently 'portamento'ed note started.

  Controller numbers from 102 to 119 are undefined

- **Controller 120**: All sounds off

Table 7.2.1 Numbers

14 bit MSBs	
0	Bank Select
1	Modulation Wheel
2	Breath Control
4	Foot Controller
5	Portamento Time
7	Volume
8	Balance
10	Pan
11	Expression
12	Effect Control 1
13	Effect Control 2
7 bit	
64	Hold, Damper, Sustain
65	Portamento
66	Sostenuto
67	Soft Pedal
91	Effects 1 (ex external effects depth)
92	Effects 2 (ex tremolo depth)
93	Effects 3 (ex chorus depth)
94	Effects 4 (ex celeste depth)
95	Effects 5 (ex phaser depth)
Non-Registered & Registered	
6	Data entry MSB
38	Data entry LSB
96	Increment data
97	Decrement data
98	Non-Registered parameter number lsb
99	Non-Registered parameter number msb
100	Registered parameter number lsb
101	Registered parameter number msb
Mode	
121	Reset All Controllers
122	Local Control On/Off
123	All Notes Off
124	Omni Off
125	Omni On
126	Mono On
127	Poly On

Controller 120 is the all sounds off message. This is an improved version of the all notes off message, since it forces any sounds which are currently playing to stop as quickly as possible.

- **Controller 121**: Reset all controllers
  Controller 121 is used for the reset all controllers message, a sort of all notes off but for controllers instead of notes. Although this message can be interpreted as meaning just

the MIDI controllers, it actually indicates that the current state of all controllers, pitch-bend, modulation wheel and pressure should be returned to the reset state (typically zero for most controllers, with the exception of the pitch band wheel, which resets to its centre position (no pitch bend), and the volume control (controller number 7), which returns to full volume (127)).

- **Controller 123**: All notes off
  The all notes off message is designed for use by sequencers when a sequence playback is stopped whilst some notes are still playing, although it often seems to be used to indicate when all the keys on a keyboard have been released.

Finally, remember that the MIDI controllers are channelized, i.e. the data is sent using a specific MIDI channel.

7.2.3 System exclusive

The system exclusive message is unusual because it is the only MIDI message which has special bytes which indicate the beginning and the end. All other MIDI messages start with a special byte which indicates the type of message which follows, and also includes the MIDI channel number. The length of a message can normally be determined by knowing what type of message it is, but system exclusive (normally abbreviated to sysex) messages can be of any length, and so the start and stop bytes are used to indicate the limits of the message.

System exclusive provides a carefully designed 'loophole' to the MIDI specification which allows manufacturers to have extra information in their own specific formats embedded in MIDI messages whilst still retaining full MIDI compatibility.

The byte which immediately follows the sysex start byte is used for the identification of the owner of the message, and it is commonly called the manufacturer's ID number. This byte is rather like the address on an envelope: the envelope is ignored by everything except the correct addressee, who can open it and see what is inside.

The actual contents of the sysex message are entirely up to the individual manufacturer, but typically it consists of sound data, editing or control functions. The format of these messages is left entirely to the manufacturer and although most manufacturers maintain a reasonably consistent standard within their own products, there is almost no commonality between manufacturers. Often, the format includes a few bytes which indicate the intended destination device – this ensures that only editing or control messages which are applicable to a specific device are acted upon. Most sysex formats have one or two bytes which indicate which parameter value is being changed, and then one or two bytes for the value of the parameter itself. To ensure that the message has been correctly received, some formats also include a checksum byte which can be used by the receiving instrument to determine if an error has occurred during transmission of the MIDI data.

Sysex editing and control messages are normally short: less than ten bytes in length. Longer messages are normally complete sets of data bytes for a sound, or a set of sounds. Information on the contents of the sysex messages for a specific synthesizer or sampler is often included with the owner's manual.

7.2.4 ZIPI

ZIPI (CMJ, 18, No. 4) is an alternative interface for connecting a performance controller to a synthesizer. It provides much more detail and flexibility in the way that pitch and performance information are conveyed. In particular, it avoids having any assumptions about the type of controller that will be used – so there are no keyboard-specific limitations. ZIPI does not seem to be intended as a replacement for MIDI, more a means of transporting performance control from a player's instrumental controller to a remote synthesizer. The exact format of the player's controller is not defined: it could be a keyboard, stringed instrument or wind instrument.

7.3 Keyboards

The music keyboard has a distinctive appearance that is widely used as a visual metaphor for music or pianos. The interlaced black and white keys are widely used for note selection, as well as starting and ending envelopes.

Keyboards provide two major outputs: discrete pitch information about the notes which are being played; and event information about the start and end of note events. Both of these are produced by pressing down on the keys and releasing them. The convention is that pressing down on the keys produces both a note pitch indication and a note start event. Releasing the key allows it to return to its rest position and produces a note end event.

Some keyboards also provide additional information on the velocity with which the key is pressed or released – this is known as either 'attack velocity' or 'release velocity', although sometimes the word dynamics is used as a synonym for velocity. When a key is held down and extra pressure is applied, some keyboards will produce information about this 'after-touch' or 'key pressure'. These are both covered in the next section.

7.3.1 Polyphony

The simplest keyboards are monophonic. A chain of resistors connected in series with a voltage applied across them can be used to output a voltage whose value is proportional to the position on the chain. By using the keys of a keyboard to switch a contact wire onto the chain of resistors, then a keyboard can be used to provide a monophonic note output. By using a second switch contact to produce a note status indication, then note event information can also be produced. This is the type of keyboard which was used in early analogue synthesizers.

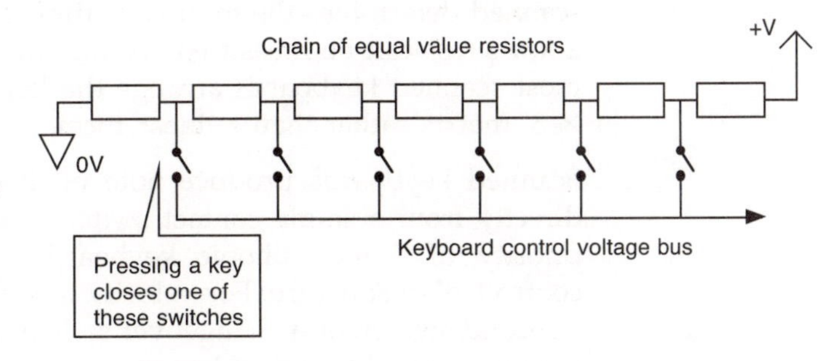

Figure 7.3.1 An analogue synthesizer keyboard uses a chain of resistors, and pressing a key closes a switch which connects a bus to the chain of resistors. Only six switches are shown – a real keyboard would have 30 or more keys. The position in the chain determines the output voltage. A separate switch (not shown here) produces the gate signal and a trigger pulse.

By extending the basic design of the monophonic 'analogue' keyboard, it is possible to produce a keyboard which will produce two voltages: one for the highest note which is being pressed down, and the other representing the next lowest note which is held down. Such 'duophonic' keyboards are often found on analogue performance synthesizers.

Keyboards based around chains of resistors are normally designed so that they have so-called 'top-note' priority. This means that if several keys are held down, then only the highest voltage will be produced. This allows a two-handed performance technique to be used where notes are held with the left hand whilst additional notes are played staccato with the right hand. Each time the right hand is released the pitch jumps down to the note being played by the left hand. Duophonic keyboards provide separate storage for these two voltages and allow two-note chords to be played instead.

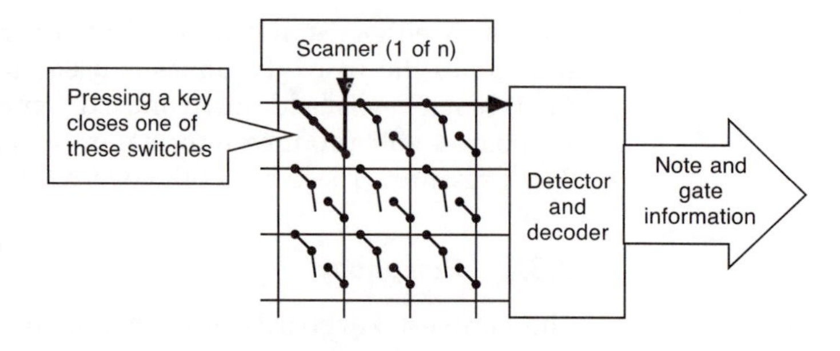

Figure 7.3.2 A scanned keyboard uses a matrix of switches which connect one of the outputs of a scanner (it cycles round each of its outputs, putting a logic one signal on each in turn) to a detector and decoder circuit. One switch can be used to produce note and gate information. Velocity information would require an additional switch or a changeover switch.

Polyphonic keyboards can be produced by extending the 'resistor' chain method, but in practice, digital scanning techniques are used. A counter is used to send a pulse to each of the key contacts in turn, and any keys which are held down will allow the pulse to be switched onto a common bus-bar contact, and the keys which are held down can be determined by examining their relationship to the counter. The keyboard scanning is thus based on time-division, and the rate at which the keyboard is scanned determines the minimum timing resolution with which a key-press can be measured. To minimize the required wiring, most scanned keyboards arrange the keys in the form of an 'N × 8' matrix rather than a linear form.

Scanned keyboards produce note pitch and event information directly from a single contact switch. The polyphony and note priority (if a monophonic keyboard is being emulated) are controlled by software. For velocity information to be produced, a second switch, or a changeover switch are required so that the time between the start of key movement and the end of the key movement can be determined.

7.3.2 Types of keyboard

Keyboards come in two major forms: organ-type and piano-type. Although both are fitted to synthesizers, the organ type tend to be used in low-cost and mid-range products, whilst the piano type are more often found in 'master' controller keyboards and 'top-of-the-range' or 'flagship' products.

Organ-type keyboards have light, hollow, plastic keys and the black keys normally slope downwards away from the front of the keyboard. Organ keys tend to be about 140 mm from the front of the keyboard to the back of the playable key surface. The key action is fast and very light, with the key being returned to the rest position by a spring. Organ-type keyboards sometimes have small weights attached to the underside of the front of each key to provide additional weight when they are used as master keyboards – this is intended to emulate some of the feel of a piano-type of keyboard. Organ-type keyboards as used on synthesizers are normally either five-octave (61 keys) or just over six octaves (76 keys) – frequently referred to as an 'extended' keyboard.

For use in small portable 'fun' keyboards and synthesizers, a variant of the organ-type keyboard is often found. This has narrower keys (an octave fits into the space of an eleventh on a normal keyboard) and the playable surface of the keys is only about 80 mm from front to back.

Piano-type keyboards have heavier keys which are normally wood covered by a plastic moulding. The black keys normally have flat tops. Piano keys are about 150 mm from the front of the keyboard to the back of the playable key surface. The piano 'action' – a complex mechanical arrangement of levers and weights – returns the key to the rest position. Piano-type keyboards are often referred to as 'weighted' because of the heavy keys and the feel of the 'action' relative to an organ-type keyboard. Piano-type keyboards normally have at least 76 keys, and frequently more.

7.4 Keyboard control

Although the basic keyboard produces only pitch and event information, the action of pressing the key and then applying pressure to the key can be used to produce additional information.

'Velocity' is the name given to the output from the keyboard which represents the rate at which the key is pressed or released. A pair of contacts are used to measure the time that the key takes to travel from just after when it is first pressed to when it is almost fully down. Changes in velocity whilst the key is being pressed do not normally affect the velocity, since the measurement is based on just the time difference between the switching of the two contacts. Attack velocity is measured when the key is first pressed, and is used to control, the dynamics of the sound – often the level and/or the timbre – whilst release velocity is normally used to control the length of the release segment of the envelope. Release velocity is only rarely implemented in keyboards, although many synthesizer sound generators will respond to release velocity.

'After-touch' or 'key pressure' is the name given to the additional pressure which is applied to a key when it is being held down. The key requires a certain amount of pressure to overcome the spring which pulls it back to the rest position when the key is released, and the after-touch measures any additional pressure which is applied in excess of this. After-touch can be a global parameter where the highest pressure applied over the entire keyboard is used as the output value, or it can be measured individually for each key. These are called 'monophonic' and 'polyphonic' respectively.

7.5 Wheels and others

The early analogue modular synthesizers used rotary control knobs to control the majority of the non-switched functions. Knobs were thus used for the first controls for pitch-bend, and modulation controls were merely the output level controls for LFOs. But knobs were not very satisfactory for live performance use, and a number of alternative approaches were tried for providing the same sort of pitch-bend and modulation control, but in a more intuitive way. The four major contenders were:

- wheels
- levers
- joysticks
- pressure pads.

7.5.1 Wheels

Wheels are a development of the rotary edge controls which turn a rotary control onto its side and use a disk instead of the knob. Pushing the disk forwards or backwards turns the rotary control. The disks used are normally about 80 or 90 mm in diameter, and only about a third or a half of the disk protrudes above the panel surface. Some disks have a textured or grooved edge, whilst others are smooth – some manufacturers made them out of clear Perspex with polished edges. A small semicircular cut-out or detent provides a reference point for the position of the disk. The disk movement normally uses about 120 degrees of rotation – a third of a complete circle.

For pitch-bend purposes, the detent is in the centre of travel of the rotary control, and so the pitch can be changed up or down: the rotation is about ± 60 degrees away from the centre point of the pitch-bend. A spring arrangement is normally used to return the pitch-bend wheel back to the centre position, and this is reinforced by the use of a second detent and a sprung cam follower which slots into the detent to provide a tactile centre position into which the wheel clicks. Pressure on the wheel is required to overcome the detent and the springs before it will move. For modulation use, the detent is with the disk full towards the player, and modulation is added as the disk is pushed forwards. No spring return is used with most modulation wheels.

Although most wheels are located so that they move away from or towards the player, some wheels have been designed which move from side to side, so that to bend the pitch upwards the wheel is moved to the right, whilst for pitch bend downwards the wheel is moved to the left. Modulation is added by pulling the wheel towards the player, or pushing the wheel away from the player.

7.5.2 Levers

The lever replaces the disk with a short lever with a length of about 50 or 60 mm. The rotational movement is normally less than a wheel: a maximum of 90 degrees and a minimum of about 60 degrees. As with wheels, the pitch-bend lever is sprung so that it returns to the central detent position.

7.5.3 Joysticks

Joysticks combine the pitch-bend and modulation functions into a single control. Both side-to-side pitch and forwards/backwards modes have been used, with modulation using the opposite axis. Additional control has been provided on some joysticks by allowing the joystick to be rotated or moved up and down.

7.5.4 Pressure pads

Soft rubber pressure pads and metal plates fitted with strain gauges have been used to provide pitch-bend and modulation controls on some synthesizers. Separate pads are required to bend the pitch up and down, whilst the modulation pad can

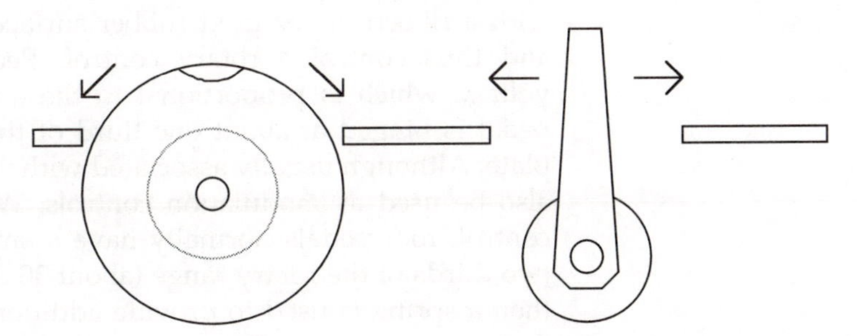

Figure 7.5.1 Wheels, levers and joysticks are all alternative ways of presenting a user interface to a rotary control. This diagram shows a cross section through a wheel and a lever.

only be used to add modulation interactively – it cannot be set to a value and left, unlike the modulation controls provided by wheels and levers.

The two main controllers that wheels and similar devices are used to control are pitch-bend and modulation.

7.5.5 Pitch-bend

Continuous control over the pitch is achieved by using a 'pitch-bend' controller. These are normally rotating wheels or levers, and usually change the pitch of the entire instrument over a specified range (often a semitone or a fifth). They produce a control voltage whose value is proportional to the angle of the control. Pitch-bend controls normally have a spring arrangement which always returns the control to the centre 'zero' position (no pitch change) when it is released. This central position is often also mechanically detented, so that it can be felt by the operator, since it will require force to move it away from the centre position.

7.5.6 Modulation

Modulation is controlled using rotary wheels or levers, where the control voltage is proportional to the angle of the control. Modulation controllers are not normally sprung so that they return to the centre position. Some instruments allow pressure on the keyboard to be used as a modulation controller.

There have been several attempts to combine the functions of pitch-bend and modulation into a single 'joystick' controller, but the most popular arrangement remains the two wheels: pitch-bend and modulation.

7.6 Foot controls

Foot controllers or foot pedals are rotary controls which are operated by the player's feet. A flat rectangular plate is covered with a ribbed or textured rubber surface and allowed to rotate and thus control a rotary control. Pedals provide a control voltage which is proportional to the angle of the pedal – the pedal is hinged at about one third of the way along the upper plate. Although usually associated with volume control, they can also be used as modulation controls. When used as a volume control, foot pedals normally have a smooth travel over about two thirds of the rotary range (about 30 degrees total travel) and then a spring is used to provide additional resistance to further

Figure 7.6.1 Foot controllers have a restricted angle of rotation – about 30°. The rotation is sensed by any of a number of methods: mechanical linkage to a potentiometer; optical sensors; or magnetic sensors. The spring provides a 'soft' end point which can be used to provide additional volume for expression purposes by using additional pressure on the foot control.

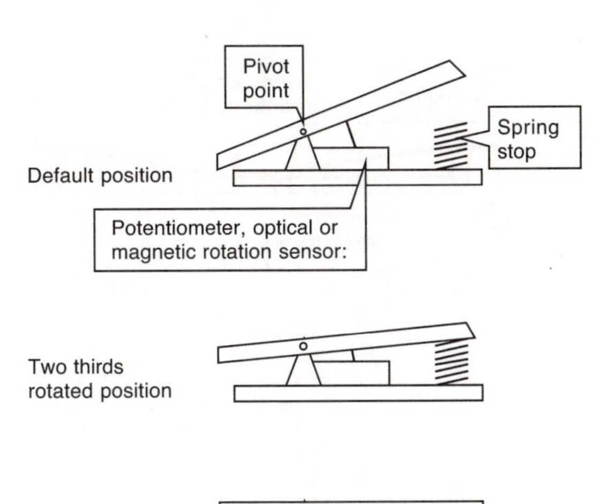

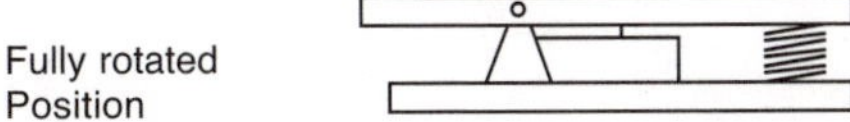

movement. This allows extra volume to be added for specific expression purposes.

Not all foot pedals use rotary potentiometers to sense the angle of rotation. Opto-electronic and magnetic rotation sensors can also be used.

Some pedals can also be used as pitch-bend controls, with springing to return the pedal to a central default position where no pitch-bend is produced.

7.6.1 Foot switches

Foot switches are foot-operated switches. They can be produced in several forms. Some operate like the sustain pedals on a piano, where a short metal lever is pressed downwards to activate the pedal. Others are like small variants on foot pedals, whilst a third type are merely push button switches. Whereas foot pedals are continuous controllers, foot switches normally only have two values, although there are some rare multi-valued variants used to control sustain on pianos. Foot switches are used to control parameters such as sustain, portamento, etc., but they can also be used to select sounds on synthesizers and to start or stop drum machines.

7.7 Ribbon controllers

The ribbon controller is a variation on the pitch-bend wheel. Instead of a rotating wheel, a flexible conductive material is held

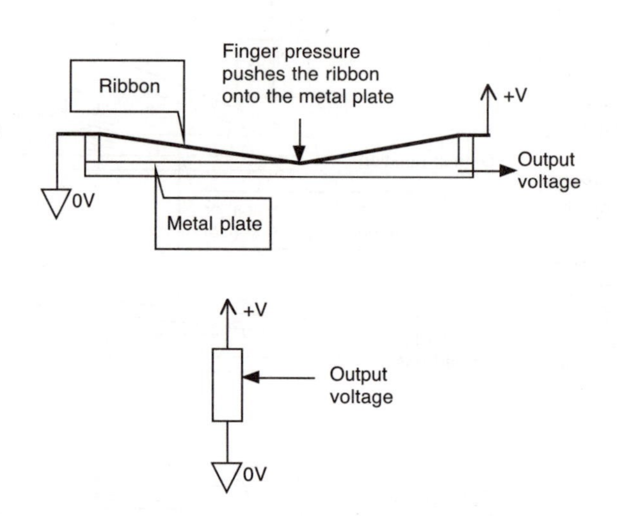

Figure 7.7.1 Ribbon controllers are actually a variation on a potentiometer or slider volume control on a mixing desk. Instead of having a slider which moves up and down a resistive track, the ribbon is moved onto a metal plate and so produces a voltage which is proportional to its position.

above a metal plate, and when pressed with a finger, the material touches the plate and a voltage is produced. Moving the finger on the material changes the voltage, and so the controller can be used as a pitch-bend control. Unlike a pitch-bend wheel, it is possible to jump in pitch by pressing a finger onto the ribbon, and then removing the finger causes a jump back to the default pitch. Ribbon controllers have the advantage of mechanical simplicity and small size, but the material does tend to wear out.

There are two main types of ribbon controller:

- 'Moog'-type ribbon controllers have a silver-coloured cloth-like appearance, are about 100 mm long and 15 mm wide, and have a central raised indication.
- 'Yamaha'-type ribbon controllers have a black flocked plastic appearance, and are about 300 mm long and 10 mm wide. The pitch changes relative to the starting point and the amount of movement along the ribbon by the finger. They are thus more suited to producing portamento and glide effects rather than pure pitch bend.

Some ribbon controllers have incorporated additional movement by mounting the ribbon on a 'log'-like wide modulation wheel, as with the joystick. This combined two-axis control remains the exception rather than the rule.

7.8 Wind controllers

Wind controllers come in two forms: breath controllers and wind instrument controllers.

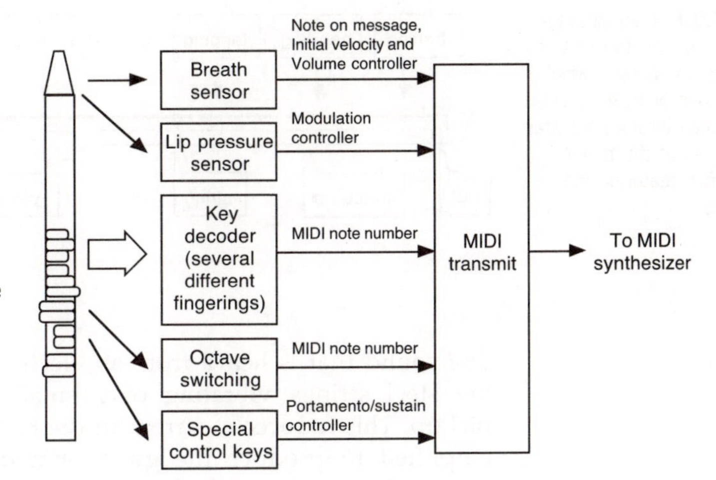

Figure 7.8.1 Wind controllers provide sensing for breath pressure and lip pressure (which are converted into MIDI controller messages for volume or modulation), and convert the keying into MIDI note numbers. Special octave switching keys and control keys (portamento/sustain) are also normally provided.

Breath controllers are simple pressure measuring devices which are blown into by the player. The output voltage is used to control either modulation or the envelope of a sound by replacing the volume control.

Wind instrument controllers use a controller which is based on a wind instrument like a flute, clarinet or saxophone and convert the finger presses on the keys into pitch information, and the blowing into additional modulation and volume information. One or more selectable modified forms of Boehm fingering are normally used, with extensions to cope with the larger pitch range provided by a synthesizer sound generator. Additional control keys allow octave switching, portamento control and sustain effects to be controlled from the wind controller.

Wind controllers are particularly effective when they are used to play lead line and solo melody sounds.

Although wind controllers are normally monophonic, some provide the ability to hold notes and then play additional notes on top, thus allowing the building up of sustained chords.

7.9 Guitar controllers

Given the enormous interest in using electronics to change the sound of the electric guitar, with a huge range of effects pedals being available, using a guitar as the controller for a synthesizer seems like an obvious development. Unfortunately, using a guitar for this purpose is not easy. There are a number of technical problems with using a guitar as the interface between a player and a synthesizer.

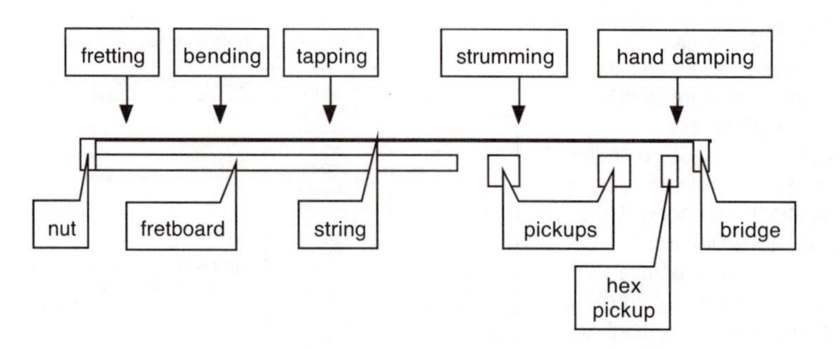

Figure 7.9.1 A guitar player can pluck, strum, tap or bend any of the six strings, which may be open or fretted. This cross-section through a guitar shows some of the major performance features and techniques.

The sound that is heard from an electric guitar is produced by the steel strings vibrating over small magnetic coils in the pickup. This induces a current in the pickup coils which is then amplified to produce the guitar sound. Normal pickups are designed to 'pick up' the vibrations from all the strings, but in order to process each string's pitch separately, the pickup needs to produce individual signals for each string. This is normally achieved by using a 'hexaphonic' or hex pickup, which has separate coils for each string.

Even with a separate signal for each string, extracting the pitch that a string is playing is far from straightforward: the pitch is dependent on the position at which the string was fretted, as well as any string bending or use of the tremolo arm, and it may also depend on how the string was plucked or strummed – the player may have deliberately produced a harmonic rather than the pitch which might be expected, for example.

In addition, the player can control the volume of each string by the amount of effort which is used in plucking or strumming the strings. This needs to be determined by the guitar controller so that appropriate control signals can be transmitted to the synthesizer. But the player also has additional control like string damping, which is achieved by placing the hand lightly against the strings near the bridge. The timbre that the strings produce can be varied by altering the position at which the string is plucked or strummed: the closer the position is to the bridge, the brighter the sound. Strings may also be tapped onto the fretboard to produce a sound, in which case the position is nowhere near the bridge at all. All of these performance techniques which are available to the guitarist make the task of converting a normal electric guitar into a suitable controller for a guitar synthesizer very difficult.

Approaches which have been tried include the hexaphonic pickup with sophisticated signal processing to try and extract the pitch; wiring up the frets so that they can be used as

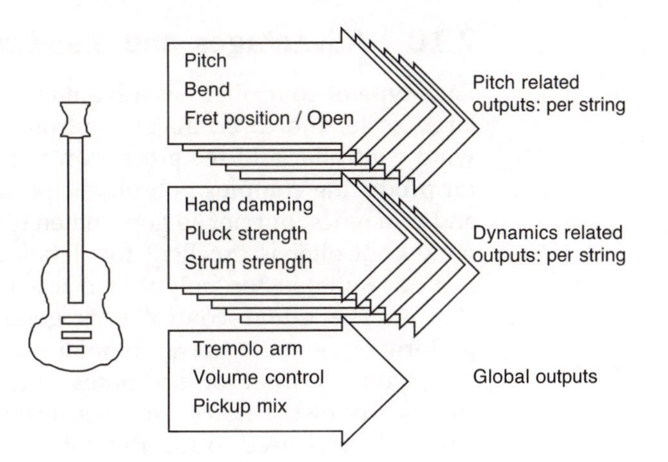

Figure 7.9.2 A guitar controller produces a large quantity of performance information.

electronic switches so that the fretting can be monitored; and even using acoustic radar signals along the strings so that the position of the plucking and fretting can be determined. None of these methods offers a complete solution, but development work is ongoing, and the limitations of the guitar as a controller seem to be gradually diminishing as the complexity of the solutions rises.

Rather than use a normal guitar as the controller, some manufacturers and experimenters have chosen to use a controller which has some elements of a guitar, but deliberately modified so that they suit the conversion process. As well as the wiring of the frets so that the fret position can be determined, some of these designs separate the strings which are fretted from those which are plucked or strummed, and others provide keys to trigger the string events instead. Most of these solutions are also technically complex, and none has been commercially successful.

One of the most useful results of guitar synthesizer research is arguably the hex pickup. By taking the six separate audio signals and applying distortion effects to them individually, a range of effects can be produced which do not have the large amounts of inter-modulation distortion present when a normal guitar pickup signal is distorted. Instead the result is a bright, synthetic sound which has many of the expressive capabilities of the guitar, but without the problems of chords producing too much distortion. Some guitar synthesizers have used this hex signal to drive the strings via a head similar to the record heads on a tape recorder, which allows the vibration of the string to be captured and amplified, opening up possibilities for increased sustain.

7.10 Advantages and disadvantages

Each type of controller has advantages and disadvantages. On a large scale, the three major controllers are the keyboard, the wind controller and the guitar controller. The keyboard is good for producing complex polyphonic performances based on notes and dynamics, but not so good when expression is required. The wind controller is excellent for detailed monophonic melodies, but not suitable for polyphonic use beyond simple sustained chords. The guitar controller is good for simple polyphonic performances using notes, dynamics and expression, but it has a polyphony limit of six notes, and has limitations in the complexity of the notes used because of the span of the single hand which is used to fret the notes.

Drum controllers are normally used to provide information on dynamics, triggering and pitch. Each drum pad is normally assigned to a fixed pitch, although this can sometimes be altered by using the velocity to change the pitch. Drum controllers are good for controlling percussion sounds, but the limitations of two hands restrict the polyphony to two or four notes at once, and so the complexity of the note patterns is limited by the handling of the drum sticks.

On the small scale, the individual controllers are the wheels, levers, pressure pads and pedals. Wheels and levers are largely similar, although wheels are much more common than levers on commercial synthesizers. Joysticks seem to be less popular than wheels, at least, manufacturers seem to prefer fitting wheels – joysticks tend to be used for mixing between sounds using a vector mixing approach. Pressure pads are rare and require an artificial separation of the pitch-bend into two pads. Pedals suffer from problems of response – the pedal is much heavier than the wheel or lever and so this tends to limit the speed at which the foot can move it. The control of a pedal with a foot does not seem to have the same degree of precise control as with a hand on a wheel or lever.

After-touch or key pressure has severe limitations as a pitch or modulation controller because it is limited for use only when the key is pressed down. This limits its use and the speed with which notes can be played whilst attempting to use after-touch. One technique which can be applied when monophonic after-touch is used is to use the right hand to play the melody notes, and use a finger on the left hand to play a related bass root note or a pedal bass note and use this finger to control the after-touch. Since monophonic after-touch applies to the entire keyboard, this allows expression to be added to the right hand, even though the right hand is apparently moving too quickly to be able to apply after-touch. Breath controllers or foot pedals can be used instead of the left hand to produce similar effects.

7.11 Front panel controls

Controllers have had a marked influence on synthesizer design. As well as the performance controllers described in the larger part of this chapter, a synthesizer also has front panel controls. The knobs, switches and sliders which are used to control the synthesizer in non-real-time performance are just as important to the function of the instrument, and they have changed significantly in the lifetime of the synthesizer.

Early analogue modular synthesizers used knobs and switches to set the operating parameters of the modules. Interconnections were made using patch-cords which were plugged into front panel sockets. This imposes a number of constraints on the use of such a synthesizer: the patch-cords can obscure the front panel, making it difficult to see the settings of the knobs and switches; and it also makes it awkward to change the patching rapidly.

As synthesizers developed, both monophonic and polyphonic versions streamlined the patching, reducing it down to the VCO/VCF/VCA format. Switches could now be used to control the routing of controllers like LFOs and envelopes, and the front panel could be designed so that it represented the signal flow through the synthesizer. The result was the front panels of the late 1970s and early 1980s, where the front panel used a large number of knobs and switches, and occasionally some sliders. In most cases, each knob or switch controlled a single parameter, which meant that learning the front panel layout told you how the synthesizer produced the sounds, and what to adjust to change a sound.

A much more minimalistic arrangement was introduced by the first digital synthesizers. The Yamaha DX7 provided one knob and a lot of buttons. The buttons were assigned so that there was usually one button per parameter, although this was complicated slightly by having several different modes, where the buttons had different meanings. Even so, for most editing, you pressed a button to select the parameter with one hand (the right), and adjusted the value with the other hand (the left). This two-handed approach to editing can be very fast, but it requires the co-ordination of both hands and the eyes, and a considerable amount of concentration.

Figure 7.11.1 Front panels have developed with technology. In the 1970s the layout reflected the structure of the synthesis method. In the 1980s the use of a single slider control with parameter selection buttons introduced a minimalistic interface. The 1990s saw a move towards soft keys and displays, with the displays increasing in size and more soft keys being added as the decade progressed.

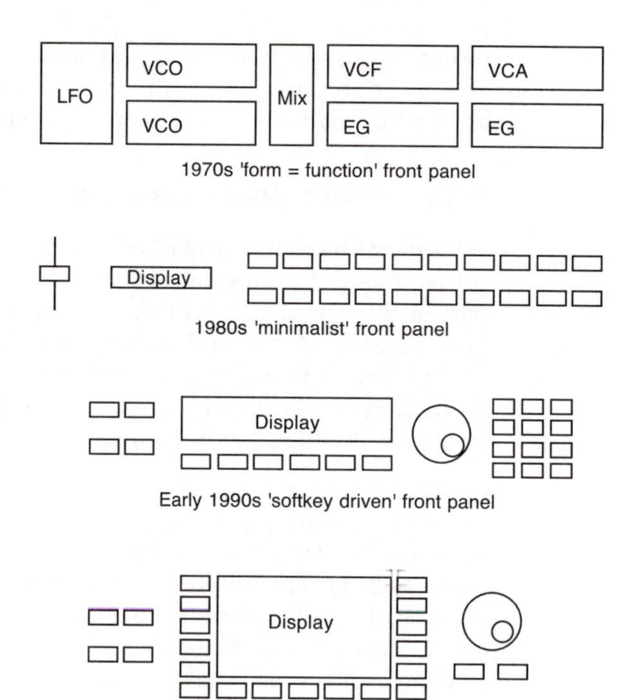

Computer-based editors can be used to replace this minimalistic interface, but screen redraws with large numbers of displayed parameters tend to be slow. The advantage of using this type of interface is that it is easy to scale to a rack mounting expander module with a limited front panel area since few controls are required and the multiple selection buttons can be replaced by a single parameter selection knob or slider. As this two-handed editing interface has been developed, the slider has been augmented by additional controls like a rotary dial and increment/decrement buttons.

When larger LCD displays became available, the emphasis changed from the button selection of parameters to using the display itself. By arranging a row of assignable buttons or soft keys underneath the display, the selection of parameters could be achieved by re-using the soft keys. LCDs gradually developed from character-based to full graphics capability, which allowed the screens to become more and more the focus of the editing – with an increasing number of soft keys clustered around the increasingly large display.

With large displays, the front panel controls became less specific to the synthesizer, and sound selection keys, numeric key pads and play/edit mode buttons replaced the named parameter buttons. This placed an increasing load on the display, but allowed the front panel to stay static whilst the software could be changed and updated. As less and less controls were required when the display became dominant, front panel space was then available for additional performance controllers: track-balls and joysticks are two examples of methods used to permit rapid real-time control of parameters from the front panel of a synthesizer.

The mid-1990s saw the first introduction of touch-screens with sophisticated graphical interfaces. Although novel, and stylistically impressive, the sparsely populated front panels tend to force the use of the touch-screen with only limited alternative methods of controlling the synthesizer. Early touch-screens were slow and awkward in their response to touch. Front panel design continues to evolve, and has moved away from hardware to solutions which increasingly depend almost entirely on software: thus mirroring the developments in synthesis techniques.

7.12 Questions

1. Describe some alternative synthesizer controllers to the music keyboard.
2. Give examples of performance parameters being produced by controllers.
3. Compare and contrast a guitar controller a wind controller and a keyboard.
4. What are the two main types of keyboard controller?
5. What is the difference between attack velocity and release velocity?
6. What limits the use of after-touch by a performer?
7. What are the differences between a wheel designed for pitch-bend and one designed for modulation?
8. Why is a guitar a difficult instrument to use as a synthesizer controller?
9. What are the limitations of drum controllers when they are used to control a synthesizer?
10. Outline the history of synthesizer front panel controls.

Time line

Date	Name	Event	Notes
1974	Lyricon	The Lyricon was the first commercially available wind-controlled electronic instrument	
1976	Lol Creme & Kevin Godley	The Gizmotron – a mechanical 'infinite sustain' device for guitars	A variation on the 'bowing with steel rods' technique
1977	ARP	ARP Avatar Guitar Synthesizer	Monophonic synthesizer with Hex pickup (for 'Hex Fuzz')
1977	Roland	GR-500 Guitar Synthesiser	Hex pickup and string drivers gave 'infinite' sustain
1979	Realton	Variophon launched – simple electronic wind instrument	
1980	Roland	GR-300 Guitar Synthesizer	
1982	Roland	Roland SH-101, a monophonic synthesizer with an add-on hand-grip performance oriented pitch bend and modulation controller	Notable for its range of colour finishes: red, blue and grey
1983	Yamaha	BC1 breath controller launched as add-on for CS01 monosynth	A small silver box that you grasped with your teeth!
1984	SynthAxe	MIDI-based guitar controller with separate strings, trigger switches and fretboard switches	Nice design, but very expensive
1985	Yamaha	FC7 Foot Pedal launched. Uses optical rotation sensor	
1987	Stepp	DG1 digital guitar – ambitious guitar synthesizer/controller	
1988	Casio	DH100 Digital Horn launched. Very low cost wind controller	
1988	Yamaha	WX11 Wind controller launched. Designed to be used with WT11 synthesizer expander	Low cost replacement for the WX7 predecessor
1991	Yamaha	BC2 Breath Controller launched. Improved version of BC1 built into a headset	Significantly improved appearance over the BC1
1995	Roland	GI-10 Guitar-oriented hex pickup pitch-to-MIDI converter	

8 The future of synthesis

Apart from minor refinements, the development of sound synthesis is nearly complete. The technology has now reached the point where the quality of the sounds that can be produced is close to the physical limits of the human ear. Apart from minor improvements to the details of the synthesis techniques, the only area which still requires large amounts of work is the interfacing between the performer and the instrument – and this is probably also true for many conventional instruments as well as electronic ones. Digital technology enables software models of real and imaginary musical instruments to be combined with purely synthetic mathematically-derived instruments – with almost no foreseeable limit. The future lies in combining sequencing, recording, composition and arranging tools into an integrated whole.

The current split into solo and accompaniment instruments actually makes a great deal of sense – after all, real world instruments naturally divide into monophonic solo instruments and polyphonic accompaniment ones. But as synthesis becomes incorporated into equipment which is concerned with producing multimedia, then these divisions will become different software routines within a much larger framework, and the individual differences will become less relevant to the end user.

The future does not depend on larger ROM samples and ever more clichéd sample sets, or even SCSI-connected CD-ROMs of prepackaged sounds. It lies with synthesis, and physical modelling seems to be in pole position, although it is likely to incorporate multiple methods of synthesis combined in software rather than the instrument simulation which the early implementations concentrate on. Just as all-digital synthesis has been used to emulate even analogue synthesis, so advanced physical modelling will be able to produce sounds which combine the best of real, analogue and digital instruments.

The powerful DSP-based engines which are used for producing sounds can also be used to process sounds, and so sound synthesis and hard-disk recording will converge. As storage prices continue to fall, hard-disk recording will become more and more attractive, and will move from the top end of the market, to the lower end.

MIDI seems to have reached a stable point where its performance is adequate for the requirements of most of its users, and ZIPI shows that special-purpose interfacing protocols can be used without having to replace MIDI. MIDI is such a powerful enabling standard that it is unlikely to be replaced – instead the problem areas like sample transfer and timing accuracy may well utilize ZIPI-type enhancements whilst still retaining backwards compatibility.

The end result will be a DSP-engined computer, which could well be a multimedia PC – since Intel's aim seems to be to mix DSP functionality with conventional processors in a composite machine which does some processing on-chip, and the numerically intensive calculations on-DSP. Since much of the software is already being written for synthesizers and hard-disk-recorders, bringing them together with a powerful PC is not going to be difficult, and it could well mark the end of the stand-alone 'studio' synthesizer, except for live performance. Since most of the R&D costs for the software will be paid off by the sales of the synthesizers using that technology, then the costs for the software add-ons to a PC will be low. This may bring the incremental cost of adding sound capabilities to a PC to the point at which it may well be added into the basic operating system.

So, by the early twenty-first century, the synthesizer will have changed into something much more like the 'orchestra in a box' of public perception. In fact, it is likely to be much more than that, since the audio will be part of a larger 'multimedia' facility which can work with pictures and moving images with the same ease as audio. This brings together all of the types of synthesis which can be imagined, and some which may not be expected. Software 'expert' systems are likely to be able to do most of the work of composing and arranging in the widest sense, which means that the role of the end-user may well become more akin to that of a mixture of movie director and screenplay writer. The creative input from a human being at the beginning and the end of making music (and moving pictures) will still be essential – the processes in-between may well be devolved to computers.

The 'synthesizer' will thus no longer be a simple musical instrument which requires large amounts of additional equipment to produce music, but will be a smaller and much more capable creative assistant. Future editions of this book may well be very different in their content!

References

Books

Buick, Peter and Lennard, Vic (1995) *Music Technology Reference Book.* PC Publishing (technology).

Clayton, George B. (1975) *Experiments with Operational Amplifiers.* MacMillan (hybrid).

Capel, Vivian (1988) *Audio and Hi-Fi Engineer's Pocket Book.* Butterworth-Heinemann (Newnes) (technology).

Cary, Tristram (1992) *Illustrated Compendium of Musical Technology.* Faber and Faber (technology).

Chamberlin, Hal (1987) *Musical Applications of Microprocessors.* Hayden Books (digital).

Chowning, John and Bristow, David (1986) *FM Theory and Applications for Musicians.* Yamaha Music Foundation (digital).

Colbeck, Julian (1985–) *Keyfax 1, 2, 3, 4, 5, ...* (ongoing) Virgin/Music Maker/Making Music (general).

De Furia, Steve (1986) *The Secrets of Analog and Digital Synthesis.* Ferro Productions/Hal Leonard Books (analogue, digital).

Forrest, Peter (1994) *The A–Z of Analogue Synthesizers, Part 1: A–M.* Susurreal Publishing (analogue).

Mellor, David (1992) *How to set up a Home Recording Studio.* PC Publishing (general).

Newcomb, Martin (1994) *Guide to the Museum of Synthesizer Technology.* The Museum of Synthesizer Technology (analogue).

Pellman, Samuel (1994) *An Introduction to the Creation of Electro-acoustic Music.* Wadsworth (general).

Pierce. John R. (1992) *The Science of Musical Sound.* Scientific American/Freeman (general).

Roland Corporation (1978) *A Foundation for Electronic Music*. Roland Corporation (analogue).

Roland Corporation (1979) *Practical Synthesis for Electronic Music*. Roland Corporation (analogue).

Rumsey, Francis (1994) *MIDI Systems and Control*. Focal Press (digital).

Tooley, Michael (1987) *Computer Engineer's Pocket Book*. Butterworth-Heinemann (Newnes) (digital).

Vail, Mark (1993) *Vintage Synthesizers*. GPI/Miller Freeman (general).

Watkinson, John (1988) *The Art of Digital Audio*. Focal Press (digital).

Other sources

Greenwald, Ted (magazine article) Samplers laid bare. Keyboard Magazine, March 1989 (digital).

The Musical Museum (museum) The Musical Museum, 368 High Street, Brentford, Middlesex, TW8 0BD. Telephone 0181 560 8108 (history).

The Museum of Synthesizer Technology (museum) The Museum of Synthesizer Technology, PO Box 36, Ware, Hertfordshire SG11 2AP. Telephone 01279 771619 (history, analogue).

The Computer Music Journal (quarterly journal) The MIT Press, Fitzroy House, 11 Chenies Street, London WC1E 7ET. Telephone 071 306 0603 Fax 071 306 0604 (history, analogue, digital, sampling).

Sound On Sound (monthly magazine) SOS Publications Limited, Media House, Burrel Road, St Ives, Cambridgeshire PE17 4LE. Telephone 01480 461 244 (history, analogue, digital, sampling).

Keefe, Douglas H. (journal paper) Physical Modelling of Wind Instruments. Computer Music Journal, Vol 16, Number 4, Winter 1992 (digital).

Karplus, K. and A. Strong, (journal paper) Digital. Synthesis of Plucked String and Drum Timbres. Computer Music Journal, Vol 7, Number 2, Summer 1983 (digital).

Cook, Perry R. (journal paper) SPASM, a Real-Time Vocal Tract Physical Model Controller. Computer Music Journal, Vol 17, Number 1, Spring 1993 (digital).

Sullivan, Charles R. (journal paper) Extending the Karplus–Strong Algorithm to Synthesise Electric Guitar Timbres with Distortion and Feedback. Computer Music Journal, Volume 14, Number 3, Fall 1990 (digital).

Julius O. Smith III (journal paper) Musical Applications of Digital. Waveguides. Stanford University Center for Computer Research in Music and Acoustics, STAN-M-39 (digital).

Julius O. Smith III (journal paper) Physical Modelling using Digital. Waveguides. Computer Music Journal, Vol 16, Number 4, Winter 1992 (digital).

White, Paul (magazine article) General MIDI. Sound On Sound, August 1993 (using synthesis).

ZIPI (journal papers) Computer Music Journal, Vol. 18, Number 4 (digital).

Glossary

AC

Abbreviation for alternating current: current which changes its value cyclically.
See also DC, Chapter 1 Background.

Accuracy

Repeatable measurements.
See also Precision, Chapter 2 Analogue.

Acoustics

The science of sound.
See also Chapter 1 Background.

Acronym

A word whose letters represent the initial letters of a sentence or phrase. MIDI = Musical Instrument Digital Interface.
See also Synonym, Chapter 1 Background.

ADBDR

Abbreviation for a type of envelope generator which has attack, two decay, and release segments.
See also EG, Chapter 2 Analogue.

ADC

Analogue-to-Digital Converter. Changes continuous values into discrete digital samples.
See also DAC, Chapter 1 Background.

Additive synthesis

Synthesis using many simple parts combined. Usually lots of sine waves.
See also Subtractive synthesis, Chapter 5, Digital.

ADR

Abbreviation for a type of envelope generator which has attack, decay and release segments.
See also EG, Chapter 2 Analogue.

ADS

Abbreviation for a type of envelope generator which has attack, decay, and sustain segments.
See also EG, Chapter 2 Analogue.

ADSR — Abbreviation for a type of envelope generator which has attack, decay, sustain and release segments.
See also EG, Chapter 2 Analogue.

After-touch — A synonym for channel or poly pressure: pressing on the key once it has reached the bottom of its travel.
See also Pressure, Chapter 7 Controllers.

Algorithm — A series of operations to carry out a process. Used by Yamaha for the arrangement of modules in an FM synthesizer.
See also FM, Chapter 5 Digital.

Aliasing — The 'reflection' of frequencies around the half sampling frequency. If a 750 Hz signal is sampled at 1000 Hz, then a 250 Hz alias frequency is the result.
See also Sampling, Chapter 1 Background.

AM — Amplitude modulation. A change in the amplitude of a signal (usually repetitive).
See also FM, Chapter 2 Analogue.

Amplitude — The loudness, level, amount or volume of a signal or waveform.
See also Level, Chapter 2 Analogue.

Analog — American spelling of 'Analogue'.
See also Analogue, Chapter 2 Analogue.

Analogue — Continuous values used to represent parameters.
See also Digital, Chapter 2 Analogue.

Anti-aliasing filter — A low-pass filter which prevents frequencies greater than half the sampling frequency being sampled. It thus prevents aliasing.
See also Reconstruction filter, Chapter 1 Background.

Antinode — The part of a vibrating object which has the largest range of movement.
See also Node, Chapter 1 Background.

AR — Abbreviation for a type of envelope generator which has only attack and release segments.
See also EG, Chapter 2 Analogue.

Arpeggiator — Cyclic repetition of a group of notes. Often used on the notes which are held down on a keyboard.
See also Sequence, Chapter 7 Controllers.

ASCII — American Standard Code for Information Exchange: computer coding system used for representing character symbols.
See also Keyboard, Chapter 5 Digital.

Assignment — The process whereby incoming notes are allocated to resources inside a synthesizer using an algorithm.
See also Keyboard, Chapter 2 Analogue.

Attack — The initial (start) segment of an envelope. Often a rapid rise to the maximum level.
See also ADSR, Chapter 2 Analogue.

Attack velocity The speed with which a keyboard note is initially pressed as it travels from the up to the down position.
See also Pressure, Chapter 7 Controllers.

Attenuation A reduction in the amplitude, level, loudness or volume of a signal or control voltage. Usually expressed as a ratio in dB.
See also Decibel, Chapter 2 Analogue.

Attenuator An electronic device or component which reduces the amplitude, level, loudness or volume of a signal or control voltage.
See also Volume, Chapter 2 Analogue.

Auto-bend Pitch chirp from AD envelope generator, triggered by start of note.
See also Pitch EG, Chapter 7 Controllers.

Auto-correct Another word for quantization. Used specifically for time quantization, where note values are adjusted to the nearest value within a range set by the quantization amount.
See also Quantization, Chapter 6 Using synthesis.

Auto-glide Synonym for portamento.
See also Portamento, Chapter 7 Controllers.

Auto-tune Retuning and calibration of VCOs, often at power-up.
See also DCOs, Chapter 3 Hybrid.

Balanced modulator Also known as a ring modulator. Takes two inputs and produces sum and difference frequencies.
See also Ring modulation, Chapter 2 Analogue.

Band A range of frequencies.
See also Bandwidth, Chapter 2 Analogue.

Band-pass A filter which passes all the frequencies between two values (the pass-band) unchanged, but attenuates all other frequencies (the stop bands).
See also Filter, Chapter 2 Analogue.

Band-reject filter A filter which does not pass the frequencies which are in a band of frequencies.
See also Filter, Chapter 2 Analogue.

Band-stop filter Synonym for band-reject filter. A filter which does not pass the frequencies which are in a band of frequencies.
See also Band-reject filter, Chapter 2 Analogue.

Bandwidth The range of frequencies which are passed by a circuit or device. Normally implies an attenuation of less than 3 dB (the 3dB points set the bandwidth).
See also Filter, Chapter 2 Analogue.

Bank A set of memories, sounds, patches, programs, etc.
See also Patch, Chapter 1 Background.

Bar (1) Unit for the measurment of pressure. (2) A number of beats.
See also Phon, beat, Chapter 1 Background.

Baud rate The serial information rate. MIDI operates at 31 250 Baud = 31 250 bits per second.
See also Serial, Chapter 6 Using synthesis.

Beat A unit of musical time. Beats are grouped into bars.
See also Bar, Chapter 2 Analogue.

Bend A change of pitch or frequency. As used in the term 'pitch-bend'.
See also Pitch-bend, Chapter 7 Controllers.

Bender An alternative to the pitch-bend wheel controller. Benders are normally either side-to-side levers, or joysticks.
See also Pitch-bend, Chapter 7 Controllers.

Bias High-frequency signal applied to tape recorder record heads to improve the frequency response of the head.
See also Tape recorder, Chapter 2 Analogue.

billion One million million. (In the UK, one billion used to be one thousand million, but the usage has now commonly come into line with the US usage.)
See also Million, Chapter 1 Background.

Binary System of counting which uses base two. Only two values: one and zero. Used in computers.
See also Decimal, Chapter 1 Background.

bit A binary digit. Bits can have one of two values: on or off = 1 or 0 = true or false. The smallest representation of a number in a computer – roughly equivalent to a unit in the decimal counting system.
See also byte, Chapter 1 Background.

Block diagram A diagram used to show high-level relationships between functional elements of a design.
See also Chapter 1 Background.

BPM Beats per minute. The tempo of a piece of music is expressed in BPM.
See also Clock, Chapter 6 Using synthesis.

Break point A place where a change takes place: applied to scaling and envelopes.
See also Chapter 5, Digital.

Buffer A circuit which has a gain of 1, and serves to isolate subsequent circuits from the input.
See also Chapter 3 Hybrid.

Bus A carrier of information, usually within a computer, although sometimes used when a set of signals are used outside the computer: the SCSI bus is one example.
See also SCSI, Chapter 5, Digital.

byte Eight bits. Sometimes arranged in two four-bit 'nibbles'.
See also kilobyte, Chapter 1 Background.

Card A plug-in printed circuit board-like device which can be used to store sounds.
See also cartridge, Chapter 5, Digital.

Carrier The output part of an FM synthesis process. The carrier waveform is the one which is frequency modulated by the modulator. Usually related to the fundamental pitch of the resulting sound.
See also FM, Chapter 5, Digital.

Cartridge A plug-in device which is used to store sounds.
See also Card, Chapter 5, Digital.

CD Compact disc: digital storage medium for up to 75 minutes of stereo digital audio.
See also CD-I, Chapter 1 Background.

CD-I Compact Disc – Interactive. A variation on CD-ROM, designed for use with a dedicated stand-alone console rather than a general purpose computer.
See also CD-ROM, Chapter 1 Background.

CD-ROM Compact disc – read only memory. A compact disc used as a means of storing digital data – not necessarily audio data, but any computer data.
See also CD, Chapter 1 Background.

Cent Sub-division of pitch. There are 100 cents in a semitone. The ratios of two frequencies which are one cent apart is 1.00057779.
See also Semitone, Chapter 1 Background.

Centre frequency The middle frequency of a notch or band-pass filter.
See also Filter, Chapter 2 Analogue.

Chain Short series of patterns or sequences, arranged into an order. Often used specificaly with drum machines.
See also Pattern, Chapter 1 Background.

Channel The complete path through a recording system.
See also Stereo, Chapter 6 Using synthesis.

Channel (MIDI) A MIDI channel is a means of splitting messages about musical events into separate streams, numbered from 1 to 16.
See also MIDI, Chapter 7 Controllers.

Chip Synonym for 'integrated circuit', although often used to mean specifically a microprocessor.
See also IC, Chapter 1 Background.

Chorus Notch-like cancellations caused by detuning of two audio signals.
See also Phasing, Chapter 6 Using synthesis.

Clamped sync Synchronization of two VCOs where the frequencies are locked to integer multiples.
See also Chapter 2 Analogue.

Clock Basic timing signal used in digital circuitry: rather like a heartbeat in an animal.
See also Chapter 6 Using synthesis.

CMOS Complimentary metal oxide semiconductor: type of digital logic chip made from FETs. Features low power consumption.
See also FET, Chapter 1 Background.

Coarse tune Semitone steps in the tuning of a VCO.
See also Chapter 2 Analogue.

Collages Assemblies of sounds which overlap.
See also Chapter 1 Background.

Colour Usually refers to the tone or timbre of a sound. Noise can be described as white (all frequencies), pink (high frequencies are attentuated) or blue (low frequencies are attentuated).
See also Filter, Chapter 1 Background.

Companding The process of compressing an audio signal before storage, and expanding in on retrieval.
See also PCM, Chapter 1 Background.

Compression A reduction in the dynamic range of an audio signal.
See also Chapter 6 Using synthesis.

Computer music Music produced by, with or from computers.
See also Chapter 1 Background.

Continuous Producing a constant output – no gaps or missing values.
See also Discrete, Chapter 5, Digital.

Contour Synonym for the 'amount of envelope applied to a VCF or VCA': Moog-speak.
See also Envelope, Chapter 2 Analogue.

Contour generator Synonym for envelope generator: Moog-speak.
See also Envelope, Chapter 2 Analogue.

Control input The input to a synthesizer module for signals which vary internal parameters.
See also Chapter 2 Analogue.

Control output The output from a synthesizer module which can be used to vary other parameters.
See also Chapter 2 Analogue.

Control sink Synonym for a control input – used in matrix modulation descriptions.
See also Control input, Chapter 2 Analogue.

Control source Synonym for a control output – used in matrix modulation descriptions.
See also Control output, Chapter 2 Analogue.

Control voltage Signals carried by patch cables in a modular synthesizer – used to vary parameters.
See also Chapter 2 Analogue.

Controller A general purpose performance control. MIDI controllers are numbered: 1 is the modulation wheel.
See also Wheel, Chapter 7 Controllers.

Counter A circuit which increments or decrements each time a specific 'clock' input is received.
See also Chapter 3 Hybrid.

Cross modulation Connecting the outputs of a pair of oscillators to the other oscillator's control input.
See also Chapter 2 Analogue.

Cross-point matrix A patching system where inputs are presented on one edge of a rectangular matrix of cross-points, and outputs are presented on the other edge. Pins, connectors or jack plugs placed at the cross-point connect the respective inputs and outputs at that joint.
See also Matrix panel, Chapter 2 Analogue.

Cut-off frequency The frequency at which a filter starts to have an effect: the 3 dB point.
See also Chapter 2 Analogue.

Cut-off Synonym for cut-off frequency.
See also Cut-off frequency, Chapter 2 Analogue.

Cut-off slope The rate at which attenuation occurs outisde of the filter pass-band.
See also Pass-band, Chapter 2 Analogue.

CV Control voltage.
See also Chapter 2 Analogue.

CV-to-MIDI A device which converts from control voltages to MIDI messages.
See also MIDI-to-CV, Chapter 2 Analogue.

DAC Digital-to-analogue converter.
See also ADC, Chapter 1 Background.

Damp Colloquialism for a signal which has been processed by a effects unit (often just an audio signal with some reverb).
See also Chapter 6 Using synthesis.

Data rate The number of bits or bytes of information transmitted in a given time period.
See also Baud rate, Chapter 1 Background.

dB Unit symbol for decibel. A measure of relative level, using a logarithmc scale for audio power level.
See also Chapter 1 Background.

DC Abbreviation for direct current – a voltage which stays at the same value.
See also AC, Chapter 1 Background.

DCO Digitally controlled oscillator.
See also VCO, Chapter 3 Hybrid.

Decay A fall in amplitude or level. Often the second segment of an envelope.
See also Attack, Chapter 2 Analogue.

Decibel A measurement unit used in audio. It represents the ratio between a reference level and the measured level. Unit is the dB.
See also Reference, Chapter 2 Analogue.

Delay An event which does not immediately follow another event is said to be delayed in time.
See also Chapter 2 Analogue.

Destination Synonym for control input.
See also Chapter 2 Analogue.

Detent A physical marker used to indicate a position on a rotary control by a change of force required to move the control.
See also Chapter 7 Controllers.

Deterministic Synonym for predictable. Many waveforms are deterministic, since each cycle is very similar to the previous cycle – and the next one!
See also Stochastic, Chapter 2 Analogue.

Detune Amount of pitch difference between two sounds or oscillators.
See also Chapter 2 Analogue.

Detuning Deliberate pitch difference between two sounds or oscillators.
See also Chapter 2 Analogue.

Digital Using numbers (often binary numbers) to represent values.
See also Logic, Chapter 1 Background.

DIN Deutsche Industrie Norm: the German Industry Standard, rather like BSI in the UK. Often used to describe connectors, as in the 5 pin DIN plug used for MIDI.
See also MIDI, Chapter 1 Background.

Discrete Split into separate elements – opposite of continuous.
See also Sample Chapter 5, Digital.

Distortion A non-linear process that changes the waveshape and adds extra frequencies.
See also Non-linear, Chapter 1 Background.

Dither Slight variation in the timing of a clock – used to improve the performance of analogue-to-digital conversion systems.
See also Clock, Chapter 1 Background.

Double Synonym for stack or layer: playing the same notes using more than one sound at once.
See also Chapter 6 Using synthesis.

Dry Unprocessed audio signal. Not processed by an effects unit. Opposite of Wet.
See also Wet, Chapter 6 Using synthesis.

DSP Abbreviation for digital signal processor. A computer processing chip which is optimized for dealing with computations on numbers.
See also Computer Chapter 5, Digital.

Duophonic An instrument that can play only two notes at once. Some analogue synthesizers have this feature.
See also Monophonic, Chapter 2 Analogue.

DVA Dynamic voice allocation or dynamic voice assignment.
See also Assignment, Chapter 7 Controllers.

Dynamic range The ratio between the loudest and softest parts of a piece of music.
See also Compression, Chapter 1 Background.

Dynamics (1) Synonym for MIDI velocity. (2) The ratio between the loudest and softest parts of a piece of music.
See also Compression, Chapter 2 Analogue.

Echo A repeat of an audio signal, delayed in time. Normally at a lower level than the original signal.
See also Delay, Chapter 6 Using synthesis.

Edit (v) To change or modify, particularly a parameter in a synthesizer. (n) An edited sound.
See also Chapter 6 Using synthesis.

Edit buffer A temporary store, often used to hold the current changes to a sound. Some synthesizers use this as a buffer for incoming MIDI sounds.
See also Chapter 6 Using synthesis.

Editor (1) A computer program which assists in editing a sound. (2) Person who edits.
See also Chapter 6 Using synthesis.

Editor/librarian Computer program which edits, categorizes and stores sounds for later recall.
See also Chapter 6 Using synthesis.

Effects Modifications of signals. Audio effects include reverberation, echo, phasing and equalization. Video effects are also possible. 'Special effects' are techniques used to produce 'the impossible' in films.
See also FX, Chapter 6 Using synthesis.

Effects unit An audio processing unit which modifies an audio signal. Multi-effects units can produce more than one type of effect at once.
See also Chapter 6 Using synthesis.

EG Envelope generator.
See also ADSR, Chapter 2 Analogue.

Electro-acoustic music Music produced by electronic methods.
See also Chapter 1 Background.

Electronic music Music produced by electronic methods.
See also Chapter 1 Background.

Electronics The study of devices which use electrons.
See also Chapter 1 Background.

Emphasis (1) Performance controller: usually a foot pedal which is used to increase the volume. (2) Boosting the audio level of a specific band of frequencies.
See also Chapter 7 Controllers.

Enharmonic Notes which can have two names depending on the context. A flat and G sharp are one example: they have different names, but are often the same pitch.
See also Chapter 1 Background.

Envelope The shape of the change of volume or level of an audio signal.
See also ADSR, Chapter 2 Analogue.

Envelope follower A circuit which outputs a voltage which represents the envelope shape of an audio signal.
See also Chapter 2 Analogue.

Envelope generator A circuit which produces a segmented sequence of segments which represent the stages of an envelope. Used as a control voltage source.
See also Chapter 2 Analogue.

Envelope tracking A parameter which controls the variation of the length of envelope segments – often the controller is keyboard note position or key velocity.
See also Chapter 2 Analogue.

EPROM Electrically programmable read only memory. A ROM memory which can be erased using UV light and reprogrammed. Often used to hold the operating system software in computers and synthesizers.
See also ROM, Chapter 1 Background.

EQ Abbreviation of equalization. Tone control: altering the level of specific bands of frequencies.
See also Filter, Chapter 2 Analogue.

Equal temperament A keyboard scale where the ratios between the frequencies of adjacent notes are the same for all notes.
See also Chapter 1 Background.

Equalization Changes to the relative level of different frequencies (or frequency bands) in an audio signal.
See also EQ, Chapter 1 Background.

Equalizer A device which changes the relative level of different frequencies (or frequency bands) in an audio signal.
See also EQ, Chapter 6 Using synthesis.

Event editing Part of a sequencer. It allows the time location and type of musical events to be altered.
See also Sequencer, Chapter 6 Using synthesis.

Excitation The signal which is connected to the input of a vocoder.
See also Vocoder, Chapter 2 Analogue.

Expander A device which increases the dynamic range of an audio signal.

See also A non-linear amplifier.
See also Chapter 2 Analogue.

Exponential A relationship between an input and an output where the output increases more rapidly (or less rapidly) than the input. A power law relationship: $y = a^x$.
See also Chapter 2 Analogue.

Feedback The connection of the output signal of a system to the input. A total gain of greater than 1 often results in oscillation at the frequency with the highest gain.
See also Chapter 2 Analogue.

FET Field effect transistor. A semiconductor device which can act as a voltage controlled switch.
See also Transistor, Chapter 2 Analogue.

FFT Abbreviation for fast Fourier transform. A mathematical process which calculates the frequency content of a sample or waveform.
See also Chapter 5, Digital.

Filter A circuit which has a gain that depends on the frequency of an audio signal.
See also Chapter 1 Background.

Filter tracking Changes to the cut-off frequency of a filter caused by a control signal: often the keyboard note position or the key velocity.
See also Chapter 2 Analogue.

Final level The final (quiescent) voltage level of an envelope generator.
See also EG, Chapter 2 Analogue.

Fine tune Sub-semitone steps in the tuning of a VCO, often providing cent resolution.
See also Chapter 2 Analogue.

Flanging Sound effect produced by mixing together a time delayed audio signal with the original.
See also Effects, Chapter 6 Using synthesis.

Flutter echo Echoes with a short time delay – from the point at which separate echoes become apparent (about 10 ms) to the point at which they become 'full' echoes (about 100 ms).
See also Echo, Chapter 6 Using synthesis.

FM Frequency modulation. A method of synthesis that uses a modulator waveform to frequency modulate a carrier waveform. Typically uses sine waveforms.
See also Carrier Chapter 5, Digital.

Foot controller A foot pedal that is used to produce a control voltage.
See also Chapter 7 Controllers.

Foot pedal A hinged flat plate which is rotated by the foot of the operator. Often used to control volume.
See also Chapter 7 Controllers.

Foot switch A hinged metal plate which is pressed down by the foot of the operator and which provides two positions: on and off.
See also Chapter 7 Controllers.

Formant A peak in a frequency spectrum. Formants are responsible for much of the characteristic timbre of a specific instrument.
See also Chapter 1 Background.

Found sounds Sounds which are literally 'found' rather than deliberately made, normally recorded in specific locations.
See also Chapter 1 Background.

Four-pole A filter with four sections. The combination produces a 'sharper' filter with a narrower pass-band or a sharper cut-off slope.
See also Chapter 2 Analogue.

Fourier analysis Conversion of an audio signal into its consitituent sine waves.
See also Chapter 5, Digital.

Fourier synthesis Construction of an audio signal by combining sine waves.
See also Additive synthesis, Chapter 2 Analogue.

FPU Floating point unit. A computer co-processor which is optimized for carrying out mathematical operations on numbers.
See also DSP Chapter 5, Digital.

Frequency Repetition rate. Can apply to waveforms or events.
See also Hertz, Chapter 2 Analogue.

Frequency response The bandwidth between the two 'half power' points, or 3 dB points. A measure of the frequency range.
See also Chapter 1 Background.

FSK Abbreviation for frequency shift keying.
See also Chapter 6 Using synthesis.

FX Abbreviation for effects.
See also Effects, Chapter 6 Using synthesis.

Gain The ratio between the input and the output of a circuit. Gains of less than 1 are normally called attenuation.
See also Chapter 1 Background.

Gate A switching action. A control signal or voltage can be used to open or close a 'gate'.
See also Chapter 2 Analogue.

Gate pulse A control signal which indicates when a key is held down on a synthesizer keyboard.
See also Chapter 2 Analogue.

Gate signal Control signal which indicates when a key is held down on a keyboard. The end of the gate usually initiates the release portion of an envelope.
See also Trigger, Envelope, Chapter 2 Analogue.

Gated A signal whose level is controlled by another control signal.
See also Gate, Chapter 2 Analogue.

Generic editor An editor which can be used to change the parameters of a number of devices – and whose characteristics can be easily varied to suit any new devices.
See also Universal editor, Chapter 6 Using synthesis.

Generic librarian A librarian program which can be used to store and recall infomation from a number of devices and which can be reconfigured to cope with new devices.
See also Chapter 6 Using synthesis.

Gigabyte One billion bytes (1 million million).
See also Chapter 1 Background.

Glide Synonym for portamento.
See also Chapter 2 Analogue.

Glissando The playing of all the notes which are present in-between two notes.
See also Portamento, Chapter 3 Hybrid.

Global A control voltage which is available throughout a snthesizer. Often used as a synonym for generic or universal.
See also Generic, Chapter 1 Background.

Graphic equalizer An equalizer which is controlled by a number of slider controls arranged to provide an approximation of the frequency response which is produced by the equalization filters.
See also Chapter 6 Using synthesis.

Graphical Shown in pictures rather than words. Many computers have both text and graphical modes.
See also Chapter 1 Background.

Growl Low frequency vibrato, or vibrato applied to low notes.
See also Chapter 2 Analogue.

Hard disk A storage device for digital information which uses a rigid rotating disk coated with magnetic material.
See also Floppy disk, Chapter 1 Background.

Hard sync Synchronization where the two oscillators are locked together at integer multiples of their frequencies.
See also Chapter 2 Analogue.

Hard-wired Either permanently wired, or available via a switch.
See also Chapter 1 Background.

Harmonic A frequency which is above the fundamental and an inverse integer multiple of the fundamental frequency.
See also Overtone, Chapter 2 Analogue.

Headroom The gap between the signal which is being recorded and the maximum allowable signal level, 'dBs below distortion'.
See also Dynamics, Chapter 1 Background.

Hertz Unit of measurement for 'how many times an event occurs in a given time (usually a second)'.
See also Chapter 1 Background.

Hex Colloquialism for hexadecimal.
See also Hexadecimal, Chapter 1 Background.

Hexadecimal Numbers expressed in base 16. The first few hex digits look like decimal digits, whilst the extra digits use alphabetic characters: 0 1 2 3 4 5 6 7 8 9 A B C D E F. Often indicated by a preceding $ or H, or a trailing subscript:$_{16}$.
See also Decimal, Chapter 1 Background.

High-note The topmost played note on a keyboard. Often used in the context of portamento or key assignment.
See also Portamento, Key assignment, Chapter 2 Analogue.

High-pass A filter whose pass band is above the cut-off frequency. Low frequencies are attenuated, whilst high frequencies are passed through.
See also Filter, Chapter 2 Analogue.

Hold (1) Latch. (2) Sustain or delay segment of an EG.
See also Chapter 2 Analogue.

Hybrid A mixture of two technologies.
See also Chapter 3 Hybrid.

Hz Unit symbol for Hertz: measure of frequency or repetition rate.
See also Chapter 1 Background.

IC Abbreviation for integrated circuit. A semiconductor device which contains several components (transistors, gates, etc.).
See also Transistor, Chapter 1 Background.

In Abbreviation for Input. One of the three MIDI ports.
See also MIDI, Chapter 1 Background.

Inharmonic Frequencies which are not related to the fundamental frequency by whole number ratios. Non-harmonics.
See also Harmonic, Chapter 2 Analogue.

Initial level The start voltage level of an envelope generator output.
See also EG, Chapter 2 Analogue.

Input The port where signals are presented to a circuit.
See also Chapter 2 Analogue.

Interface The input and output ports of a device or system.
See also Chapter 2 Analogue.

Interpolate Calculating in-between values from existing values.
See also Chapter 5. Sampling.

Intervals The spacing between notes. Often expressed in semitones.
See also Chapter 2 Analogue.

Invert To turn 'upside down'. In digital terms, to change 1s into 0s and vice-versa. In analogue terms, to change the sign of the signal.
See also Chapter 2 Analogue.

Inverter A circuit which inverts anything presented at its input.
See also Chapter 1 Background.

IRCAM French experimental music research centre in Paris: Institute for Research into Co-ordination of Acoustics and Music.
See also Bell Labs, Chapter 8 Other types.

Jack Abbreviation for jack connector – based on a telephone exchange connector. Available in mono and stereo versions in a variety of sizes. For synthesizers the quarter inch jack connector is almost a standard for line level audio interconnections.
See also Connectors, Chapter 1 Background.

Jitter Variation in the timing of an edge in a digital circuit – often clock signals.
See also Chapter 3 Hybrid.

Joystick A two-axis controller consisting of a stick and two orthogonal potentiometers. Normally produces a control voltage or MIDI controller messages.
See also Wheels, Chapter 7 Controllers.

k Prefix for kilo, meaning one thousand.
See also Chapter 1 Background.

K Prefix meaning 1024, as used in computer terminology, e.g. 1 Kbyte = 1024 bytes.
See also Chapter 1 Background.

Kb A kilobyte: 1024 bytes.
See also bit, Chapter 1 Background.

Kbyte
1024 bytes: abbreviated to Kb. (1000 bytes is a kilobyte, abbreviated to Kbyte.)
See also Kb, Chapter 1 Background.

Key assignment
Also known as key/note allocation. The process of routing incoming note events to voice resouces inside a synthesizer.
See also Chapter 2 Analogue.

Key follow
Synonym for yey tracking: the amount of the keyboard control voltage (determining the pitch of a note played on the keyboard).
See also Key tracking, Chapter 2 Analogue.

Keyboard
(1) Musical keyboard, consisting of a series of black and white notes. (2) Qwerty keyboard, consisting of a number of switches with alphanumeric legends.
See also Controller, Chapter 1 Background.

Keyboard amount
Synonym for key tracking: the amount of the keyboard control voltage (determining the pitch of a note played on the keyboard) which is applied to a control sink.
See also Key tracking, Chapter 2 Analogue.

Keyboard level scaling
The variation of level with keyboard position. Can be used to split sounds on a keyboard.
See also Chapter 2 Analogue.

Keyboard logic
The circuitry which converts key depressions into control voltages, gates and triggers, or digital events.
See also Chapter 2 Analogue.

Keyboard rate scaling
The variation in timing of an envelope caused by note position on the keyboard.
See also Chapter 2 Analogue.

Keyboard scaling
A change in a parameter due to note position. Can be applied to level, envelope rates and the keyboard note position control voltage.
See also Chapter 2 Analogue.

Keyboard tracking
The amount of the keyboard control voltage (determining the pitch of a note played on the keyboard) which is sent to a control sink.
See also Chapter 2 Analogue.

Keyboard voltage
The output of the keyboard controller: one or more control voltages which represent any notes which are being held down.
See also Chapter 2 Analogue.

kHz
Unit symbol for thousands of Hertz. The k is not capitalized, since this would mean 1024ths of Hertz!.
See also Chapter 1 Background.

Knob
The front panel appearance of a rotary control. Normally a small cyclinder with some means of indicating the rotation position.
See also Chapter 7 Controllers.

Lag processor
A low-pass filter with a cut-off frequency of a fraction of a Hertz. Intended to 'smooth' out rapid changes in voltages, and so can be used to extract the envelope from a sound. Portamento can be produced by using a lag processor on the keyboard voltage.
See also Chapter 2 Analogue.

Last-note The last note played on a keyboard so far. Often used to control the mode of portamento or note assignment.
See also Chapter 2 Analogue.

Latch A simple memory circuit. The state at the input is stored when a trigger input is activated.
See also Chapter 2 Analogue.

Latching A latching control is one where it maintains a state until changed. Some switches are latching, some are momentary.
See also Momentary, Chapter 7 Controllers.

Layer A composite sound made up from two or more separate sounds.
See also Stack, Chapter 6 Using synthesis.

Layering The process of making a composite sound from two or more separate sounds.
See also Stacking, Chapter 6 Using synthesis.

LCD Abbreviation for liquid crystal display. A type of display used on electronic equipment. LCDs require only low power because they use electric fields to alter the rotation of light as it passes through the LCD. Polarization effects either let the light through or stop it, depending on the rotation, and so the dots on the LCD can appear as either transparent or opaque.
See also LED, Chapter 1 Background.

LED Abbreviation for light emitting diode. A semiconducting device which produces light when a current is passed through it.
See also LCD, Chapter 1 Background.

Level The amplitude, level, loudness or volume of a signal or waveform.
See also Volume, Chapter 2 Analogue.

Lever Method of controlling pitch on some synthesizers. A performance controller – often fabricated as a 'paddle' connected to a rotary control.
See also Chapter 7 Controllers.

LFO Low-frequency oscillator.
See also VCO, Chapter 2 Analogue.

Librarian Computer application program which stores data, usually sounds from synthesizers.
See also MIDI, Chapter 6 Using synthesis.

Limit The maximum or minimum value which can be reached.
See also Chapter 1 Background.

Linear A transfer function which is a straight line. Cicruits which are linear do not cause significant distortion.
See also Chapter 1 Background.

Lock Synonym for 'phase-lock'.
See also Chapter 2 Analogue.

Log Colloquialism for logarithmic. Derived from logarithms, which are a means of carrying out multiplication and division using just addition and subtraction operations.
See also Logarithmic, Chapter 2 Analogue.

Logarithmic A relationship where the output is the power of a number that equals the input number. If $y = e^x$, then x is the log of y. Logs are normally to base e or base 10.
See also Chapter 1 Background.

Look-up A mapping between two variables. One is used as an index and points to the other number.
See also Chapter 3 Hybrid.

Loop A repeated series of events. Can be audio, MIDI messages or a physical loop of tape.
See also Chapter 2 Analogue.

Loudness (1) The volume of an audio signal. (2) Compensation of the response of the ear at low volumes.
See also Chapter 1 Background.

Low-note The lowest note which is being pressed on a keyboard. Low-note priority is a type of key assignment where the lowest note has precedence over higher notes.
See also High-note, Chapter 2 Analogue.

Low-pass Low pass filter: a circuit which passes frequencies below a special 'cut-off' frequency.
See also Filter, Chapter 2 Analogue.

LSB Least significant byte or least significant bit.
See also MSB, Chapter 1 Background.

μ Unit prefix for 'one millionth'.
See also m, Chapter 1 Background.

M Unit prefix for 'million'.
See also m, Chapter 1 Background.

m Unit prefix for 'one thousandth'.
See also M, Chapter 1 Background.

Map A relationship between two variables. It can be one-to-one, or one-to-many. A MIDI program change table is a map which connects program change messages to stored sounds.
See also Look-up, Chapter 6 Using synthesis.

Mapper A device which carries out mapping – especially changing one MIDI message to another.
See also Chapter 6 Using synthesis.

Mapping The process of noting down the map which connects two variables.
See also Chapter 6 Using synthesis.

Master A keyboard which is used as a source for MIDI messages is called a 'master' keyboard.
See also MIDI, Chapter 6 Using synthesis.

Matrix A two axis table which can be used to show the relationship between an input and an output.
See also Chapter 2 Analogue.

Matrix modulation A method of connecting control sources and sinks so that any source can be connected to any sink.
See also Chapter 2 Analogue.

Matrix panel A flexible patching system found on some modular synthesizers: ARP, EMS, Dewtron, etc. Not restricted to rectangular matrices like cross-point matrices.
See also Cross-point matrix, Chapter 2 Analogue.

Matrix switching Synonym for patching systems which use a matrix of switches.
See also Chapter 2 Analogue.

Mb Unit symbol for Megabyte: just over one million bytes.
See also Byte, Chapter 1 Background.

Mbyte Abbreviation for megabyte.
See also Chapter 1 Background.

Megabyte Just over 1 million bytes. Abbreviated to MB.
See also Byte, Chapter 1 Background.

Memories The sounds that a synthesizer makes are stored in memories. Different manufacturers use different names for memories.
See also Jargon, Chapter 1 Background.

Memory A store for analogue or digital information.
See also Chapter 1 Background.

Messages MIDI commands: sent as one or more bytes of MIDI data.
See also Chapter 1 Background.

MG Acronym for modulation generator: an LFO in Korg-speak.
See also LFO, Chapter 2 Analogue.

MIDI Acronym for Musical Instrument Digital Interface. An international standard for the interchange of information about musical events, and the inter-communication of electronic musical instruments.
See also Serial, Chapter 1 Background.

MIDI byte Eight bits from the MIDI serial stream. Also used for the seven bits which are used in a byte for carrying information.
See also MIDI, Chapter 1 Background.

MIDI clock A MIDI message which indicates timing. Sent at the rate of 24 clocks per quarter note.
See also MIDI, Chapter 1 Background.

MIDI message One or more MIDI bytes with a specific meaning as defined in the MIDI specification.
See also MIDI, Chapter 1 Background.

MIDI mode One of the four ways of operating a MIDI system: mono, 'multi', omni and poly.
See also Chapter 1 Background.

MIDI-to-CV A conversion between control voltages and MIDI messages.
See also Chapter 2 Analogue.

millisecond Unit of time measurement: one thousandth of a second.
See also ms, Chapter 1 Background.

Mixer A device which mixes together two or more audio signals.
See also Chapter 2 Analogue.

Mod Abbreviation for modulation or modifier.
See also Chapter 2 Analogue.

Mode Colloquialism for one of the four ways of operating a MIDI system: mono, 'multi', omni and poly.
See also Chapter 1 Background.

Modifier A circuit or device which changes a control voltage, audio signal or digital information.
See also Chapter 2 Analogue.

Modular Built up from simpler units. Modular synthesizers have discrete VCOs, VCFs, etc.
See also Chapter 2 Analogue.

Modulation Changing or varying a parameter. Often used to mean a cyclic variation, but not exclusively.
See also Chapter 2 Analogue.

Modulation generator Term used by Korg for an LFO.
See also LFO, Chapter 2 Analogue.

Modulator A device which imposes modulation.
See also FM Chapter 5, Digital.

Module One of the simpler component units which form a larger and more complex complete assembly.
See also Modular, Chapter 2 Analogue.

Momentary A momentary control is one which is normally in one state, and can be temporarily put into another state. Some switches are momentary, others latch in one of several states.
See also Latching, See also Chapter 7 Controllers.

Monitor (1) To note down or keep track of. (2) A TV display.
See also Chapter 6 Using synthesis.

Mono Single channel of audio. Many synthesizers have mono synthesis functions, but pan controls and stereo effects processors are used to provide a 'pseudo-stereo' output. Some samplers provide true stereo samples.
See also Stereo, Chapter 6 Using synthesis.

Mono mode The monophonic MIDI mode. Provides up to 16 separate individual monophonic channels.
See also Chapter 1 Background.

Monochord A stringed instrument with only one string.
See also Chapter 1 Background.

Monophonic An instrument that can only play one note at once. Most wind and brass instruments are naturally monophonic.
See also Polyphonic, Chapter 2 Analogue.

Monostable A circuit which has a default state, and which can be forced into an alternative state which it remains in for a fixed time interval, after which it reverts to the default state.
See also Chapter 1 Background.

Mother Synonym for 'master', as in a master keyboard.
See also Chapter 7 Controllers.

Mouse An x–y positioning device, often used as a pointing device for a computer.
See also Chapter 7 Controllers.

ms Unit symbol for millisecond: one thousandth of a second.
See also μs, Chapter 1 Background.

MSB Most significant byte or most significant bit.
See also LSB, Chapter 1 Background.

MTC MIDI time code. A simple form of SMPTE-like time code, used to provide synchronization in MIDI systems.
See also SMPTE, Chapter 6 Using synthesis.

Multi mode An 'unofficial' MIDI mode. When several channels can be used in poly modes, then the resulting multi-timbral mode is often called 'multi' mode.
See also Chapter 1 Background.

Multi trigger (1) An envelope generator which can restart the envelope. (2) A monophonic keyboard which can produce additional triggers if a note is pressed whilst another is held down.
See also Chapter 2 Analogue.

Multi-mode filter A filter whose design allows it to operate in low-pass, high-pass and band-pass modes. Sometimes also provides band-reject modes, and sometimes multiple outputs.
See also Chapter 2 Analogue.

Multi-sample Assigning several samples to notes (velocity layering) or groups of notes (keyboard layering).
See also Chapter 4 Sampling

Multi-timbral An instrument which is able to produce several different timbres simultaneously. A guitar can produce noises from the strings and from the soundboard by slapping it.
See also Polyphony, Chapter 2 Analogue.

Multiple A passive splitting module which is found in modular synthesizers. Can also be used as a pasive mixer.
See also Chapter 2 Analogue.

Multiple trigger Synonym for multi trigger.
See also Chapter 2 Analogue.

Multitimbral An instrument which can play more than one sound at once.
See also Polyphony, Chapter 6 Using synthesis.

Musique concrète Music produced using pre-prepared audio fragments.
See also Tape, Chapter 1 Background.

μs Unit of time measurement, a microsecond: one millionth of a second.
See also s, ms, Chapter 1 Background.

Natural harmonic series A sequence of harmonics produced by taking a fundamental frequency and multiplying it by integers: e.g. 55 Hz, 110 Hz, 165 Hz, 220 Hz, 275 Hz, etc.
See also Chapter 1 Background.

nibble Half of a byte: four bits.
See also bit, byte, Chapter 1 Background.

Node A position on a vibrating string where no movement occurs.
See also Antinode, Chapter 1 Background.

Noise A random mix of all frequencies.
See also Chapter 2 Analogue.

Noise generator A circuit which produces noise.
See also Chapter 2 Analogue.

Non-linear A transfer function which is not a straight line. Non-linear circuits normally distort a signal passed through them.
See also Chapter 1 Background.

Non-real-time Not happening in real-time. Either with a delay, or taking longer to process than the length of the information being processed.
See also Chapter 1 Background.

Non-sinusoidal Not a sine wave wavform or waveshape.
See also Sine, Chapter 2 Analogue.

Non-volatile Permanently stored.
See also Chapter 1 Background.

Notch Synonym for a notch filter.
See also Chapter 2 Analogue.

Notch filter A filter which rejects one frequency.
See also Chapter 2 Analogue.

Note stealing The process of re-assigning playing notes when the available polyphony has all been allocated.
See also Chapter 2 Analogue.

Nyquist frequency The highest frequency which can be coded and restored in a digital codec system which is sampled at twice the Nyquist frequency.
See also Chapter 1 Background.

Octal Numbers expressed in base 8. The first eight digits are used: 0 1 2 3 4 5 6 7.
See also Chapter 1 Background.

Omni mode A MIDI mode where messages on any channel are played by the receiving instrument.
See also Chapter 1 Background.

One-shot Synonym for monostable. A circuit which produces a short pulse when it is triggered.
See also Monostable, Chapter 1 Background.

Operating system The overall underlying software which runs a computer.
See also Chapter 5, Digital.

Operator Yamaha FM-speak for an oscillator, VCA and envelope generator.
See also FM, Chapter 5 Digital.

Orthogonal (1) Two axes or planes which are at 90 degrees to each other. (2) Two quantities which are unrelated.
See also Chapter 6 Using synthesis.

OS Operating system The overall underlying control program of a computer.
See also Computer Chapter 5 Digital.

Oscillator A circuit which produces a repetitive output.
See also Chapter 2 Analogue.

Oscillator sync Locking together of two oscillators so that the two frequencies have some relationship.
See also Hard sync, Chapter 2 Analogue.

Out Abbreviation for output. One of the three MIDI ports.
See also MIDI, Chapter 1 Background.

Overdrive Synonym for distortion.
See also Distortion, Chapter 2 Analogue.

Overflow When a digital circuit exceeds the largest number it can cope with, it has overflowed.
See also Chapter 1 Background.

Overtone Similar to harmonic: whole number multiples of the fundamental. The second harmonic is the FIRST overtone.
See also Harmonic, partial, Chapter 2 Analogue.

Paddle Method of controlling pitch on some synthesizers. A performance controller – often made from a flat textured vertical piece of plastic, with a rectangular hole in the front panel, and a fulcrum below the surface of the panel.
See also Chapter 7 Controllers.

Page User interface term, meaning a new set of controls in the same screen or panel area.
See also Chapter 7 Controllers.

Parallel In a parallel communication channel the data is sent with several bits simultaneously, using at least as many wires as there are bits.
See also Serial, Chapter 4 Sampling

Parameter An abstract value or variable quantity. Often used to represent the controls of a synthesizer.
See also Value, Chapter 5 Digital.

Part (1) One line in a piece of music. (2) One channel in a multi-timbral arrangement.
See also Chapter 2 Analogue.

Partial
(1) Synonym for harmonic, although can be used for inharmonics too. (2) Roland-speak for a 'voice'.
See also Voice, harmonic, Chapter 2 Analogue.

Patch
A sound – originally referred to the layout of the patch cords and control settings, but now generic.
See also Chapter 2 Analogue.

Patch cord
A cable with connectors at each end. Used for connecting modules together in a modular synthesizer.
See also Modular, Chapter 2 Analogue.

Patching
The process of connecting together modules using patch cords.
See also Chapter 2 Analogue.

PCM
Pulse code modulation. A method of encoding audio signals into digital form – the audio is sampled and the level is converted into a number.
See also ADC, DAC, Sampling, Chapter 4 Sampling.

Peak level
(1) The highest voltage present in an audio signal. (2) The voltage level reached by the attack segment of an envelope generator.
See also EG, Chapter 2 Analogue.

Performance
Many meanings: live performance. user controls. Complete front panel settings.
See also Memories, Chapter 6 Using synthesis.

Phase
The position in a waveform or waveshape. Normally given in degrees. One cycle contains 360 degrees.
See also Chapter 1 Background.

Phase shifter
A circuit which alters the relative phase of an audio signal.
See also Chapter 6 Using synthesis.

Phase sync
A method of locking two oscillators together so that they have a fixed phase relationship between their outputs.
See also Chapter 2 Analogue.

Phase-lock
A method of locking two oscialltors together so that they have a fixed phase relationship between their outputs.
See also Chapter 2 Analogue.

Phasing
Sound effect produced by mixing together a phase changed audio signal with the original.
See also Effects, Chapter 6 Using synthesis.

Phon
Unit of perceived loudness, which takes into account the response of the human ear.
See also Loudness, Chapter 1 Background.

Pink noise
Noise passed through a low-pass filter so that the level decreases with frequency. Produces equal energy for any bandwidth in the audio spectrum.
See also Chapter 2 Analogue.

Pitch
Synonym for the frequency of a note. Pitch is a musical term for frequency.
See also Frequency, Chapter 1 Background.

Pitch bend Changes of pitch or frequency.
See also Chapter 7 Controllers.

Pitch-to-MIDI The conversion of pitched audio signals into the equivalent MIDI note numbers. Normally only monophonic conversions are possible.
See also Chapter 5 Digital.

Pitch-to-voltage The conversion of pitched audio signals into a control voltage. Normally only monophonic conversions are possible.
See also Chapter 2 Analogue.

Pivot point Synonym for a break point in scaling.
See also Chapter 2 Analogue.

Pole One section of a complex filter. 2-pole and 4-pole VCFs are commonly found in analogue synthesizers. The number of poles affects the cut-off slope: 2-pole = 12 dB/octave, 4-pole = 24 db/octave. Poles are normally associated with peaks in the frequency response.
See also Cut-off slope, Chapter 2 Analogue.

Poly mode MIDI mode where polyphonic notes can be produced from a single MIDI channel.
See also Mode, Chapter 1 Background.

Poly-mod Synonym for cross-mod. Modulation of an oscillator by another oscillator.
See also Cross-mod, Chapter 2 Analogue.

Polyphonic An instrument that can play more than one note at once.
See also Monophonic, Chapter 2 Analogue.

Polyphony The total number of simultaneous 'notes' that an instrument can produce at once. A piano is fully polyphonic – with enough hands you can play all of the notes simultaneously!
See also Timbrality, Chapter 2 Analogue.

Port A MIDI port.
See also Chapter 1 Background.

Portamento A smooth change of frequency between two notes instead of the normal abrupt transition.
See also Glissando, Chapter 2 Analogue.

Pot Abbreviation for potentiometer: a rotary or slider control whose output is proportional to the position of the indicator knob.
See also Knob, Chapter 7 Controllers.

PPQ Acronym for pulses per quarter-note.
See also MIDI clock, Chapter 2 Analogue.

PPQN Pulses per quarter note. The number of pulses used to represent a quarter note length of time in a sequencer.
See also Sequencer, Chapter 7 Controllers.

Pre-emphasis The pre-processing of an audio signal before digitally coding it. A slight gain at high-frequencies is intended to improve the dynamic range for high frequency components.
See also Chapter 1 Background.

Precision The number of significant digits.
See also Accuracy, Chapter 2 Analogue.

Preset Synonym for patch.
See also Chapter 1 Background.

Pressure Pressing on the key with the finger, once it has reached the bottom of its travel.
See also After-touch, Chapter 7 Controllers.

Pressure sensitivity The size of the control signal produced by key pressure depends on the 'pressure sensitivity'.
See also Chapter 2 Analogue.

Priority Part of an assignment strategy.
See also DVA, Chapter 7 Controllers.

Program (1) Synonym for sound or patch. (2) A sequence of instructions to a computer.
See also Chapter 1 Background.

Program change A MIDI message which is used to select a sound.
See also Sound, Chapter 1 Background.

Programmable The behaviour can be controlled using a parameter or a script.
See also Chapter 2 Analogue.

Programmer (1) A person who writes computer programs. (2) A remote front panel for expander modules which do not have a complete front panel.
See also Chapter 2 Analogue.

Programming The process of writing a computer program, or the process of producing a sound on a synthesizer.
See also Chapter 2 Analogue.

PROM Programmable read only memory – a fixed one-time-programmable storage medium for computer programs.
See also EPROM, Chapter 1 Background.

Pulse A digital signal which consists of a change to one state for a fixed period of time, followed by a reversion to the original state.
See also Chapter 2 Analogue.

Pulse Width The ratio between the two parts of a pulse waveform. Normally expressed as a percentage.
See also Chapter 2 Analogue.

PW Abbreviation for pulse width.
See also Pulse width, Chapter 2 Analogue.

Q Synonym for resonance. A measure of the selectivity/ sharpness/ resonance of the filter.
See also Resonance, Chapter 2 Analogue.

Quantization Converting a continuous value into a series of discrete values.
See also Chapter 6 Using synthesis.

Quantization noise Noise produced by a digital coding system because of the limitations of the representation of small signals.
See also Chapter 1 Background.

Quantize The process of converting a continuous value into discrete values.
See also Chapter 1 Background.

Quantized A value which has been converted to a quantized form.
See also Chapter 1 Background.

Quantizer A circuit or device which quantizes.
See also Chapter 1 Background.

RAM Random access memory. Memory which can be written to, and read from.
See also ROM, Chapter 1 Background.

Ramp Synonym for sawtooth. Decription for a waveform shape which rises linearly and then abruptly returns to the start level.
See also Chapter 2 Analogue.

Random Unpredictable.
See also Noise, Chapter 2 Analogue.

Rate/level EG A type of envelope generator, where the shape is defined by a number of rate and level values.
See also Chapter 2 Analogue.

Rate scaling Variation of envelope generator times by a control voltage – usually derived from the keyboard control voltage.
See also Chapter 2 Analogue.

RCA Abbreviation for Radio Corporation of America. Electronics innovation company.
See also Chapter 1 Background.

Real-time Happens at the same rate as normal time. Implies a short time delay – almost instantaneous.
See also Chapter 1 Background.

Reconstruction filter Filter which removes all frequencies above the Nyquist frequency from the output of a DAC.
See also Nyquist, Chapter 1 Background.

Rectangle A waveshape which is equivalent to a pulse waveshape. The 50% pulse is also known as a square wave.
See also Square, Chapter 2 Analogue.

Rectangular Shaped like a rectangle.
See also Rectangle, Chapter 2 Analogue.

Reference A standard tone or level which is used to make comparisons against.
See also Chapter 1 Background.

Regeneration Synonym for feedback or resonance.
See also Chapter 2 Analogue.

Release The final segment of an envelope generator.
See also EG, Chapter 2 Analogue.

Release velocity The speed with which a key is released.
See also Attack velocity, Chapter 2 Analogue.

Remote keyboard A controller keyboard which is remote from the expander module (or modules) which it is controlling. Often limited in functionality with respect to a master keyboard.
See also Master keyboard, Chapter 7 Controllers.

Resolution The smallest change which can be represented. In integer mathematics, the resolution is one.
See also Precision, Chapter 1 Background.

Resonance A measure of the way that a system responds to an external stimulus. A tuning fork is resonant because it vibrates when struck. Filters can 'ring' or self-oscillate when the resonance is high.
See also Q, Chapter 2 Analogue.

Resynthesis Analysing an audio signal and extracting parameters which are then used as the basis for synthesising the sound with modifications.
See also Chapter 5, Digital.

Reverb Abbreviation for reverberation.
See also Chapter 6 Using synthesis.

Reverberation A series of echoes caused by reflections from the acoustic environment.
See also Echo, Chapter 6 Using synthesis.

Ribbon (1) Type of microphone. (2) Colloquialism for a performance controller using a flexible strip which is activated by pressing with the finger.
See also Chapter 7 Controllers.

Ribbon controller A performance controller using a flexible strip which is activated by pressing with the finger.
See also Chapter 7 Controllers.

Ring modulation Balanced modulation with two inputs and one output. Output is sum and difference of input frequencies.
See also AM, Chapter 2 Analogue.

Ringing Resonant circuits will oscillate momentarily when there is a sudden change at their input. This is called ringing. Some VCFs can be forced to ring by using high resonance settings.
See also Oscillator, Chapter 2 Analogue.

Roll-off Synonym for filter cut-off slope.
See also Slope, Chapter 2 Analogue.

Roll-off slope Synonym for filter cut-off slope.
See also Slope, Chapter 2 Analogue.

ROM Read-only memory. Memory which can only be read – the contents are fixed and so cannot be altered.
See also EPROM, Chapter 1 Background.

Routeing The path which an audio or control signal takes through a device or circuit.
See also Patching, Chapter 2 Analogue.

Running status A method of removing redundant information in repeated MIDI messages to reduce the required data bandwidth.
See also Chapter 1 Background.

s Unit of time measurement: second. One sixtieth of a minute.
See also Chapter 1 Background.

S&H Abbreviation for sample and hold. Also known as S/H.
See also Sample and Hold, Chapter 2 Analogue.

S/H Abbreviation for sample and hold. Also known as S&H.
See also Sample and Hold, Chapter 2 Analogue.

Sample Data which represents an audio signal. Samples can be analogue or digital, although they are usually digital.
See also Sampling, Chapter 4 Samplers.

Sample and hold An analogue latch which stores an analogue voltage when triggered by a clock or trigger signal.
See also Chapter 2 Analogue.

Sample rate The frequency at which analogue samples are converted to digital representations.
See also Sampling, Chapter 4 Samplers.

Sampler A device which can sample and replay sounds. Often used to mean a sample replay device.
See also Chapter 4 Samplers.

Samples Analogue or digital values which make up the information contained within an audio signal.
See also Chapter 4 Samplers.

Sampling The process of capturing part of an audio signal. Usually refers to digital sampling.
See also Chapter 4 Samplers.

Sawtooth A waveshape found in analogue synthesizer VCOs. Consists of a rising slope followed by a sudden return to the start value.
See also Ramp, Chapter 2 Analogue.

Scale The relationship between the keyboard control voltage and the VCO frequency. Normally set so that 1 volt of CV change produces an octave change in pitch.
See also Chapter 2 Analogue.

Scaling curve The shape of the curve used to modify the rate or level scaling with keyboard note position.
See also Chapter 2 Analogue.

Schmitt trigger A circuit which can be used to convert analogue signals into digital signals with two values.
See also Chapter 2 Analogue.

Scrub Moving backwards and forwards through events in a sequencer (audio, video or MIDI) or on a tape (audio or video).
See also Chapter 6 Using synthesis.

SCSI Acronym for small computer systems interface. A method of connecting peripherals (disks, CD-ROM drives, etc.) to a computer.
See also Bus, Chapter 4 Samplers.

SDS
Acronym for sample dump standard – the method defined in the MIDI specification for transferring samples using MIDI messages.
See also Chapter 1 Background.

Second touch
Synonym for after-touch or pressure.
See also After-touch, Chapter 7 Controllers.

Segment
One part of an envelope cycle. An AD envelope has two segments: attack and decay, whilst an ADSR envelope has four segments. In a rate and level envelope, one of the pairs of rates and levels.
See also Envelope, Chapter 2 Analogue.

Self-oscillation
VCOs and clock generators are designed to maintain self-oscillation, where a continuous repetitive output is produced. Often associated with resonant circuits.
See also Oscillator, Chapter 2 Analogue.

Semiconductor
A material whose electrical properties can be changed by applying voltages or currents. The basis of electronic devices.
See also Electronics, Chapter 1 Background.

Semitone
The smallest change of pitch on a conventional music keyboard. The ratios of two frequencies which are one semitone apart is 1.059463.
See also Cent, Chapter 1 Background.

Sequence
A series of musical events.
See also Chapter 6 Using synthesis.

Sequencer
A computer or hardware device which is designed to produce a series of musical events.
See also Chapter 6 Using synthesis.

Serial
In a serial communication channel the data is sent one bit at a time along a single piece of wire.
See also Parallel, Chapter 4 Samplers.

Sideband
A frequency produced above or below the carrier frequency by an FM, AM or RM modulation technique.
See also Chapter 5, Digital.

Sine
A smooth curved waveform which contains only one frequency component.
See also Harmonic, Chapter 2 Analogue.

Single trigger
A keyboard which produces only one trigger when a key is held down and other keys are subsequently pressed.
See also Multi trigger, Chapter 2 Analogue.

Single-step
Method of recording notes and other musical events into a sequencer. Each event is entered individually, in non-real-time.
See also Real-time, Chapter 6 Using synthesis.

Sinusoidal
Shaped like a sine wave.
See also Sine, Chapter 2 Analogue.

Slave
A 'slave' is a keyboard or expander synthesizer which receives MIDI messages.
See also MIDI, Chapter 6 Using synthesis.

Slope The ratio between the horizontal and vertical parts of a line. A slope of 1:1 is at an angle of 45 degrees.
See also Chapter 2 Analogue.

SMPTE Society of Motion Picture and Television Engineers. Short-hand for the timecode that is used to synchronize video, TV and audio.
See also MTC, Chapter 7 Controllers.

SNR Signal to noise ratio. A measure of the signal level against the background noise.
See also THD, Chapter 1 Background.

Song A collection of phrases of music. Typically starting with an introduction, then a verse, followed by a chorus, repeats of verse and chorus, a break for eight bars where a second theme is introduced, then more verse and chorus pairs, a key change and the closing section.
See also Chapter 6 Using synthesis.

Sostenuto pedal A performance controller which mimics the behaviour of the piano pedal: only notes which are being held down on the keyboard are sustained when the sostenuto function is activated.
See also Sustain, Chapter 7 Controllers.

Source A control voltage or control signal output.
See also Chapter 2 Analogue.

Spectrum A plot of the frequency content of an audio signal. The horizontal axis is the frequency axis, whilst the vertical axis shows the level of the frequency components.
See also Harmonic, Chapter 2 Analogue.

Speed The tape speed of a tape recorder is the speed at which the magnetic tape passes through the machine.
See also Chapter 1 Background.

Spillover Synonym for overflow.
See also Chapter 6 Using synthesis.

Splice A join between two bits of magnetic tape. This has also come to mean a join between any two bits of audio, even in software based audio editors and samplers.
See also Tape, Chapter 1 Background.

Split The allocation of two sounds to different parts of the keyboard. Named from the major usage, which is to play different parts with each hand on the same keyboard.
See also Chapter 6 Using synthesis.

Split point The point on the keyboard at which a split between two sounds occurs.
See also Split, Chapter 6 Using synthesis.

Square A waveshape which has equal time periods in two states. A special case of a rectangular or pulse waveshape.
See also Rectangle, Chapter 2 Analogue.

Stack A composite sound made up of two or more sounds.
See also Layer, Chapter 6 Using synthesis.

Stacking The process of making a composite sound from two or more sounds.
See also Layering, Chapter 6 Using synthesis.

Stage Synonym for a segment in an envelope. One of the pairs of rates and levels, or one of the attack, decay or release periods.
See also Chapter 2 Analogue.

State-variable A multi-mode filter made from a loop of op-amps, which produces several types of filter response outputs simultaneously.
See also Chapter 2 Analogue.

Step input Sequencer input mode: one note/event at a time. Non-real-time.
See also Single step, Chapter 6 Using synthesis.

Stereo Two audio channels (left and right), which can provide the illusion of a complete sound stage.
See also Mono, Chapter 6 Using synthesis.

Stochastic Synonym for unpredictable. Noise is an example of a stochastic waveform, since no two cycles are the same.
See also Deterministic, Chapter 2 Analogue.

Storage Often implies 'data storage media' – somewhere to hold information.
See also Chapter 1 Background.

Sub-octave A note one or two octaves down from the fundamental pitch. Often produced using digital divider circuits.
See also Chapter 3 Hybrid.

Subtractive Synthesis technique which uses harmonically rich waveforms and filters them to produce sounds.
See also Additive, Chapter 2 Analogue.

Subtractive synthesis Synthesis technique which uses harmonically rich waveforms and filters them to produce sounds.
See also Additive, Chapter 2 Analogue.

Sustain One of the segments or stages in an ADSR envelope. This is the level which the envelope will decay to and stay at whilst the key is held down.
See also Chapter 2 Analogue.

Switch trigger Envelope trigger signal which uses a brief pulse to initiate an envelope.
See also Chapter 2 Analogue.

Sync Abbreviation for synchronization: (1) locking a sequencer or tape recorder to another timing source. (2) Locking two VCOs together.
See also Chapter 2 Analogue.

Sync track A track on a tape recorder which is reserved for the recording and playback of synchronization signals.
See also Chapter 6 Using synthesis.

Synchronization (1) Locking a sequencer or tape recorder to another timing source. (2) Locking the frequency of two VCOs together.
See also Chapter 2 Analogue.

Synonym A word having almost exactly the same meaning as another word.
See also Acronym, Chapter 1 Background.

Synthesiser Alternative UK spelling for synthesizer. A musical instrument which produces sound from simple resources.
See also Chapter 1 Background.

Synthesizer US spelling of synthesizer. A musical instrument which produces sound from simple resources.
See also Chapter 1 Background.

Sysex Abbreviation for system exclusive: Instrument and manufacturer specific information.
See also System exclusive, Chapter 7 Controllers.

System MIDI-speak for global MIDI messages.
See also MIDI, Chapter 7 Controllers.

System exclusive Instrument and manufacturer specific information – transmitted as MIDI messages with a distinctive header and footer ($F0 and $F7).
See also MIDI, Chapter 7 Controllers.

Tape Recording tape. Polyester plastic backing with an iron oxide magnetic coating.
See also Speed, Chapter 4 Samplers.

Tape loops A piece of magnetic tape looped around and the end spliced to the beginning. Also used for looped audio in digital equipment.
See also Chapter 4 Samplers.

THD Abbreviation for total harmonic distortion. A measure of the distortion produced by an audio system. Measures the amount of additional harmonics which are added to a signal by a process. Usually given in percent.
See also Distortion, Chapter 1 Background.

Thru Abbreviation for 'through' (US). One of the three MIDI ports.
See also MIDI, Chapter 1 Background.

Timbrality The number of different 'timbres' or 'sounds' that an instrument can produce at once. Different to polyphony, which is the total number of notes which can be played at once, regardless of the 'timbre' or 'sound'.
See also Polyphony, Chapter 6 Using synthesis.

Timbre (1) The 'tone colour' of a sound. (2) Roland-speak for a sound.
See also Chapter 1 Background.

Time-code A method of attaching information about timing to an audio or video recording.
See also Chapter 6 Using synthesis.

Tone colour The timbre of a sound.
See also Timbre, Chapter 1 Background.

Tone control Synonym for equalization.
See also EQ, Chapter 1 Background.

Tone poems A piece of music which attempts to describe a place, event, emotion or person using just notes and timbre.
See also Timbre, Chapter 1 Background.

Top-note priority A key assignment algorithm which gives priority to the highest note played on a monophonic keyboard.
See also Low-note, Chapter 6 Using synthesis.

Total harmonic distortion Audio measurement which is used to determine the amount of additional harmonics which are added to a signal by a process.
See also THD, distortion, Chapter 1 Background.

Touch pad A performance controller.
See also Chapter 7 Controllers.

Touch sensitive Something which is changed by pressure. Normally additional pressure once a key has been depressed.
See also After-touch, Chapter 7 Controllers.

Track One of several paths on a piece of magnetic recording tape. This is also used in the context of digital sequencers.
See also Chapter 4 Samplers.

Tracking generator A module which is concerned with controlling the rate and level scaling functions.
See also Chapter 2 Analogue.

Transient Brief, non-sustaining, events. Can refer to sounds, harmonics or one-off occurrences.
See also Chapter 2 Analogue.

Transient generator Synonym for envelope generator.
See also EG, Chapter 2 Analogue.

Transistor Abbreviation for transfer resistor: a simple semiconducting device which can switch or amplify signals.
See also Semiconductor, FET, Chapter 1 Background.

Tremolo Periodic variation in the volume, level or amplitude of a signal (audio or control voltage). The LFO speed is usually between 5 and 25 Hz.
See also LFO, AM, Chapter 2 Analogue.

Triangle A waveform which has linear slopes which rise and fall alternately.
See also Sawtooth, Chapter 2 Analogue.

Trigger Signal which starts an event – usually an envelope.
See also Chapter 2 Analogue.

Trigger pulse (1) The initiation signal for an EG. (2) The output of a keyboard controller note detector – indicates that a key has been pressed.
See also Chapter 2 Analogue.

TTL Transistor transistor logic: type of digital logic chip made from transistors.
See also CMOS, Chapter 1 Background.

TVA Abbreviation for time-variant amplifier. Roland-speak for a VCA in its digital synthesizers.
See also VCA, Chapter 5 Digital.

TVF Abbreviation for time-variant filter. Roland-speak for a VCF in its digital synthesizers.
See also VCF, Chapter 5 Digital.

Two-pole A filter with two sections – normally with a cut-off slope of 12 dB/octave.
See also Chapter 2 Analogue.

UART Universal asynchronous receiver/transmitter: converts data into serial format (as used in a MIDI port).
See also ACIA, Chapter 5 Digital.

Unison Several events happening as one. Often used for a stack of sounds which all sound when a key is played on a keyboard.
See also Chapter 6 Using synthesis.

Universal Synonym for generic, as in the sense of globally available or usable. Often used in the context of editor/librarian software.
See also Generic, Chapter 6 Using synthesis.

VCA Abbreviation for voltage controlled amplifier. An amplifier whose gain can be controlled by a control voltage.
See also VCO, Chapter 2 Analogue.

VCF Abbreviation for voltage controller filter. A filter whose cut-off frequency can be controlled by a control voltage.
See also VCA, Chapter 2 Analogue.

VCO Abbreviation for voltage controller oscillator. An oscillator whose frequency can be controlled by a control voltage.
See also VCF, Chapter 2 Analogue.

Vector synthesis A synthesis technique where two or four separate sounds can be mixed in real-time using a joystick, and the mix recorded for subsequent replay as part of the sound.
See also Chapter 2 Analogue.

Velocity The speed with which a key is pressed. Most synthesizers merely measure the time from just after the key is depressed to just before it stops moving at the bottom of its travel.
See also Chapter 2 Analogue.

Velocity curve The relationship between the speed of depressing a key and the control voltage or control signal which is produced. This can often be mapped.
See also Chapter 2 Analogue.

Velocity sensitivity Synonym for velocity curve.
See also Chapter 2 Analogue.

Vernier A method of allowing precise measurement by having a small scale attached to a rotary or linear control. Often used for the frequency control knobs of analogue synthesizers.
See also Chapter 2 Analogue.

Vibrato Periodic variation in pitch or frequency of an audio signal. (The LFO speed is usually between 5 and 25 Hz.).
See also LFO, FM, Chapter 2 Analogue.

Vocoder A modifier which splits an audio signal into separate frequency bands for processing.
See also Chapter 2 Analogue.

Voice A complete sound producing system for one note.
See also Chapter 2 Analogue.

Voice channel Synonym for voice.
See also Voice, Chapter 2 Analogue.

Voice stealing Synonym for note stealing.
See also Note stealing, Chapter 2 Analogue.

Volatile Memory which loses its contents when the power is removed.
See also Non-volatile, Chapter 1 Background.

Voltage control Using a voltage as a means of controlling parameters.
See also CV, Chapter 2 Analogue.

Voltage pedal A performance controller which produces a control voltage as the output of a foot pedal.
See also Chapter 7 Controllers.

Voltage trigger Envelope trigger signal which uses a brief pulse to initiate an envelope.
See also Chapter 2 Analogue.

Volume The amplitude, loudness or level of a signal or waveform.
See also Loudness, Chapter 2 Analogue.

Volume pedal A performance controller which varies the attenuation (or gain) of an audio signal as it passes through the foot pedal.
See also Chapter 7 Controllers.

Wave sequencing The concatenation of two or more wavecycles, multi-cycle waveforms, or samples.
See also Chapter 3 Hybrid.

Wavecycle A single cycle of a waveform.
See also Waveform, Chapter 3 Hybrid.

Waveform The shape of a wave. Some waves only repeat over several cycles.
See also Wavecycle, Chapter 2 Analogue.

Waveform modulation Changes in the shape of a waveform under the control of a control voltage or a control signal. PWM is one example.
See also PWM, Chapter 2 Analogue.

Wavelength An alternative name for a cycle of a waveform: The time for one complete cycle.
See also Cycle, Chapter 1 Background.

Waveshape Synonym for waveform.
See also Waveform, Chapter 2 Analogue.

Wavetable An area of memory which is used to hold wavecycle and multi-cycle waveforms.
See also Chapter 3 Hybrid.

Wet A processed audio signal. Often refers to reverberation.
See also Chapter 6 Using synthesis.

Wheel Performance control where a vertical wheel is moved to change a controlled parameter. Often used for pitch or modulation parameter.
See also Chapter 7 Controllers.

Wheels Wheels are performance controllers used to control pitch or modulation functions. Usually take the form of a 50 cm disk set next to the left hand side of the keyboard.
See also Joystick, Chapter 7 Controllers.

White noise Noise which has an equal mix of all frequencies.
See also Chapter 2 Analogue.

Wobble Low frequency vibrato is sometimes called wobble.
See also Wow, Chapter 2 Analogue.

Word A set of bits or bytes containing a single sample or block of data. Often 8, 12 or 16 bits.
See also Chapter 1 Background.

Workstation (1) A powerful networked personal computer. (2) A synthesizer with a built-in sequencer and disk storage.
See also Chapter 6 Using synthesis.

Wow Low-frequency variation in the speed of a tape recorder, which results in wobble or vibrato.
See also Vibrato, Chapter 2 Analogue.

XLR Identifier name for a specific type of audio connector – also called a Cannon connector.
See also DIN, Chapter 2 Analogue.

Y leads An audio adapter which has one input and two outputs, or two inputs and one output. Cannot be used with MIDI cables.
See also Chapter 1 Background.

Zero Part of the mathematical background to filter design. Zeroes are associated with notches and dips in the filter response.
See also Pole, Chapter 1 Background.

Zero crossing The points at which the audio signal crosses the zero axis at zero volts.
See also Chapter 1 Background.

Zone Part of a keyboard. Used in describing splits. Can be a single note, or a contiguous range of notes.
See also Split, Chapter 6 Using synthesis.

Jargon

Jargon

Specialist	Used only in a given subject.
Particular	Words used in one context
Subject	Synthesizers is the current subject
Alternative	Jargon often replaces more common words
Glossary	Also in the Glossary section
Acronym	Jargon frequently uses acronyms

Jargon is a descriptive term for those specialist words which are used only in a particular subject. This entry shows how they are dealt with here. Look up words which are used throughout this book, and the alternatives will be shown. This is somewhat like a thesaurus, but for jargon. The entries include both synonyms and antonyms. Most of these words will also be in the Glossary section.

Vibrato

FM	Frequency modulation: cyclic pitch changes.
AM	Amplitude modulation: cyclic volume changes (not vibrato).
Tremolo	Cyclic changes in volume (not vibrato).
Flutter	Rapid random changes in pitch (usually on a tape recorder).
Wow	Slow cyclic pitch changes (usually on a tape recorder).
Wobble	Cyclic pitch changes.

Modulation	Can be any pitch, volume or timbre change, often vibrato.
Leslie effect	A complex combination of vibrato and tremolo effects.

Vibrato is the name applied to cyclic changes in frequency. Tremolo is the term for cyclic changes in volume. The terms vibrato and tremolo are often mistakenly used interchangeably. For example, the 'tremolo' arm on a guitar produces vibrato and pitch bending, whilst a violinist's 'vibrato' is a mixture of vibrato and tremolo.

Pulse width

Width	Duration in time of mark portion of a pulse waveform
Duty cycle	Alternative term for pulse width
Shape	Used for non-rectangular waveforms: sawtooth, sine, etc.
PW	Abbreviation for pulse width
PWM	Pulse width modulation: cyclic change of pulse width
Skew	Used for non-rectangular waveforms: sawtooth, sine, etc.
Percent	Alternative measure of PW: 50% = square, 1% = narrow
Ratio	Alternative measure of PW: 1:1 = square, 99:1 = narrow
Rectangle	Alternative term for pulse width
Symmetry	Alternative term for pulse width

The pulse width of a waveform is a measure of the duration of a specific state. Usually it refers to a rectangular waveform, and is then a measure of the length of the mark portion of the waveform. Pulse waveforms are made up of two parts: the mark (usually the most positive portion), and the space (usually the most negative portion). When the mark and space times are equal, then the waveform is said to be 'square' in shape.

Envelope

Contour	Alternative name for an envelope.
ADSR	Attack decay sustain release – commonest form.
ADR	Attack decay release – old Moog-style of envelope.
AD	Attack decay – percussive envelopes only.
EG	Abbreviation for envelope generator.

Segment	One part of an envelope: attack is the first segment.
Stage	Alternative name for segment.
Slope	The rate of change of the envelope during a segment.
Rate	The rate of change of the envelope during a segment.
Time	The length of time of a segment.
Level	The level at the start of a segment (and the sustain level).
Break Point	Indicates a change of slope at a specific time or level.
Pivot	Alternative name for break point.
Trapezoid	Function generator sometimes used as an envelope.
Follower	Extracts an envelope from an audio signal.
Scaling	Changes in rates or levels.

Envelopes are one of the major sources of control voltages and control signals in a synthesizer. Whereas most other control sources are cyclic (LFOs, VCOs) the envelope can produce complex shapes which happen only once (although some envelopes do allow looping between segments). Envelope terminology can vary considerably between instruments. The method used to indicate the time of a segment may be a time, or a slope, and may use low numbers to indicate a short segment, or high numbers!

Source

Output	The source of the control voltage = output of circuit.
Out	Abbreviation for output.
Source	The output of the circuit.
Origin	Alternative name for source.

Controls can be divided into sources and destinations. Sources provide control voltages and control signals. Matrix modulation schemes use the words source and destination for the outputs and inputs to a matrix of connection points.

Sink

Destination	The input to the circuit = destination of control signal/voltage.
Input	The input of the circuit.
In	Abbreviation for input.
Sink	Alternative name for destination.

Controls can be divided into sources and destinations. Destinations take control voltages and signals and use them to make changes to parameters. Matrix modulation schemes use the words source and destination for the outputs and inputs to a matrix of connection points.

Detent

Dead band	The centre of the control has an area where nothing happens.
Dead zone	Alternative name for dead band.
Detent	A physical click at the centre.
Notch	Alternative name for detent.
Click	Alternative name for detent.
Stop	Alternative name for detent.
Spring	Springing to return to the detented position

Pitch wheels and levers need to have a way for the performer to know when the pitch bend is centred: no pitch bend. This can be achieved by using a dead band where moving the control has no effect, or a detent for the 'no pitch bend' position, or a combination of springs and a dead band or detent.

Layer

Split	Layering of two sounds controlled from different notes.
Double	Two sounds playing the same notes.
Stack	Two sounds playing the same notes.
Layer	Two sounds playing the same notes.
Hyperpreset	The setup of two or more sounds in a layered form.
Setup	Alternative name for hyperpreset.
Performance	Alternative name for hyperpreset.
Multi	Alternative name for hyperpreset.
Combination	Alternative name for hyperpreset.
Combi	Shortened form of combination.
Function	Alternative name for hyperpreset.

A layer is two or more sounds playing the same notes, but often arranged so that the two sounds are complementary rather than very similar (unless detuning is the purpose of the layering). A split allows two timbres to be layered and to be controlled independently by allocating them to different areas of the keyboard. Layering thus requires only one-handed playing or single notes, whilst splits require two hands or two notes. The setup of two or more voices has many names. Some synthesizers enable the complete setup of the synthesizer to be stored as a hyperpreset.

Resonance

Q	The sharpness or selectivity of a filter.
Feedback	The signal which is fedback from the filter output to the input.
Emphasis	Alternative name for Q.
Quality	The full name for Q.
Regeneration	Alternative name for feedback.
Sharpness	Alternative name for resonance.

Low-pass filters are good for making tonal changes to the brightness of a sound, but for making more major changes a band-pass filter is required. By adding feedback to a low-pass filter, it is possible to turn it into a resonant band-pass filter which has a strong peak at the cut-off frequency.

Memory

Program	Used by MIDI. Computer programming term.
Patch	Analogue modular synthesizers.
Voice	Yamaha-speak.
Preset	Usually permanently stored in ROM memory.
Sound	The real world.
Tone	Electronic organs.
Waveform	Electronics term.
Wave	Electronics slang.
Combi	More than one sound layered or stacked.
Performance	More than one sound layered or stacked.
Stack	More than one sound layered or stacked.
Split	More than one sound across the keyboard range.
Timbre	Musical term for sound quality.
Store	Computer term.
Location	Computer programming term.
Bank	Several sounds.
Set	Alternative name for several sounds.

Synthesizers store sounds in 'memories'. There are many alternative names for the memories. The distinction between a sound and a combination of several sounds is becoming blurred. Many sounds are made up of simpler sub-sounds, parts or elements.

Storage

Disk	Hard or floppy: rotating magnetic/optical data memory.

Disc	Vinyl record or album: holds audio information.
RAM	Random access memory: read/write solid-state data chips.
ROM	Read only memory: permanent solid-state data chips.
Media	The physical carrier/holder of the data: disk, tape, card etc.
Floppy	Flexible circle of magnetic material in a rigid casing.
DAT	Tape based data media using rotating helical scan heads.
Card	Generic term for memory card (using ROM or RAM).
PCMCIA	Specific type of standardized card and interface.
Network	A means of connecting computers together: non-local.
Server	A remote computer used to hold data, accessed via a network.
Optical	Disk using optical rather than magnetic storage for data.
Virtual	RAM used as a hard disk replacement (faster).
Tape	Flexible plastic strip with magnetic coating for data.
Off-line	Off-line storage needs to be physically loaded by hand.

Computers use many forms of storage for holding data (information). Storage is really an shortened form of the compound word 'mass storage'.

After-touch

Touch	Additional key pressure used as a controller.
Pressure	Alternative name for touch.
Sustain pressure	After-touch is often used in the sustain segment.
Second touch	Alternative name for touch (initial touch = velocity).
Expression	Alternative name for touch.
After-touching	Alternative name for touch.

After-touch is the name for the additional pressure that some keyboards allow to be used as a controller after the key has travelled from the up to the down position. After-touch sensors vary from those of soft rubber which provides several millimetres of movement, to ones which are hard rubber and hardly move at all. After-touch sensing can be global for the entire keyboard (the hardest pressure is normally the output) or polyphonic (much rarer).

Messages

Data	The information contained in the message
Signals	Alternative name for messages.
Packets	Alternative name for messages.
Commands	Alternative name for messages.
Streams	Used for several messages in sequence.
Information	The data in the message.

MIDI messages are sent in the form of one or more bytes in sequence.

Harmonics

Partial	A single frequency component (could be an inharmonic).
Overtone	First overtone is second harmonic.
Line	A single frequency component (could be an inharmonic).
Spectral line	One frequency in the spectrum (could be an inharmonic)
Component	One frequency in the spectrum (could be an inharmonic).
Frequency	A single spectral component (could be an inharmonic).
Harmonic	Related to the fundamental.
Inharmonic	Not related to the fundamental.

Harmonics are based around multiples of the fundamental frequency. The 2nd, 4th and 8th harmonic are octaves, whilst the 3rd harmonic is a perfect fifth. Inharmonics are frequencies which are not harmonically related to the fundamental. On a spectrum, individual frequencies appear as vertical lines or peaks.

Volatile

RAM	Random access memory = read/write memory.
Battery-backed	The battery maintains the data when the power is down.
Eraseable	The memory contents can be changed.
Non-volatile	The device holds its data without any power present.

Permanent	The device holds its data without any power present.
EPROM	Erasable programmable read-only memory.
EEPROM	Electrically erasable ROM.
EAROM	Electrically alterable ROM.
Flash EPROM	EPROM that behaves like non-volatile RAM.
Flash chips	Alternative name for flash EPROM.
Flash memory	Alternative name for flash EPROM.
Flash	Alternative name for flash EPROM.

'Volatile' means that the storage medium depends on a supply of electrical power to hold data. If the power fails then the data is lost. RAM is usually volatile. 'Non-volatile' means that the storage does not depend on an external power supply. It may have its own local power supply, or use a technology which requires no power to maintain the data.

Index